Surface Science

Springer

Berlin
Heidelberg
New York
Barcelona
Budapest
Hong Kong
London
Milan
Paris
Santa Clara
Singapore
Tokyo

R. J. MacDonald
E. C. Taglauer
K. R. Wandelt (Eds.)

Surface Science
Principles and Current Applications

With 196 Figures

 Springer

Professor Dr. R. J. MacDonald
University of Newcastle
Callaghan, NSW 2308
Australia

Dr. E. C. Taglauer
Max-Planck-Institut für Plasmaphysik
Boltzmannstr. 2
D-85748 Garching
Germany

Professor Dr. K. R. Wandelt
Universität Bonn
Institut für Physikalische und Theoretische Chemie
Wegelerstr. 12
D-53115 Bonn
Germany

Library of Congress, Cataloging in Publication Data applied for

Die Deutsche Bibliothek – CIP-Einheitsaufnahme
Surface Science: principles and current applications/R. J. MacDonald ... (ed.).
Berlin; Heidelberg; New York; Barcelona; Budapest; Hong Kong; London; Milan;
Paris; Santa Clara; Singapore; Tokyo: Springer 1996
ISBN-13:978-3-642-80283-6   e-ISBN-13:978-3-642-80281-2
DOI:10.1007/978-3-642-80281-2

NE: MacDonald, Richard J. (Hrsg.)

© Springer-Verlag Berlin Heidelberg 1996
Softcover reprint of the hardcover 1 edition 1996

The use of general descriptive names, registered names, trademarks, etc. in this publication does not imply, even in
the absence of a specific statement, that such names are exempt from the relevant protective laws and regulations and
therefore free for general use.

Typesetting: Camera-ready copies from the authors
Cover design: Design & Production, Heidelberg
SPIN 10539289   57/3144-5 4 3 2 1 0 – Printed on acid-free paper

Preface

This book includes case studies and status reports from many timely research areas in surface science.

Surface science holds a key position in fundamental research in that it must provide the basis for future technologies that are to garantee the supply of sufficient and reasonably priced energy, the intelligent use of existing materials as well as the design of new ones, and the fast availability and exchange of information. Heterogenous catalysis as well as thin film technologies in conjunction with surface modification on the atomic scale are based on surface (and interface) processes per se. The exploitation of solar and renewable energy sources would be unthinkable without catalysis (e.g., in fuel cells) and thin film technology (e.g., for solar cells). Also the development of materials with new chemical and physical properties will include catalytic and thin film growth processes. Electronics, including information technology, sensorics, etc., is the obvious exponent for the development and application of nano-scale structures. As well as having unprecedented chemical and physical (e.g., quantum-size) properties, thin-film and nano-scale structures also help to reduce the consumption of materials and to economize on our limited natural resources.

These technological demands together with the continuous refinement of existing methods as well as the development of new techniques are making surface science both a steadily expanding and an increasingly relevant field of research. As a consequence it is important, but more and more difficult, for us to keep up with the current literature. The intention of this volume is therefore two-fold. Firstly, it aims at demonstrating the breadth of surface-science-related problems and at bridging the gap between the development of basic experimental methods and theoretical models, on the one hand, and the challenge of using them for solving technological problems on the other hand. Secondly, it intends to provide the reader with a comprehensive review of several major and most timely research areas of fundamental and applied surface science. The various contributions are based on invited lectures presented by the authors at the "German-Australian Workshop on Surface Science" held at Schloß Ringberg, Tegernsee, Germany, in January 1994. The papers were written and submitted by late summer 1995 and are arranged here in six parts: (1) Structure at Single Crystal Surfaces; (2) Surface Electronic Properties; (3) Surface Composition Analysis; (4) Properties and Influence of Adsorbates; (5) Thin Film Growth and Interfaces; and (6) Characterization of Catalysts. The individual papers have been selected and grouped in such a way that the different parts provide state-of-the-art overviews of timely research topics such as surface structure determination using Tensor-LEED or surface X-ray diffraction; the preparation and detection of low-dimensional electronic surface states; the further evaluation of ion scattering for quantitative surface analysis; the dynamics of adsorption and reaction of adsorbates, including kinetic oscillations; the characterization and control of thin film and multilayer growth, also under the influence of surfactants; and a critical assessment of the surface-physics approach to the characterization and "design" of catalysts.

The "German-Australian Workshop on Surface Science", which provided the stimulating basis for this volume received generous financial support from the Deutsche Forschungsgemeinschaft (DFG), the Max Planck Society (MPG), the Australian Research Council (ARC), and the Department of Industry, Technology and Commerce (DITAC) of Australia. The organizers of this workshop being also the editors of this volume are once again greatly indepted to these organizations as well as to the staff of Springer-Verlag for arranging the publication of this book.

Newcastle, Australia　　　　　　　　　　　R.J. MacDonald
München, Germany　　　　　　　　　　　　E. Taglauer
Bonn, Germany　　　　　　　　　　　　　　K. Wandelt
April 1996

Contents

IV. Properties and Influence of Adsorbates

V. Thin Film Growth and Interfaces

VI. Characterization of Catalysts

I.

Structure at Single Crystal Surfaces

Tensor-LEED, Diffuse LEED, and LEED Holography

K. HEINZ

Lehrstuhl für Festkörperphysik, Universität Erlangen–Nürnberg,
Staudtstr. 7, D-91058 Erlangen, Germany

Abstract. Low energy electron diffraction (LEED) has developed powerful new features and extensions in recent years. Intensity analyses were dramatically speeded up by the development of Tensor LEED and structural search procedures allowing access to structures of increased complexity. The restriction to long range order has largely been lifted by the development of diffuse LEED and patterns from disordered adsorbates could be interpreted in a holographic way resulting in atomically resolved images of surfaces.

1. Introduction

For a long time the applicability of low energy electron diffraction (LEED) for the quantitative analysis of surface structures was restricted to the presence of long range order and to structures of only very limited complexity. On the one hand, long range order seemed to be absolutely necessary because spectra of LEED spots have to be measured. It seemed trivial that this requires the existence of more or less sharp spots. On the other hand, the structural complexity of surfaces tackled by LEED was limited by the immense amount of computational work required by full dynamical calculations and the necessary variation of numerous structural parameters.

The present paper shows that these restrictions could be largely lifted by the introduction of diffuse LEED (DLEED) and tensor LEED (TLEED) in recent years allowing to approach new frontiers. To illuminate this exciting development we resume the standards of conventional LEED analysis first and describe the break through in theory and experiment by TLEED and DLEED, respectively, in paragraphs 3 and 4. Section 5 addresses a recent holographic interpretation of DLEED intensities, i.e. the reconstruction of real images of surfaces with atomic resolution.

2. Conventional LEED Intensity Analysis

It was only in the late sixties and early seventies that impressing theoretical efforts resulted in a full dynamical theory which allowed the reliable calculation of intensities [1]. The corresponding computer programs provided by theorists [1,2] represented the state of the art until recently. The codes follow some hierarchical sequence, i.e. scattering by atoms building up the diffraction of layers which in turn stacks to produce the full surface diffraction. For simple structures the accuracy of both theory and experiment has reached a remarkable level. Experimental and theoretical data can come so close that additional to visual comparison it needs a quantitative measure (R-factors [3]) to compare spectra and to find the best fit model.

The computer time necessary for a set of intensity spectra for a given structure varies with its structural complexity. For a medium complex structure, e.g. a p(2x2)/fcc(100) superstructure with some substrate reconstruction, the necessary CPU time for spectra in a range 50 - 250 eV amounts to about 10 minutes on a fast main frame computer. This seems to be tolerable when only a few parameters have to be varied. However, if many parameters must be fixed as e.g. in the case of large unit cells and/or complex substrate reconstructions, the situation changes dramatically. So, with N atoms in the unit cell with 3N coordinates and M values to be checked for each, the number of trial structures $n = (3N)^M$ scales in a non-polynomial way. Even for modest values, say $N = M = 5$, there are already $n = (15)^5 \sim 7 \cdot 10^5$ trials necessary to scan the parameter space. Even if one considers that some computer time can be saved by intermediate storage and multiple use of layer diffraction matrices, the necessary amount of computer time is practically unacceptable.

Some way out of the dilemma is opened by the introduction of search procedures [4,5]. They do not simply trace the full parameter space but – at a certain starting point in parameter space – probe the structural environment and determine in which direction the R-factor improves. By some method of steepest descent a local minimum is found. If different starting points lead to the same minimum one can assume that it is global rather than only local, i.e. that the best fit structure is found. However, still a huge number of intensity calculations must be carried out. So, inevitably there is need for a considerable speed up of full dynamical calculations.

3. Tensor LEED

The break through for speeding up LEED intensity calculations came in the mid 80ies by the development of tensor LEED [6-8] and the later combination with search procedures [9]. The basic idea is demonstrated in fig.1 where curves calculated for the 11 beam of Ni(100) are displayed for different surface relaxations (full lines). Apparently, the spectra change only smoothly with small changes of the structural parameter. This feature gave birth to the idea that starting from the spectra of a certain structure (*reference structure*) all other spectra may be obtained by perturbation of the reference. Mathematically this can be expressed by

$$A = A_0 + {}_i\Sigma T_i \, \delta t(\delta R_i) \tag{1}$$

whereby A_0 is the diffraction amplitude of the reference, δR_i is the geometrical displacement of an atom at original position R_i and A is the new diffraction amplitude resulting with the displacements. The quantitiy δt describes the change of the atomic scattering due to the displacement δR_i. In good approximation kinematic calculation yields

$$\delta t = G(-\delta R_i) \, t \, G(+\delta R_i) - t \tag{2}$$

whereby $G(\pm \delta R_i)$ are propagators linking the old and new atomic positions. The quantity T_i couples $\delta t(\delta R_i)$ to the total wave field to yield the corresponding change of the diffraction amplitude $\delta A = A - A_0 = {}_i\Sigma T_i \, \delta t(\delta R_i)$. In angular momentum representation T_i is a tensor which explains the term *Tensor LEED*, or as geometrical displacements are involved, *Geoemtrical TLEED*. It depends only on the reference and so has to be calculated only once. Spectra for structures deviating from the reference result easily and fast by use of the simple equs. (1,2). Depending on the structural complexity this can save orders of magnitude of CPU time.

Geometrical TLEED is valid for displacements $0.2\text{Å} \leq \delta R \leq 0.5\text{Å}$ depending on the atomic scattering strength. So, for e.g. Ni atoms the approximation is good up to $\delta R = 0.4$ Å. The quality of the approximation is demonstrated by the broken line spectra in fig. 1 which were calculated via TLEED with the bulk terminated surface as reference. Of course TLEED proves its real power only for much more complex structures. As an example fig. 2 shows an ultrathin epitaxial film of 4 monolayers fcc-Fe on Cu(100). Different from current asumptions the Fe layers are not simply the pseudomorphic continuation of Cu(100). Iron atoms shift in-plane and buckle

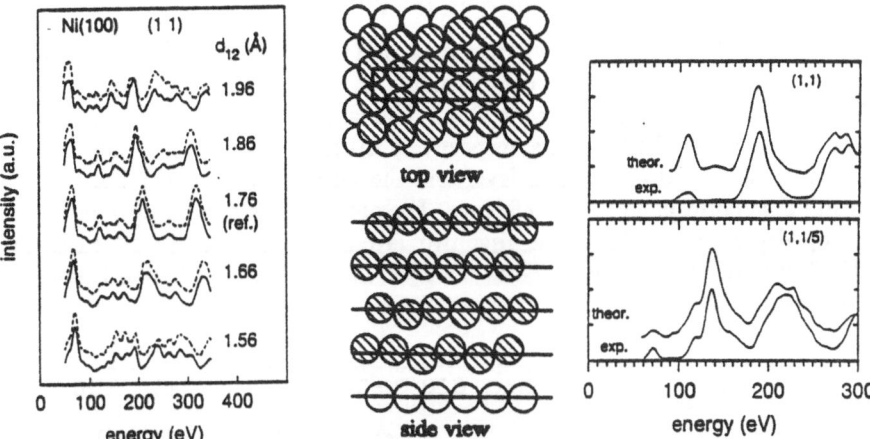

Fig. 1: Exact (full lines) and TLEED (broken lines) for various relaxations of Ni(100)

Fig. 2: Structure model of a 4 ML fcc Fe film epitaxially grown on Cu(100) and experimental and calculated best fit spectra for two selected beams.

substantially (up to 0.4 Å [10]). The reconstruction produces a 5×1 superstructure with 5 atoms in the layer unit cell. In-plane and vertical shifts in the 4 layers demand the adjustment of as much as 40 parameters. This was successfully done recently applying TLEED yielding a minimum Pendry R-factor [11] of R_p= 0.14. Measured and best fit spectra are displayed in fig. 2 for two selected beams.

It is worth to inspect equ. (1) once more. If we rewrite it without considering that δt may be caused by some geometrical displacement δR, it becomes more general,

$$A = A_0 + {}_i\Sigma T_i \, \delta t_i \qquad (3)$$

whereby δt_i can be due to *any* change in the scattering of atom i. This change may also come by a chemical substitution of that atom (*Chemical TLEED* [12]) or by its atomic vibration (*Thermal TLEED* [13]). Of course, both changes can be combined with additional geometrical displacements. Test calculations show that e.g. in a reference calculation for p(2×2)O/Ni(100) oxygen atoms can be replaced by sulphur and shifted to appropriate new positions. So, simply by perturbation spectra for p(2×2)S/Ni(100) result with an accuracy mirrored by an R-factor R_p<0.1 [12]. The same holds for *Thermal TLEED*, details can be found in the literature [13].

An ideal field of application of *Chemical TLEED* is the structure determination of substitutionally disordered alloys [12]. The disorder can be accounted for by the average t-matric approximation (ATA) [14] which considers the j-th layer built up by average scatterers

$$t_j = c_j t^A + (1-c_j)t^B \qquad (4)$$

whereby c_j and $(1-c_j)$ are the concentrations of atoms of type A and B, respectively and t^A and t^B their scattering matrices. In a conventional determination of the layer specific stoichiometry of a sample one has to carry out full dynamical calculations for each set of $\{c_j\}$. With *Chemical TLEED*, however, one can start from a certain reference (e.g. bulk concentrations in each layer) and subsequently use

$$\delta t_j = \delta c_j \, (t^A - t^B) \qquad (5)$$

as input to equ. (3). This allows a fast scan of the stoichiometry space. The validity range is rather large as demonstrated in fig. 3. with spectra for the substitutionally disordered alloy Mo_cRe_{1-c}. The labels of the curves correspond to the ATA and the TLEED results, respectively. At the left the spectra refer to bulk concentrations $c = 0.75$ in each layer and the curves are identical because the TLEED data correspond to the reference. The other graphs display spectra for increasing depletion of Mo in the first two layers as indicated. In all cases the TLEED spectra compare very well to the exact data as mirrored by the Pendry R-factors given.

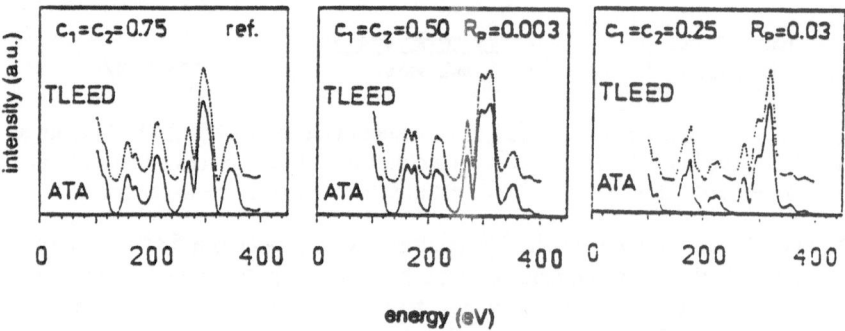

Fig. 3: Spectra of the 10 beam of Mo_cRe_{1-c} calculated by the conventional ATA method and via TLEED for varying concentrations c_1 and c_2 in the 1. and 2. layer.

Even with TLEED available the conventional way to find the best structural model requires to scan the full parameter space. Therefore, search procedures were developed similar to that mentioned above but using the TLEED approximation [9]. An alternative solution comes by *direct* inversion of equ.(3) combined with equ.(5). Using intensities rather than amplitudes and neglecting quadratic terms in δc_j the layer dependent stoichiometry $\{c_{0j} + \delta c_j\}$ results *directly*, by

$$\delta c_j = {}_g\Sigma (M_{gj})^{-1} (I_g - I_{0g})$$

whereby g labels independent data points and $M_{gj} = 2(t^A - t^B) \, Re[(A_{0g})^* T_{gj}]$. Unfortunately, errors in theory and experiment enforce to use more data points $\{g\}$ than unknown variables $\{c_j\}$ and to minimize the corresponding misfit between theory and experiment. The minimization procedes iteratively and the correct stoichiometry results after only a few steps [15]. This *direct* approach saves another considerable amount of computer time. It can be extended also to geometrical structure parameters though the range of convergence is limited to 0.1 Å [16,17].

4. Diffuse LEED

So far we have dealt only with ordered structures. However, it is known from various adsorption experiments that not all species made to adsorb on a crystalline surface order to show a long range superstructure. In many cases depending on temperature, coverage and adsorptives only diffuse intensities develop superimposed on the sharp substrate spots. Figure 4 (left, bottom) displays the example of K/Ni(100) at a coverage of $\Theta=0.04$ with potassium deposited at 90 K [18]. At this coverage K atoms have an average distance of 5 substrate lattice constants prohibiting mutual interaction and consequently ordering. The assumption is reasonable that locally all adatoms reside in the same adsorption position and produce the same intensity distribution as indicated in fig. 4 (left, top). The lack of correlations between adatoms means that the observed intensity distribution is a multiple of that formed locally. It can be calculated full dynamically using a cluster approach [19] or large unit cells [20] which include the multiple scattering between adatom and adsorbate. For the measurement parallel detection is best suited as e.g. by a video camera [21,22] or a channel plate followed by a position sensitive device [23]. Computational variation of the local structural parameters and comparison to the experiment allows the retrieval of the local adsorption structure very much in the same way as for ordered structures whereby TLEED again can be of much help [6]. A number of disordered adsorption systems were solved in this way (for a review see ref. [24]).

So far, diffuse intensity distributions and patterns with sharp spots qualitatively seem to be different things. However, inspection of fig. 4 (right, top) tells that there is an intrinsic connection: Ordering means that intensities cancel in most directions (broken line arrows) and remain only in directions allowed by symmetry (full line

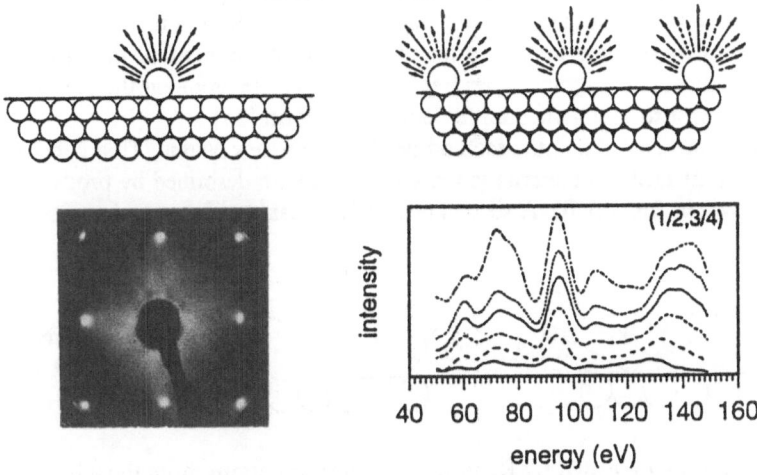

Fig. 4: Local formation (schematic) of diffuse intensities (left, top) and diffuse pattern for K/Ni(100) at 90 eV and coverage $\Theta = 0.04$ (left, bottom). Also, the formation of Bragg spots by ordering is indicated (right, top). DI(E) spectra (right, bottom) for the ½¾ position for various coverages ($\Theta = 0.04, 0.08, 0.14, 0.17, 0.22$ from bottom) and the I(E) curve for the ordered c(4×2) phase (top curve) are compared.

arrows). Consequently, the correponding spot intensities must be multiples of the diffuse intensities (if intralayer scattering is weak). This means that the I(E) spectrum of a sharp spot must – apart from the intensity level – have the same structure as the dependence of diffuse intensities DI(E) if both are recorded for the same constant surface parallel momentum transfer. In fact, this is confirmed by measurement at the ½¾ position of the K/Ni(100) system. Figure 4 (right, bottom) displays DI(E) spectra for various coverages of the disordered phase and the I(E) spectrum for the ordered c(4×2) phase. The spectra practically show the same structure, small peak shifts may come by intralayer scattering or structural modifications varying slightly with coverage [18]. So, there is no physical difference between I(E) and DI(E) spectra both with respect to the measurement and the full dynamical analysis. In this sense both LEED and DLEED are local probes of surface structure, i.e. the intensities are formed locally.

5. DLEED - Holography

It is well known that Gabor in the first place wanted to reconstruct 3-dimensional images by the use not of light waves but of electrons. His efforts, however, were in vain due to the very limited coherence length of electrons, which is of the order of 100 Å. So, with macroscopic beam splitters the reference and object waves are incoherent prohibiting any interference nececsarry to form a hologram. A way out seemed to come by a recent proposal according to which DLEED patterns can be interpreted as holograms from which atomic real space images can be reconstructed [25]. The basic idea is illuminated by fig. 5 (left) where the adatom is shown to act as a *microscopic beam splitter*. The wave R_0 immediately scattered back serves as *reference wave* because its spatial structure is relatively homogeneous. The wave scattered to the substrate (= object) and diffracted back from there is interpreted as the *object wave* O. Interference of the waves at the position of the detector in the far field produces a hologram. A real space image of the atoms in the surface should result by reconstruction via simple Fourier transform [25].

It soon turned out that the method works not as easy as that. This is because the formation of DLEED patterns is more complex than described by processes of the type shown in fig. 5 (left). A second class of processes of in general comparable

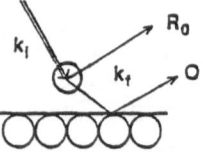 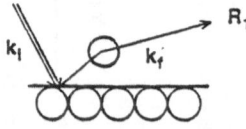

Fig. 5: Different types of DLEED scattering processes

weight adds as indicated by R_1 in fig. 5 (right). Emerging from the adatom they become part of the reference wave and so generally make the latter ill defined [26]. Only in cases where this process is negligible (e.g. at certain energies) or can be considered computationally, a direct holographic reconstruction seems to be possible [27-30]. So, after a first period of exciting research in DLEED holography the method seemed turning out to be a flop.

However, very recently a new idea came up. Instead of using single energy DLEED patterns it was proposed that diffuse intensities should be measured at many energies and averaged in such a way that the disturbing processes indicated in fig. 5 (right) cancel [31]. This can be done by phase modulated averaging, i.e. by introduction of a constant phase condition which favours the process R_0, i.e.

$$A(\mathbf{k_f}, \mathbf{k_i}, \mathbf{r}) = \int I(\mathbf{k_f}, \mathbf{k_i}) \exp\{-ikr + ik_f\ \mathbf{r}\}\ k^2 dk$$

whereby k is the modulus of the electron momentum. $I(\mathbf{k_f}, \mathbf{k_i})$ is the measured diffuse intensity for an incident wave vector $\mathbf{k_i}$ and the final state wave vector $\mathbf{k_f}$ as defined by the detector's position. It was shown in ref. [31] that this averaging not only favours processes R_0 but simultaneously makes all other unwanted processes largely cancel. The final image function results by integration over all final states

$$A(\mathbf{k_i}, \mathbf{r}) = \int A(\mathbf{k_f}, \mathbf{k_i}, \mathbf{r})\ d\mathbf{k_f}.$$

The image $|A(\mathbf{k_i}, \mathbf{r})|^2$ by dependence on $\mathbf{k_i}$ predominantly contains atoms positioned in direction $\mathbf{k_i}$ seen from the adsorbate. However, with data measured for a sufficient set of directions of incidence each substrate atom should show up [31].

The reconstruction procedure described was applied for the disordered adsorption of K on Ni(100) [32]. At a coverage $\Theta = 0.05$ data were taken for two angles of incidence ($\theta = \phi = 0°$ and $\theta = 35°$, $\phi = 22°$) and for multiple energies entering the reconstruction process in the range 100 - 300 eV with a constant step width $\Delta k = 0.075$ (atomic units). Figure 6 displays the results for different cuts through the

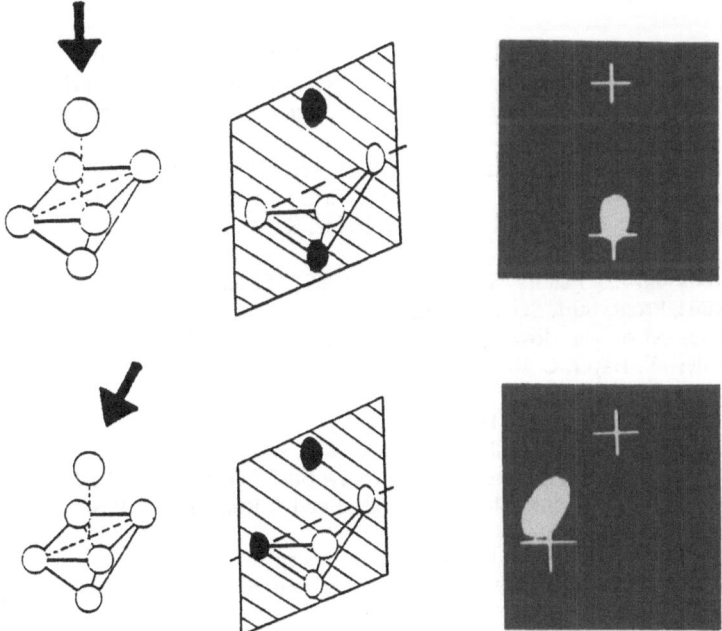

Fig. 6: Holographic image reconstruction in the planes indicated by hatched areas for K/Ni(100) from normal (top) and oblique (bottom) incidence DLEED data.

surface. Upper crosses in each case mark the position of the reference (ad)atom, the others give the exact positions of substrate atoms as known from conventional analysis [18]. Apparently, bright spots with a FWHM of less than 1 Å appear very near the correct positions. Potassium appears to reside in four fold symmetric positions with an adsorption height of 2.2–2.4 Å above the first and 3.9–4.1 Å above the second substrate layer. These values are within about 0.5 Å of the correct values. Probably, the error comes from the phase shift introduced by atomic backscattering. As demonstrated in ref.[33] this can be easily corrected for as will be done in future work. Though the accuracy of atomic positions is much lower than typical for conventional LEED, the resulting image gives a good idea of the underlying structure. It could serve as reference for structural refinement via TLEED.

6. Conclusion

As demonstrated, LEED recently has developed powerful new features and extensions. Tensor LEED and structural search routines considerably speed up the intensity analysis and allow access to structures of increasing complexity. Restrictions to the existence of long range order are lifted by the development of diffuse LEED. The intelligent use of adatoms as microscopic beam splitters seems to open the field of a special holographic technique which yields real space images of surfaces with atomic resolution. So, LEED may look forward to a future of exciting applications.

References

[1] J.B. Pendry, *Low Energy Electron Diffraction*, Academic Press (1974)

[2] M.A. Van Hove and S.Y. Tong, *Surface Crystallography by LEED*, Springer (1979)

[3] M.A. Van Hove, W.H. Weinberg and C.-M. Chan, *Low-Energy Electron Diffraction*, Springer (1986)

[4] P.G. Cowell, M. Prutton and S.P. Tear, Surf. Sci. **177** (1986) L915

[5] G. Kleinle, W. Moritz, D.L. Adams and G. Ertl, Surf. Sci. **219** (1989) L637

[6] P.J. Rous, J.B. Pendry, D.K. Saldin, K. Heinz, K. Müller and N. Bickel, Phys. Rev. Lett. **57** (1986) 2951

[7] P.J. Rous and J.B. Pendry, Surf. Sci. **219** (1989) 355 and 373

[8] P.J. Rous, Progr. Surf. Sci. **39** (1992) 3

[9] P.J. Rous, M.A. Van Hove and G.A. Somorjai, Surf. Sci. **226** (1990) 15

[10] S. Müller , P. Bayer, C. Reischl, K. Heinz, B. Feldmann, H. Zillgen and M. Wuttig, Phys. Rev. Lett. submitted

[11] J.B. Pendry, J. Phys. C **13** (1980) 239

[12] R. Döll, M. Kottcke and K. Heinz, Phys. Rev. B. **48** (1993) 1973

[13] U. Löffler, R. Döll, K. Heinz and J.B. Pendry, Surf. Sci. **301** (1994) 346

[14] R. Baudoing, Y. Gauthier, M. Lundbergand and J. Rundgren, J. Phys. C **19** (1986) 2825

[15] R. Döll, M. Kottcke and K. Heinz, Surf. Sci. accepted

[16] J.B. Pendry, K. Heinz and W. Oed, Phys. Rev. Lett. **61** (1990) 2953

[17] J.B. Pendry and K. Heinz, Surf. Sci. **230** (1990) 137

[18] H. Wedler, M.A. Mendez, P. Bayer, U. Löffler, K. Heinz, V. Fritzsche and J.B. Pendry, Surf. Sci. **293** (1993) 47

[19] J.B. Pendry and D.K. Saldin, Surf. Sci. **145** (1984) 33

[20] D.K. Saldin, J.B. Pendry, M.A. Van Hove and G.A. Somorjai, Phys. Rev. B **31** (1985) 1216

[21] K. Müller and K. Heinz, in *The Structure of Surfaces*, Eds. M.A. Van Hove and S.Y. Tong, Springer (1985) p.105

[22] K. Heinz, Progr. Surf. Sci. **27** (1988) 239

[23] D.F. Ogletree, G.S. Blackman, R.Q. Wang, U. Starke, G.A. Somorjai and J.E. Katz, Rev. Sci. Instr. **63** (1991) 104

[24] K. Heinz, Vacuum **41** (1990) 328

[25] D.K. Saldin and P.L de Andres, Phys. Rev. Lett. **64** (1990) 1270

[26] K. Heinz, R. Döll, M. Wagner, U. Löffler and M.A. Mendez, Appl. Surf. Sci. **70/71** (1993) 367

[27] M.A. Mendez, C. Glück and K. Heinz, J. Phys. Cond. Matt. **4** (1992) 999

[28] M.A. Mendez. J. Guerrero, C. Glück, P.L. de Andres, K. Heinz, D.K. Saldin and J.B. Pendry, Phys. Rev. B **45** (1992) 9402

[29] P. Hu, C.J. Barnes and D.A. King, Chem Phys. Lett. **183** (1991) 521

[30] P.L. de Andres, Surf. Sci. **269/270** (1992) 1

[28] C.M. Wei and S.Y. Tong, Surf. Sci. Lett. **274** (1992) L577

[29] C.M. Wei, S.Y. Tong, H. Wedler, M.A. Mendez and K. Heinz, Phys. Rev. Lett. submitted

[30] S.Y. Tong, H. Li and H. Huang, Phys. Rev B **46** (1992) 4155

3-D Surface Structure Analysis by X-Ray Diffraction

W. Moritz and H. L. Meyerheim
Institut für Kristallographie und Mineralogie, Universität München
Theresienstr. 41, 80333 München, Germany

Abstract.

Surface x-ray diffraction is by now a well established technique for the structure analysis of clean and adsorbate covered crystal surfaces. In the present article we briefly review the most important features of surface x-ray diffraction in combination with some examples 3-D surface structure determinations. The structures of the clean and alkali covered Ge(100)-(2x1) surface as well as the (3x3) reconstruction of InSb($\overline{1}\overline{1}\overline{1}$) have been studied. For the uncovered Ge(100)(2x1) surface we have obtained clear evidence for the asymmetry of the dimers. Adsorption of K and Cs is found to take place in the large grooves between the Ge- dimer rows. At about half monolayer coverage the high coordinated site above the third layer Ge atom is occupied, at lower and higher coverage occupation of an asymmetric site close to the Ge-dangling bonds is observed. For the InSb($\overline{1}\overline{1}\overline{1}$) surface a new type of reconstruction has been found. It is characterised by adatom rings located at the corners of the (3x3) unit cell above a slightly relaxed InSb double layer. Two types of six-atom rings have been found, an elliptic ring occurring in three different orientations and an trigonal ring occurring in two different orientations. The rings are randomly distributed and oriented. The analysis of this structure demonstrates the importance of the out-of-plane diffracted intensity with a large momentum transfer q_z normal to the sample surface. Only the consideration of these reflections allow a complete three dimensional analysis of the surface structure including subsurface relaxation and anisotropic thermal vibrations. The specific advantages and disadvantages of grazing incidence x-ray diffraction are compared with low energy electron diffraction.

1. Introduction

One of the main advantages of x-ray diffraction (XRD) for surface structure determination is the applicability of the kinematical diffraction theory. It allows to analyse complex surface structures which could not be solved with low energy electron diffraction (LEED) though this technique has made considerable progress in the last years. For X-rays the theory and the computational methods are well developed for three dimensional structure determination and the application to surface structure analysis is straightforward. Existing program packages can be used to a large extent for surface structures allowing the application of the usual Fourier methods such as the calculation of the Patterson function, the Fourier and the difference Fourier synthesis. The techniques and applications of surface x-ray diffraction have been reviewed in detail [1,2] recently. We therefore only briefly discuss in chapter 2 some basic experimental and theoretical aspects of surface XRD which are different from the conventional 3-D diffraction methods. These are the so called truncation rods arising from the lack of periodicity along the surface normal and the specific features of the Patterson function obtained when using partial data sets. The latter is a result that part of the reflections originate solely from the superstructure of the surface and part of the reflections, originate from both, the superstructure and the bulk, where an interference of both parts occurs. The latter are also called integer order reflections following the language in the LEED literature. Frequently only superstructure reflections are included in the data analysis and integer order reflections related to the (1x1) bulk structure are excluded or analysed separately because the conventional structure refinement programs cannot treat the interference between bulk and surface part properly. Moreover, the interpretation of the Patterson function of all reflections (excluding bulk Bragg reflections, of course) has not been worked out so far. The Patterson function therefore has to be calculated from the superstructure reflections alone. The special features related to the partial data sets are briefly discussed below.

In most experiments reported so far the data sets are confined 'in-plane' diffraction data, $q_z \approx 0$, allowing only a two dimensional structure determination of the z-projected structure or to small vertical momentum transfer with a low resolution normal to the surface. This shortcoming can be removed by including high q_z reflections. It requires, however, the measurement of diffraction data at high exit angles of the diffracted beam and thus results in a large data set and long measuring times. We demonstrate the capabilities of the method with two recent results presented in chapter 3, the structure analyses of the clean [3] and alkali covered Ge(100)-(2x1) surface [4] and the (3x3) reconstruction of the InSb($\overline{1}\overline{1}\overline{1}$) surface [5].

The main technique for surface structure analysis has been LEED in the past and the development of surface XRD at grazing incidence rises the question which technique should be applied for which specific problem and what are the advantages and disadvantages of either method. These questions are discussed in the last chapter.

2.1 Experiment and Theory

The experiments were performed using a 4-circle UHV diffractometer which has been especially designed for the measurement of high q_z reflections [6,7]. The surface can be cleaned by standard techniques and controlled by LEED and Auger electron spectroscopy (AES). The data collection is performed using the so called 'Z'-axis geometry with two independent angles for the detector setting [1,6]. The scattering geometry in reciprocal space is shown in the lower part of fig. 1.

The sample surface is illuminated by the primary beam k_i under grazing incidence angle α_i which is close to the critical angle of total external reflection. The diffracted beam k_f associated with the scattering vector $q=(q_\parallel,q_z)$ is directed to the detector at an exit angle α_f relative to the sample surface. For the measurement of the intensity at a given position q_z along the rod the sample is rotated around its surface normal which is equivalent to the rotation of the rod around the origin (O*) of the reciprocal

14

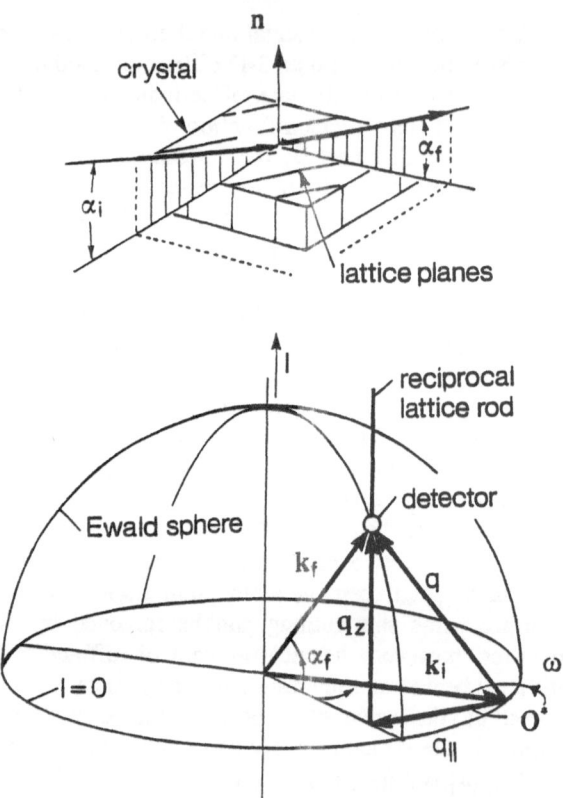

Fig. 1
Diffraction geometry at grazing incidence X-ray diffraction. a: At in-plane diffraction the angles α_i and α_f are near the critical angle of total external reflection. b: Illustration of the z-axis geometry. The intensity along the rods in reciprocal space is scanned by rotating the crystal around the surface normal.

lattice. The intensity along the rod is measured in this way in equidistant steps in q_z, where the detector position has to be readjusted for each scan. Due to the continuous intensity distribution along q_z the detailed knowledge of the detector resolution function is a prerequisite for the proper derivation of the experimental structure factor intensities $|F^{obs}|^2$ [7]. The present diffractometer allows a maximum exit angle of about 70° which corresponds to an momentum transfer q_z of about 0.7-0.8 Å^{-1} (omitting the factor 2π).

The fig. 2 shows in the upper panel a crystal model in an a section normal to the surface. We assume the bulk crystal structure to extend to infinity along the a- and b-direction, but not along the c- direction, where the periodicity is truncated by the surface. In addition, the surface structure is reconstructed by dimerisation of the top layer atoms as it is found for Si(001) and Ge(001) leading to a doubling of the bulk lattice periodicity a_0. The lower panel shows the corresponding reciprocal space. In

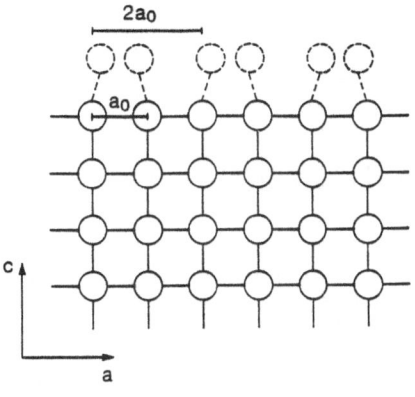

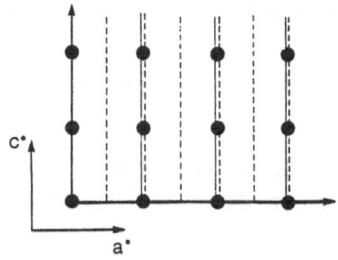

Fig. 2
a: Side view of a surface model with a 2x1 reconstruction in the topmost layer. b: Corresponding reciprocal lattice. The dots represent Bragg points of the volume structure, the dotted lines represent the lattice rods of the 2-D surface structure and the solid lines represent the crystal truncation rods (CTR).

comparison with the analysis of three dimensional periodic crystals, where only the intensity at the bulk reciprocal lattice points (filled circles) is measured, two important modifications arise from the truncation of the bulk crystal and the surface reconstruction. Due to the crystal truncation there are rods of intensity along the c*-direction. They pass through the bulk reciprocal lattice points and usually are called crystal truncation rods (CTR). They are not related to the surface reconstruction which itself gives rise to superlattice rods along c* shown as dashed lines. Due to the $2a_0$ superperiodicity of the surface structure the superlattice rods are found at half order positions along a*.

The intensity $\left| F^{CTR}(hkl) \right|^2$ along the CTRs can be calculated analytically by summing up the scattering amplitudes of the unit cells. We assume the bulk crystal structure to extend to infinity along c, layer relaxation is not included here for simplicity. With an absorption coefficient μ we obtain:

$$F^{CTR}(hkl) = \sum_{n_3=0}^{-\infty} F(hkl) \cdot e^{2\pi i n_3 \cdot c - \frac{n_3 \cdot c}{\mu}} \tag{1}$$

where $F(hkl)$ is the structure factor of the unit cell. For the limiting case of negligible absorption and extinction corresponding to the large penetration depth of the x-rays ($\mu \approx 1\,\mu m$) one obtains after summation [8,9]:

$$\left|F^{CTR}(hkl)\right|^2 = \left|F(hkl)\right|^2 \cdot g(h) \cdot g(k) \cdot \frac{1}{4\sin^2(\pi l)} \qquad (2)$$

where g(h) and g(k) are the Laue functions which may be represented by delta functions. It should be noted that equation 2 is only valid away from the bulk reflections given by $l \neq 0, \pm 1, \pm 2,...$ Along the CTRs there is a steep intensity variation due to the interference between the amplitudes scattered by all crystal layers. A simple estimation shows that the ratio between a bulk Bragg-reflection intensity and the intensity at the so called 'anti'-Bragg point at $l=(2n+1)/2$ is in the order of 10^5. In contrast, along superlattice rods only a few atomic layers along the c-axis contribute to the scattering amplitude, $F^{SURF}(hkl)$ leading to a slowly varying intensity along c*. At integer order positions along a* there is interference between the scattering amplitude of the substrate, $F^{CTR}(hkl)$, and the reconstructed surface structure, $F^{SURF}(hkl)$ as illustrated in fig. 1. The intensity analysis of the truncation rods therefore allows the determination of the registry between the substrate and adsorbate layer by

$$F^{TOT}(hkl) = F^{CTR}(hkl) + F^{SURF}(hkl) \cdot e^{i\varphi} \qquad (3)$$

where the phase factor $e^{i\phi}$ contains the relative position of the surface structure relative to the origin of the perfectly truncated (1x1) substrate crystal [8,9]. The possibility to extract from the known bulk structure factors phase information about the surface structure factors has been recently investigated [10]. The analysis of the superstructure reflections alone does not give any information about the position of the surface unit cell relative to the bulk and the Patterson function shows only those features deviating from the mean (1x1) structure.

2.2. Patterson function and Fourier synthesis of partial data sets

The Patterson function shows the inter atomic vectors and their weight [11,12]:

$$\begin{aligned} P(u,v,w) &= \int \rho(x,y,z) \cdot \rho(x+u, y+v, z+w) \cdot dx \cdot dy \cdot dz \\ &= \sum_{hkl} 2\left|F(hkl)\right|^2 \cdot \cos 2\pi(hu + kv + lw). \end{aligned} \qquad (4)$$

The first step in the analysis is usually the calculation of the Patterson function with in-plane superstructure reflections to derive a model for the projection of the structure. The use of superstructure reflections only results in a partial Patterson function which is the autocorrelation function of the difference structure to the average (1x1) structure [12]. Writing the electron density as

$$\rho(x,y,z) = \langle \rho(x,y,z) \rangle_{1x1} + \Delta\rho(x,y,z) \tag{5}$$

the Patterson function can be devided into two parts because the cross terms in the convolution function vanish:

$$
\begin{aligned}
P(u,v,w) &= \int \langle \rho(x,y,z) \rangle_{1x1} \cdot \langle \rho(x+u,y+v,z+w) \rangle_{1x1} \cdot dx \cdot dy \cdot dz \\
&+ \int \Delta\rho(x,y,z) \cdot \Delta\rho(x+u,y+v,z+w) \cdot dx \cdot dy \cdot dz \\
&= \sum_{h_i k_i l_i} 2 \cdot |F(h_i k_i l_i)|^2 \cdot \cos 2\pi(h_i u + k_i v + l_i w) \\
&+ \sum_{h_s k_s l_s} 2 \cdot |F(h_s k_s l_s)|^2 \cdot \cos 2\pi(h_s u + k_s v + l_s w) \\
&= P_i(u,v,w) + P_s(u,v,w)
\end{aligned} \tag{6}
$$

where h_i, k_i, l_i denote integer order reflections and h_s, k_s, l_s denote superstructure reflections. In the partial Patterson function derived from the superstructure reflections also negative maxima occur as is schematically explained in Fig. 3 with an one dimensional example. The total electron density $\rho(x)_{tot}$ shown in panel (a) is composed of the sum of the electron density $\rho(x)_{1x1}$ of the underlying (1x1) bulk substrate, panel (b), and the electron density $\rho(x)_{2x1}$ representing the (2x1) reconstructed surface. The latter is shown in panel (c) assuming the dimerisation of the substrate atoms and an adsorbed atom in the centre as hashed column. The analysis of the integer order reflections only would lead to an averaged electron density with (1x1) periodicity shown in panel (d). The superstructure reflections alone lead to the difference structure $\Delta\rho(x)_{2x1}$ shown in panel e. It has a (2x1) translational symmetry with no (1x1) contribution and there are positive and negative maxima observable. In the Patterson function positive and negative maxima appear as well, where positive are due to the correlation of two positive (++) or two negative (--) difference densities and negative are due to the correlation (+-) or (-+).

In simple cases the positive maxima can still be interpreted as vectors between atoms in the superstructure unit cell. We illustrate this with the example of the Patterson functions of the clean and Cs covered Ge(100)-(2x1) surface. Two z-projected Patterson functions P(u,v) are shown for the clean (top) and Cs covered (centre) Ge(001) surface in fig. 4. In both cases a (2x1) superstructure is observed and 3 positive maxima (solid lines) are determined which are labelled 1, 2 and 3. The negative maxima appear as dashed lines. The lower part of the figure shows in a projection along the surface normal the structure model of the Ge(001)-(2x1) surface. It is characterised by the dimerisation of surface atoms (large solid circles) and by a small shift of second layer atoms (small solid circles). Further details are neglected. The vectors in the Patterson function of the clean Ge(001) surface can be correlated with the inter atomic vectors 1, 2 and 3.

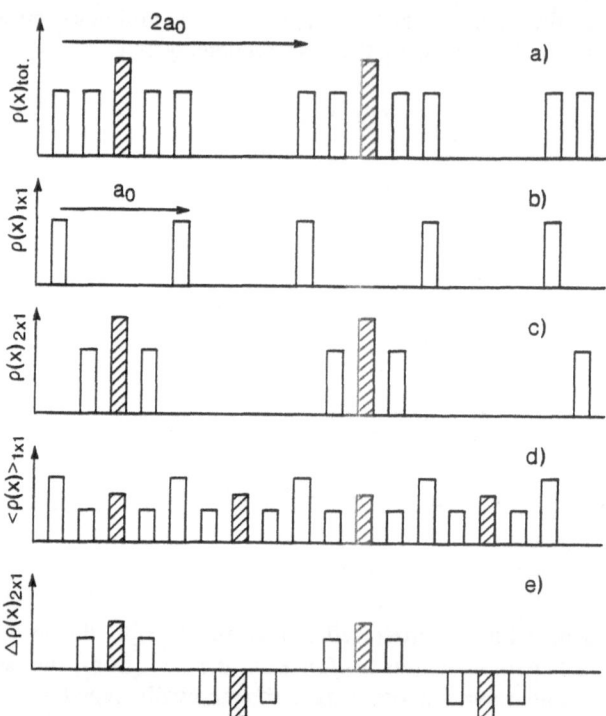

Fig. 3

Schematic drawing of the electron density along the a axis of a model surface with a 2x1 superstructure. a: total electron density, b: electron density with (1x1) periodicity from the substrate layer, c: electron density of the surface layer with (2x1) periodicity, d: averaged (1x1) structure. d: difference of the total electron density to the averaged (1x1) structure.

For the Cs covered Ge(001)-(2x1) surface the same vectors are observed, however with different relative weight. Qualitatively, the weights of the vectors 1 and 3 are enhanced relative to the clean surface, the vector 2 is nearly unchanged. For a starting model we can assume Cs to be adsorbed in the centre of the (2x1) unit cell as shown by the large dashed circle. Another hint for the correctness of this model comes from the calculation of the difference Fourier synthesis which for the projected structure is given by:

$$\delta\rho(x,y) = \sum_{hk} \left(\left| F^{obs}(hk0) \right| - \left| F^{calc}(hk0) \right| \right) \cdot e^{(2\pi i(hx+ky) - \alpha^{calc})} \tag{7}$$

In equ. 7 F^{obs} and F^{calc} represent the observed and calculated structure factors, respectively. Both, F^{calc} and the phase α^{calc} are based on a model structure. For example we show in the upper row of Fig. 5 the model structure that is used for the

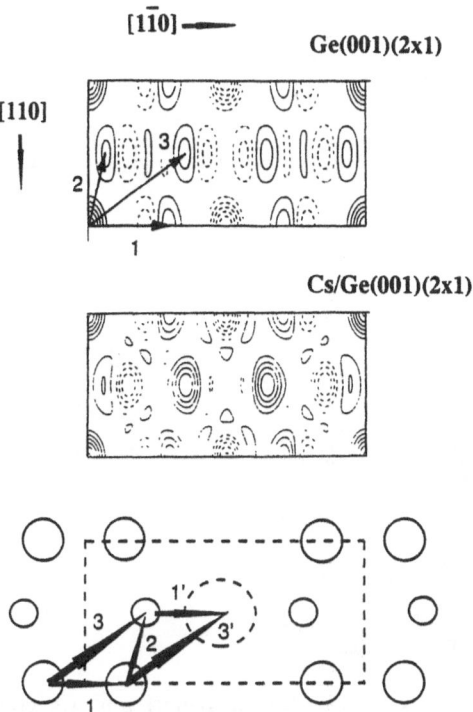

Fig. 4
Patterson function of the projection of the clean Ge(100)-(2x1) surface (upper panel) and the Cs covered surface (middle panel). Structure model of the Cs covered surface. Dimer atoms: large circles, second layer Ge-atoms: small circles, Cs: dashed circle.

Cs covered Ge(001)-(2x1) surface. Since the Cs atom is completely neglected in the model structure the agreement between F^{obs} and F^{calc} is bad which is expressed by the weighted residuum (R_W) of 41.3% [4]. The corresponding difference Fourier synthesis $\delta\rho(x,y)$ is shown on the right side. Here, an intense positive peak at (1/2,1/2) within the reconstructed unit cell is observed which indicates that electron density has to be added to correctly model the surface structure. Including Cs into the model structure (large hatched circle in the second row) leads the agreement parameter R_W to drop to 4.6% indicating a good fit between observed and calculated structure factors. Correspondingly, $\delta\rho(x,y)$ does no longer show significant maxima above the noise level even if the distance between contour lines is reduced by a factor of five as compared to the first calculation of $\delta\rho(x,y)$. Finally, in the lower part of fig. 5 the Fourier synthesis defined by

$$\rho(x,y) = \sum_{hk} F^{obs}(hk0) \cdot e^{2\pi i(hx+ky)}$$

(8)

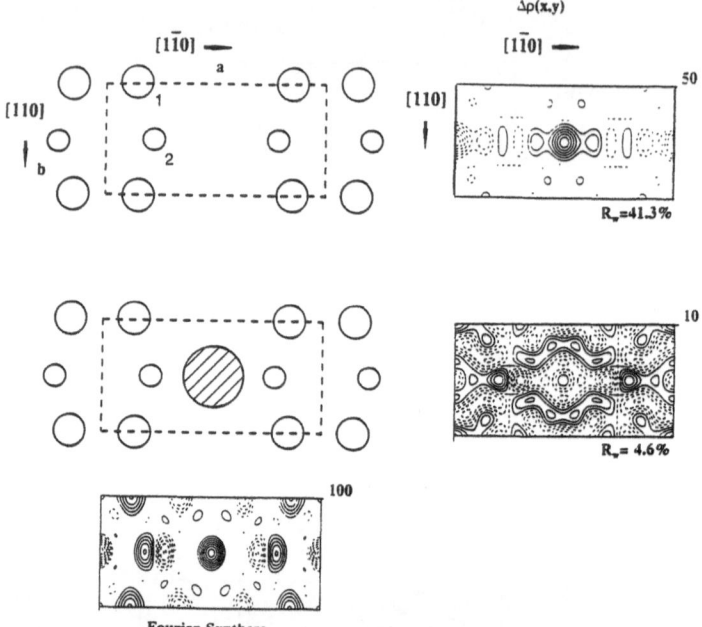

Fig. 5
Left side: top view of the clean (upper panel) and Cs covered (middle panel) Ge(100)-(2x1) surface. Large circles: dimer atoms, small circles: second layer Ge-atoms, hatched circle: Cs. Lower panel: Fourier synthesis of superstructure reflections. Right side, upper panel: difference Fourier map with calculated intensities of the clean surface; lower panel: difference Fourier map (5 times enhanced) with calculated intensities of the final model.

is shown, where the positive maxima can be clearly identified with the Ge-dimer atoms, the Ge atoms in the second layer and the adsorbed Cs atom. It should be noted that apart from truncation errors the positions of the atoms contributing to the (2x1) superstructure are given by the positive maxima in the difference density. This is also evident by comparison with Fig. 5.

However, in more complicated structures like the InSb($\overline{1}\overline{1}\overline{1}$)-(3x3) surface [5] with a large number of atoms contributing to the superstructure positive and negative difference densities may overlap which makes the analysis of the projected Patterson function P(u,v) impossible. Therefore in these cases the measurement of reflections with q_z as large as possible is required in order to obtain a reasonable model for the structure refinement. The 3-D surface structure analysis with the out-of-plane data is discussed in the next chapter using the Cs/Ge(001)-(2x1) and the InSb($\overline{1}\overline{1}\overline{1}$)-(3x3) surfaces as examples.

3. Analysis of the out-of-plane data

3.1. Structure refinement of Cs/Ge(001)-(2x1)

As a first example we consider the Cs/Ge(001)-(2x1) superstructure mentioned above. Starting from the projected structure the z-coordinates of the atoms are refined with the out of plane structure factor intensities $|F_{hkl}|^2$. The Fig. 6 shows as filled circles the structure factor intensities measured along the (1/2 0), (3/2 0), (1/2 1) and the (3/2 1) rod. The oscillating intensity distribution along q_z is indicative that several atomic layers contribute to the superstructure. This has also been observed for the clean Ge(001)-(2x1) surface [3]. The solid lines represent the best fits to the measured data. A weighted residual of $R_w=10.2\%$ and a GOF of 1.1 is obtained, the latter indicating that the agreement between observed and calculated intensities is within the experimental uncertainty.

The Fig. 7 shows the Fourier Synthesis of the superstructure in two sections perpendicular to the [110] direction (see Fig. 5), the first at y=0 (left) containing the

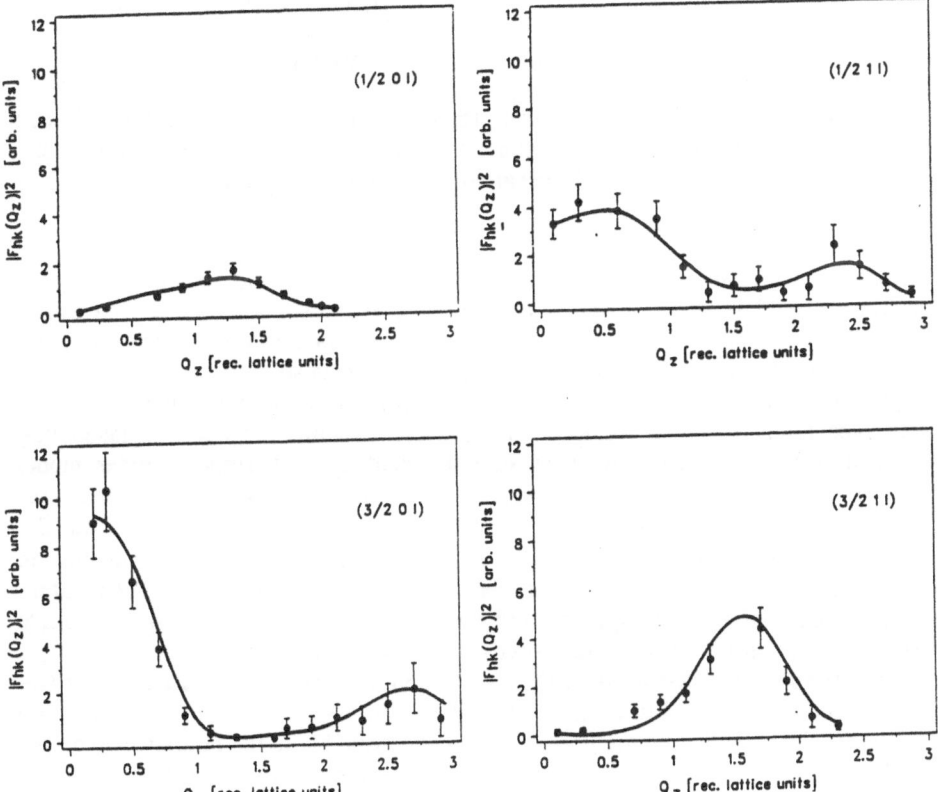

Fig. 6
Structure factor intensities of 4 superlattice rods versus $q_z=lc^*$. The solid line is the calculated intensity of the best fit model.

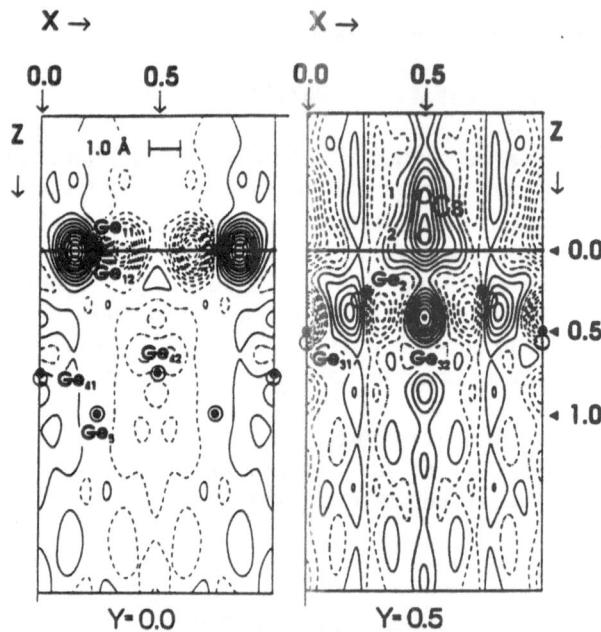

Fig. 7
Electron density map in the X-Z plane for two sections at Y=0 and Y=1/2. Open circles indicate the position of the atoms as determined in the structure refinement, full circles give the position of the Ge atoms in the bulk (1x1) structure.

Ge-dimer atoms and the second at y=0.5 (right) showing the adsorbed Cs- atom above the z=0 level which is indicated by the horizontal solid line. The open circles indicate the Cs- and Ge-atoms as derived by the least square fit of the data, the solid circles represent the positions of the bulk unreconstructed structure.

Several results are important. At first, large disorder along the surface normal is observed for both, the Ge-dimer atoms and the Cs-atom. This can be taken into account with about the same level of agreement by a dynamic disorder model allowing for anisotropic vibrations or by a static disorder model using split atoms. The latter is shown in Fig. 8 and corresponds to a structure model, where the Ge-dimer axes are randomly inclined by about ±17° to the surface and the Cs-atoms are adsorbed in two different height levels separated by about 1.4 Å. Temperature dependent measurements might rule out the unlikely possibility of a large amplitude thermal vibrations of the Ge-dimer atoms and the Cs-atoms.

The second important result is the observation of large shifts of deeper layer Ge-atoms in the second and third layer labelled by Ge_2, Ge_{31} and Ge_{32}. The shifts are directly evident by the intense positive maxima observable in the Fourier-map on the right side of Fig. 8. The atoms Ge_2 and Ge_{32} are the nearest neighbour atoms to the Cs adsorbate, within the dynamical model the (average) bond lengths were determined to 3.61(20) Å and 4.01(15) Å which can be compared with Cs-Ge bond lengths found in bulk CsGe-structures (3.58 Å) [13] and with the sum of the covalent radii of Cs and Ge (3.91 A).

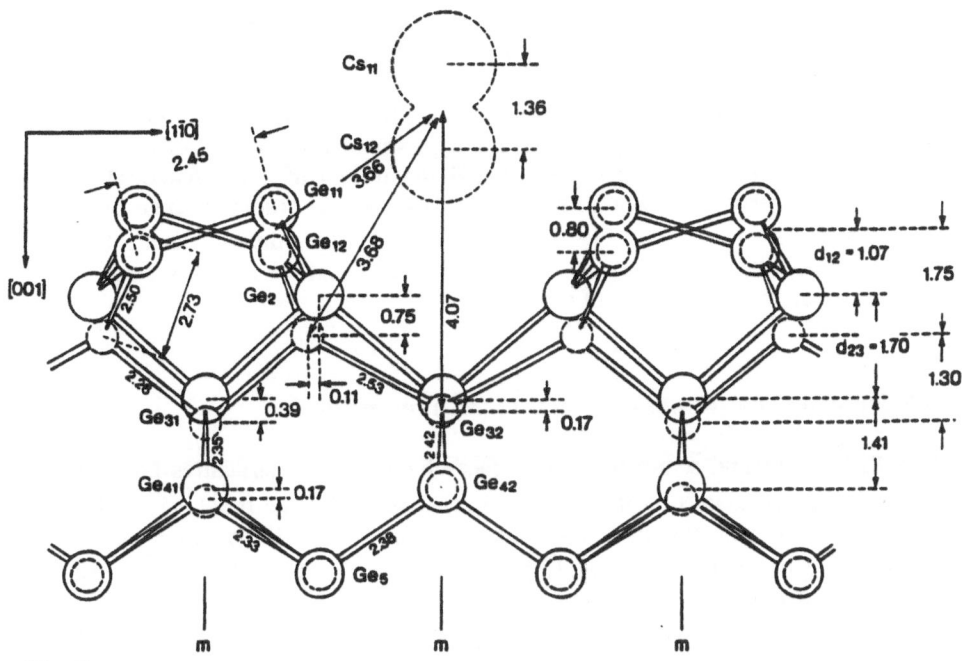

Fig. 8
Side view of the structure model. The random orientation of the asymmetric dimers
is taken into account by split positions of the dimer atoms and in the substrate layers.
Cs is adsorbed in two different heights probably due to the asymmetry of the dimers.

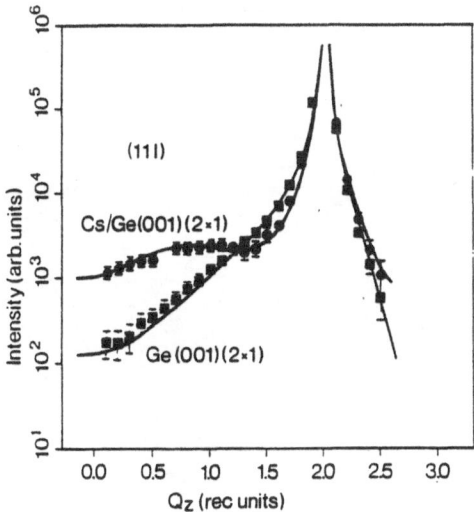

Fig. 9
Intensities of the truncation rod (11l) of the clean and Cs covered surface. The solid
line is the calculated intensity of the best fit model.

The analysis of the integer order CTR's must lead to results consistent with that of the superstructure reflections. To the intensity of the CTR contributes, according to chapter 2.1, the whole semi-infinite crystal. In the present analysis for the clean and the Cs-covered Ge(001)-(2x1) surface three Ge-layers above the truncated bulk structure were allowed to relax. Within the experimental uncertainty of the analysis of the clean Ge(001)-(2x1) surface [3] and the results of the Cs/Ge(001)-(2x1)-structure the intensity variation along the (111)- and the (10l)-rod could be fitted consistently. This is shown in Fig. 9 for the (11l) rod, where the calculated $|F_{hkl}|^2$ are represented by the solid lines. It should be noted that the intensity range to be fitted extends over three orders of magnitude.

3.2. Structure analysis of the InSb($\overline{111}$)-(3x3) reconstruction

The III/V compound semiconductor surfaces have acentric structures and exhibit different reconstructions on the (111) and ($\overline{111}$) surfaces. The InSb(111) surface is terminated by In and exhibits a (2x2) reconstruction for which the vacancy buckling model could be confirmed using LEED [14] and XRD [15]. The ($\overline{111}$) surfaces

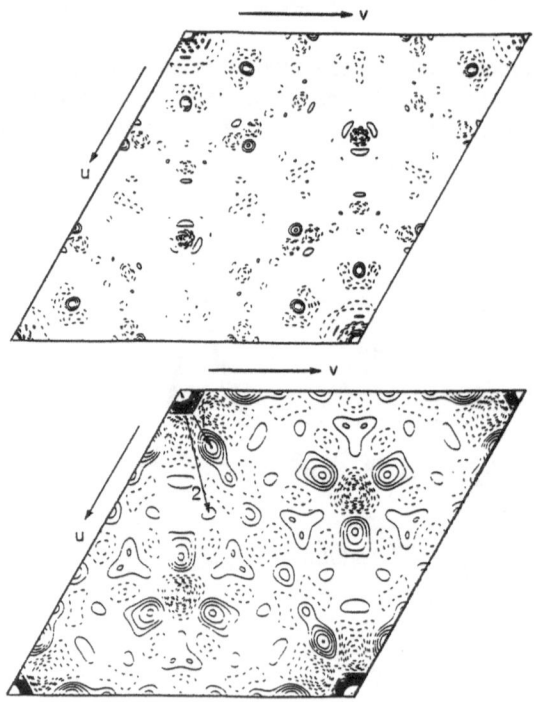

Fig. 10

Upper panel: Patterson function of the projected structure of InSb($\overline{111}$)-(3x3) obtained from the 'in-plane' superstructure reflections. Lower panel: section of the Patterson function at w=0 obtained from the complete data set.

exhibit more complicated diffraction patterns. For InSb a (3x3) reconstruction is observed, for GaAs a number of different structures depending on the As coverage and the preparation conditions have been found [16,17]. None of these structures could be determined up to now.

Careful preparation by Ar^+-ion bombardment and annealing to 400 K followed by slow cooling to room temperature is necessary to obtain a well ordered (3x3) superstructure. In total 70 in plane and 207 out of plane symmetry non-equivalent reflections were collected at the wiggler beam line W1 of the Hamburger Synchrotron Strahlungslabor (HASYLAB). It turned out that the measurement of a large out of plane data set was essential to obtain a starting model for the structure refinement. The importance of the out of plane data set is demonstrated in Fig. 10 which compares the Patterson function of the projected structure P(u,v) shown in the upper part with the Patterson section P(u,v,w=0) shown in the lower part, where two inter atomic vectors are indicated by the labels 1 and 2. The interpretation of these vectors which do not appear in the projected Patterson function together with STM measurements [5] was the key to solve the structure. The in plane vectors 1 and 2 can be related to inter atomic vectors within a six member ring shown in fig. 11. The assignment of the atoms to In and Sb was made on the basis of the bonding angles (sp^2 hybridisation of In and p-type bonding of Sb). The six-member rings are located above a complete InSb-double layer. The rings are centred around an Sb atom and are randomly oriented. This is shown schematically in Fig. 12. A small fraction ($\approx$7 %) of trigonal rings is also present. The random orientation and the fraction of elliptic and trigonal rings is consistent with STM observations [5]

4. Comparison with LEED

LEED and x-ray diffraction are both diffraction techniques operating in the same range of wave lengths and should provide similar information in principle. Each method has it specific advantages and disadvantages and there is a large overlap in subjects of interest in surface science where both methods can be applied. There are, nevertheless, specific fields of application of LEED and X-ray diffraction resulting from the difference in cross sections of electrons and photons. Of course, the main difference is that a dynamical calculation is required in the case of LEED while for X-rays the kinematical theory is sufficient. The dynamical theory leads to lengthy calculations and this has limited the structure analysis by LEED to relative simple cases in the past. The recent progress made by the development of Tensor LEED [18,19] and optimisation procedures [20] has overcome these problems to a large extent and there are now fairly complex structures solvable by LEED.

The difference to X-ray diffraction therefore lies mainly in the methods to derive a suitable structure model from the experimental data. With X-rays the Patterson function can be used, while for LEED a comparable procedure has not yet been developed. As we have shown above the interpretation of the Patterson function of the 'in-plane' data is straight forward for simple structures, but for complex structures the full data set and the 3-D Patterson function is required. It may be even then difficult to derive a structure model from the Patterson function. The interpretation of the Patterson function is further complicated by the use of partial

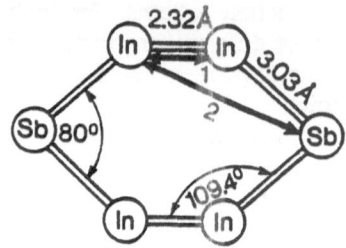

Fig. 11

Model of the elliptic ring of 4 In and 2 Sb atoms. Two inter atomic vectors observed in the section of the Patterson function but not in the projected structure are indicated in the figure.

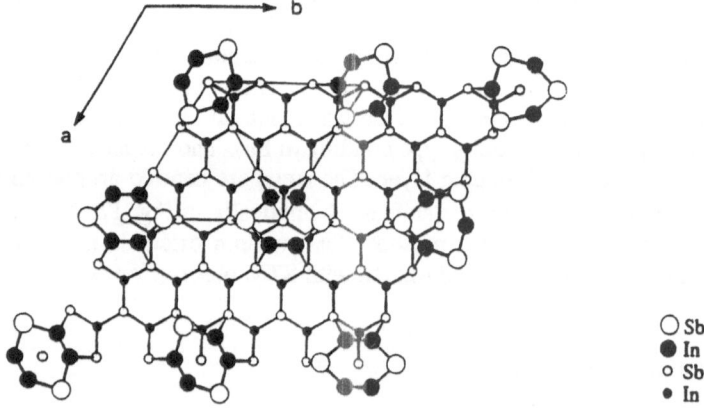

○ Sb
● In
○ Sb
• In

Fig. 12

Structure model of the (3x3) reconstruction of InSb($\overline{1}\overline{1}\overline{1}$). One (3x3) unit cell is indicated by solid lines. Two types of 6-atom rings occur in random orientation. The rings are centred above an Sb atom.

data sets as discussed above. Other methods of structure determination routinely used in 3-D X-ray crystallography are the so called direct methods. These methods are not yet fully developed for the application with surfaces but there does not seem to exist a principal limitation.

While in the data analysis X-rays have a clear advantage over LEED there are experimental requirements which gives preference to one method in certain cases. Besides the fact that XRD is not bound to UHV conditions and that isolator surfaces can be studied without the charging effects occurring with LEED there exist some specific differences in the experimental requirements for both methods:

Sample quality: With LEED usually a back scattering geometry is applied which makes it less sensitive to surface defects because the integral intensity is required for structure analysis. The beam profiles are narrower in the back scattering geometry and the integral intensity can be easier measured. With grazing incidence X-ray diffraction the diffraction geometry is comparable to RHEED and large flat and homogeneous surfaces are required. The terrace sizes should be large because the reflections become broad at rough surfaces which lowers the peak intensity. With X-rays the illuminated area of the surface is usually in the order of 10 mm^2 though intensive X-ray sources at the ESRF in Grenoble provide now diffractometers with a much smaller focus. LEED does not require a large area, typically 1 mm^2 is illuminated.

Measuring time: With X-rays the duration of the measurements is typically from several hours to one or two days. During this time the surface should be stable and inert against contamination. The duration of the x-ray measurements sets certain limits to the class of adsorption systems which can be studied. The measurement with LEED can be very fast, in the order of minutes when the time to prepare the surface is not counted. The problem of contamination is, on the other hand, less severe with x-rays than with LEED as no electron gun is required which is usually a source of CO. Furthermore, photon induced desorption can usually be neglected while desorption induced by electrons is a severe problem for many adsorbate systems. By using low primary beam currents and channel plates this problem can be overcome but requires special equipment.

Penetration depths: Electrons have a elastic cross section in the order of 1 Å^2 while the cross section for x-rays is 10^{-4} to 10^{-6} times smaller. Therefore the penetration depth for x-rays is large which allows to study interfaces. There exists also the possibility to investigate liquid-solid interfaces with x-ray diffraction. With LEED the penetration depth is limited to about 10 Å.

Disordered surfaces: Besides the strong multiple scattering effects dominant in LEED which is usually considered as disadvantage there results a specific application from the large cross section in LEED. The local geometry of point defects can be studied by the intensity analysis of the diffuse background (DLEED). For X-rays the background from the bulk is superimposed on the diffuse scattering from the surface which make it impossible to analyse the diffuse intensity. There is, however, for XRD the possibility to analyse the intensity of the truncation rods which allows to study disordered adsorbate layers as well. The condition for DLEED as well as for XRD is that localised adsorption sites exist.

Resolution: The lateral resolution in the analysis of defect distributions is in the order of μm for X-rays when 'in-plane' reflection profiles are measured. For out-of-plane measurements the resolution is lower and becomes comparable to that obtained with a high resolution LEED system (SPALEED). The momentum transfer normal to the surface is for X-rays limited to $|k|$, see fig. 1, due to the fact that either the incident or the diffracted beam should be grazing to the surface in order to minimise the background from the bulk. From that follows that the resolution in z-direction is in principle smaller than in LEED where the momentum transfer limit is $2|k|$.

References

[1] R. Feidenhans'l, Surf. Sci. Rep. 10, 105, (1989).

[2] I. K. Robinson, in: Handbook of Synchrotron radiation, Vol. 3, Eds.: G. S. Brown and D. E. Moncton, Elsevier, Amsterdam (1991).

[3] R. Rossmann, H. L. Meyerheim, V. Jahns, J. Wever, W. Moritz, D. Wolf, D. Dornisch and H. Schulz, Surf. Sci. 279, 199-209 (1992)

[4] H. L. Meyerheim and R. Sawitzki. Surf. Sci. Letters, 301 (1994) L203.

[5] J. Wever, H. L. Meyerheim, W. Moritz, V. Jahns, D.Wolf, H. Schulz, L. Seehofer and R. L. Johnson. Surf. Sci. Letters, submitted.

[6] F. Kretschmar, D. Wolf, H. Schulz, H. Huber and H. Plöckl, Z. Krist. 178 (1987) 130.

[7] C. Schamper, H. L. Meyerheim and W. Moritz, J. Appl. Cryst. 26, 687-696 (1993).

[8] I. K. Robinson, Phys. Rev. B33 (1986) 3830.

[9] E. Vlieg, J. F. Van der Veen, S. J. Gurman, C. Norris, J. E. Macdonald, Surf. Sci. 232 (1990) 417.

[10] S. Ferrer, Surf. Sci. Letters 286 (1993) L564.

[11] B.E. Warren, "X-ray diffraction", Addison Wesley, Reading, Mass. (1969).

[12] M. J. Buerger, "Vector space and its application in crystal structure determination", Wiley, New York (1959).

[13] J. Witte, H.G. v. Schnering and W. Klemm, Z. anorg. Chemie 327 (1964) 260.

[14] S. Y. Tong, G. Xu and W. N. Wei, Phys. Rev. Letters 52 (1984) 1693.

[15] J. Bohr, R. Feidenhans'l, M. Nielsen, M. Toney, R. L. Johnson, and I. K. Robinson, Phys. Rev. Letters 54 (1985) 1275.

[16] K. Jakobi, C. v. Muschwitz, and W. Ranke, Surf. Sci. 82 (1979) 270.

[17] R. D. Bringans, R. Z. Bachrach, Phys. Rev. Letters 53 (1984) 1954.

[18] J. B. Pendry, K. Heinz and W. Oed, Phys. Rev. Letters 61 (1988) 2953.

[19] P. J. Rous, Progress in Surface Science, 39 (1992) 3.

[20] M. A. van Hove, W. Moritz, H. Over, P. J. Rous, A. Wander, A. Barbieri, N. Materer, U. Starke ans G. A. Somorjai, Surface Science Reports, 19 (1993) 191.

Surface Structural Determination by VLEED Analysis

S.M. Thurgate, Chang Sun and G. Hitchen*

School of Mathematical and Physical Sciences
Murdoch University
Western Australia, 6150
Australia

* CSIRO Floreat Park Laboratories
Underwood Avenue, Floreat Park
Western Australia, 6014
Australia

Abstract LEED intensity curves contain fine structure at low energies due to the interaction of pre-emergent beams with the surface potential barrier. Generally these regions of the spectra are avoided due to the difficulty in obtaining reliable data and fitting the spectra. We have recently developed computer programs that model the effect of the surface barrier potential on the I/V curves. These differ from our previous programs in that they interface to the Van Hove-Tong LEED package and so can be used to model the I/V curves from surfaces with more than one atom per unit cell. We have used these to model the surface reconstructions that occur when oxygen chemisorbs on Cu(001). We have found good agreement with the missing row model of Zeng et al [1]. We have also found strong evidence to suggest a c(2×2) structure intermediate between the clean surface and the missing row reconstruction.

1. Introduction

The VLEED region, below 40 eV, is not often used in determining surface structure. This is despite a number of advantages that accrue to analysis of data from this energy range. In general, fewer beams and fewer phase shifts must be considered and peaks occur with a greater density [2], hence computation time is considerably reduced. The scattering from light adsorbates is often stronger at these energies, and this may led to improved knowledge of their position. These advantages are offset to some extent however by a number of difficulties. The imaginary part of the inner potential is typically -0.5 eV compared to -5.0 eV in the regular LEED range. This can cause convergence problems in some parts of the LEED calculations. Changes may also occur in the inelastic potential that are due to the excitation of electrons between bands. Such effects have been studied in Ru [3]. The surface potential barrier can induce rapidly varying fine structure to appear in the spectra that are not easily accounted for. This range of energies is also difficult to make measurements in because the electron beam is easily deflected by any stray field.

In this work, we have chosen a system where the I/V curves are dominated by fine structure induced by the surface barrier potential. Such fine structure is a series of rapidly varying intensity oscillations that are seen in LEED I/V curves. These were first

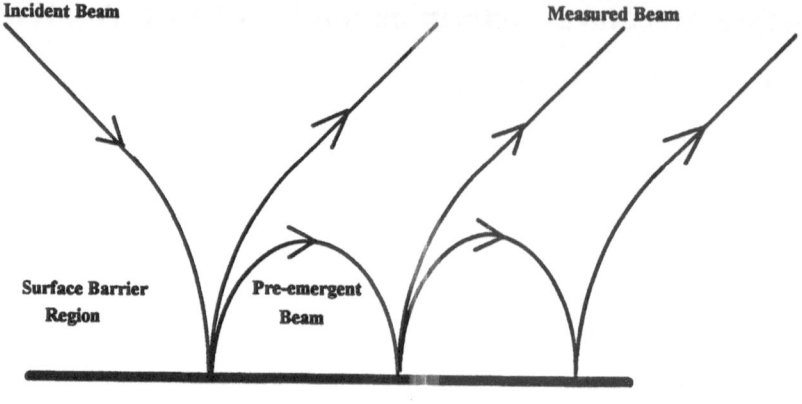

Figure 1. The mechanism by which fine structure peaks are formed is represented
diagrammatically. The incident beam is diffracted into a pre-emergent beam,
which has insufficient momentum perpendicular to the surface to overcome the
surface potential barrier and so diffracts from the substrate again. Part of this
amplitude can diffract back into the direction of the measured beam, giving rise
to the fine structure fringes.

correctly identified by McRae [4] as being due to the interaction of pre-emergent beams
with the surface potential barrier. Pre-emergent beams are those that have emerged from
the crystal but have insufficient momentum perpendicular to the surface to escape to the
vacuum. The surface potential turns them back to the substrate where they diffract
again. Some amplitude can diffract into the direction of the measured beam where it can
interfere with the measured beam amplitude. This process can repeat itself, with many
substrate scatterings adding to the amplitude of the measured beam. The process is
illustrated diagrammatically in figure 1. The long range dependence of this potential on
1/z (the image potential) means that the peaks caused by single emergence converge to a
limit, forming a Rydberg like series. The same physical process is responsible for image
like states seen in inverse photoelectron spectra [5]. Each pre-emergent beam is
potentially capable of contributing to the observed fine structure. The degree to which it
will produce visible modulations in the observed intensity and the exact form of these
modulations is dependent on a number of factors. These include:

- The relative amplitudes of the pre-emergent beam and the observed beam (usually
  the specular beam). These beams must have similar amplitudes for there to be any
  observable modulation in the intensity [6].
- The "emergence speed". This term was proposed by Gaubert et al [7] to describe
  the rate at which a pre-emergent beam emerged from the crystal with increasing
  energy. This is a function of the crystal geometry. In general, those beams that are
  anti-parallel to the observed beam have the slowest emergence speed. These beams
  produce fine structure that is spread over the largest energy range.
- The emergence speed and the shape of the potential barrier determine the observed
  peak spacing. The 1/z dependence produces the Rydberg like series of fine fringes.
  The width of each peak, and so the degree of overlap between peaks, depends on

the lifetime of each state, which is determined by the inelastic scattering in the barrier region. An analysis of these effects has been done by Echenique et al [8]. The situation is clearly complex as the inelastic potential is also a function of distance from the substrate.

- The distance between the surface layer atoms and the classical turning point in the potential barrier. This determines the phase change between the measured beam and the pre-emergent beam. This phase change determines the position of peaks within the range of the fine structure.
- The emergence energy of the pre-emergent beam. This is a function of the two dimensional crystal geometry and is independent of the barrier shape or height. The point here is that whatever energy the electron gains as it enters the crystal, it looses as it leaves. The dependence of the emergence energy on the two dimensional crystal geometry can be assessed from the Ewald construction. The emergence energy corresponds to the point where the sphere first touches the rod in reciprocal space. The fine structure features converge in a Rydberg like series to this energy. They extend to energies less than the emergence energy by an amount determined by the emergence speed and the height of the potential barrier.

It is clear from the above discussion that these features are well understood. The question is then to what extent we can learn something new from the measurement of these features. We can only measure the variations in intensity with energy and these are determined by the geometry of the surface and the spatial variation in the surface potential barrier. In principle, if we had an adequate theory describing the scattering from the substrate and the potential barrier then a knowledge of one of these might make it possible to determine the other.

To this end, there have been a number of attempts to determine the shape of the surface potential barrier from a number of clean surfaces where the surface geometry is well known. These include Cu(001) [9], Cu(111) [10], W(001) [11] and W(110) [12]. There have also been several attempts to reconcile measured I/V fine structure profiles from reconstructed surfaces using known surface structures and assumed surface potential barriers. Jones and Jennings [13] have reviewed fine structure analysis of clean and chemisorbed surfaces.

In this work, we have attempted to determine the surface structure of O on Cu(001) by fitting a theoretical model to the observed data. The effect of oxygen on Cu has been extensively studied for the past three decades. There have been a number of dramatic advances in the past five years, and now it seems that this system is understood. Besenbacher and Norskov [14] have recently produced an excellent review that explains the similarities between oxygen adsorption on Cu, Ni and Ag. Our analysis confirms the missing row model of Zeng et al [15], but also suggests a phase that lies between the clean surface and the $(\sqrt{2} \times 2\sqrt{2})R45°$ O surface. This seems to be the c(2×2) which has been largely discounted by investigators following the work of Mayer et al [16] which showed that the increase in intensity with oxygen exposure of the 1/4 order spots matched the rate of increase in the 1/2 order spots.

2. Experiment

In general, fine structure occurs at very low energies, typically less than 40 eV and often less than 10 eV. The spacing between peaks can be as little as 20 meV. Hence gathering this data requires special care. One must ensure that the diffraction conditions are well known because at such low energies it is likely that the electron paths will not be straight lines due to stray fields entering the field free regions. To take this data, we developed a specialised electron spectrometer. This had moderately high resolution (50 meV) but could be rotated about the crystallographic axes, allowing the azimuthal angle to be varied continuously [17]. From this variation, the symmetries of the crystal could be exploited to unambiguously determine the incident direction of the electron beam and the contact potential difference between the spectrometer and the sample [18]. This instrument has been described previously. We used it to measure the I/V spectra shown. The spectrometer was housed in a UHV system with a base pressure of 5×10^{-10} Torr. High purity oxygen was admitted via a leak valve into the system during oxygen exposures. A conventional LEED system was used to judge the order of the surface. The Cu sample was cut from a single crystal, aligned to with 0.5 degrees with a Laue diffractometer and mechanically polished and electro polished before entry into the vacuum system. The sample was prepared by repeated ion bombardment (Ar^+ Ions at 500 eV) and annealing until Auger scans revealed no surface contamination and LEED indicated a well ordered surface.

3. Theory

We have previously published a number of papers where we have determined the shape of the surface potential barrier by fitting the I/V spectra to a theoretical model using the known surface geometry of several simple clean surfaces [9,10,13]. The code used to do this was developed for a single atom per unit cell and so was unsuitable for systems where the surface had reconstructed or where there was more than one atom per unit cell. In order to extend our analysis to such complex systems we have modified the multi-atom LEED package of Tong and Van Hove [19]. Two modifications were necessary to do this. The first was to modify the existing code that calculated the layer scattering matrices to make it suitable for such low energies. This had been done previously by Lindroos [2] and we made use of his modified code. He used Kambe's method for summing the scattering from atoms in the layer. We wrote our own code for calculating the reflection and transmission matrices for the barrier and made it seamless with the Van Hove package. Malmstrom and Rundgren [20] have previously published a package for calculating these reflection and transmission coefficients but it used different conventions for the zero of potential and different external representations of energy and distance which made it somewhat cumbersome to use with the Van Hove package.

The package by Malmstrom and Rundgren calculates the reflection and transmission coefficients of a barrier by integrating the image like part of the potential analytically and numerically integrating a selvedge potential that smoothly joins the image like potential to the flat inner potential of the substrate [21]. The analytical integration involves the numerical evaluation of Whittaker functions. We chose to follow the techniques developed by Jennings, Read [22], and McRae [23] and numerically integrated an analytic form of the potential to large distances from the substrate. It should be noted here that all these attempts to find the effect of the surface potential barrier have assumed that the potential could be represented as a one dimensional function of distance from the substrate. We used the Runga-Kutta method to integrate the Schrodinger equation from the surface layer out to distance from the substrate that depended on the energy of the electron compared to the height of the potential barrier. When the electron was unbound and had an energy of just greater than the barrier, then we integrated out to a distance of 300 au. This had to be reduced in order to keep the integration convergent when the electron was bound, or had an energy much greater than the height of the barrier. Having calculated the wave function and its derivative after propagation through the surface potential barrier in both directions, we calculated the reflection and transmission coefficients making use of the well known analytic forms that connect these quantities. The code we used was based on code supplied with the Van Hove package.

Once the reflection and transmission coefficients were calculated, we produced reflection and transmission matrices for the surface barrier layer. In the Van Hove package, the scattering of a single layer from all incident beams into all exiting beams is represented by a square ($n \times n$) matrix, where n is the number of beams in the current set of active beams. In this convention, the barrier matrices must be diagonal as the barrier is one dimensional so a beam incident in one direction cannot be scattered into a beam in another direction. In order to save space, we represented the reflection and transmission matrices as vectors and wrote a subroutine to add the effect of the barrier to the substrate scattering. This was done using the sum to infinity that describes the effect of an additional surface layer on a substrate [19]. This sum to infinity includes all multiple scattering effects between the substrate and the barrier.

4. Analysis

4.1 The Clean Surface
In order to establish the validity of our code, we calculated the scattering from the clean surface of Cu(001), using a barrier of the form:

If $z < z_0$

$$(z) = \frac{1}{4(z - z_0)}\{1 - \exp[\lambda(z - z_0)]\}$$

else

$$(z) = \frac{-V_0}{A \exp[-B(z - z_0)] + 1} \tag{1}$$

where A and B are constants given by $B = V_0/A$ and $A = -1 + 4V_0/\lambda$, as originally proposed by Jennings and Jones [24]. We found that we were able to satisfactorily describe the fine structure features found on this surface. A comparison between the measured curve and the calculated curve is shown in figure 3. Interestingly, we found that we were able to achieve a good match with experiment using the sum to infinity rather than the double diffraction model suggested by Deitz et al [23] and previously

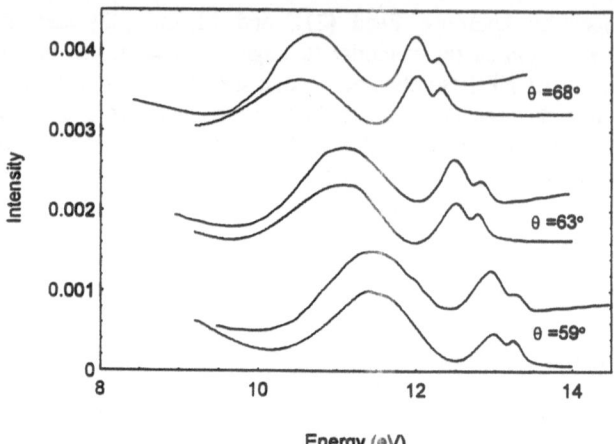

Energy (eV)

Figure 2. Comparison between the experimental curves (top) and the theoretical
curves (bottom) for the clean surface at a number of different angle sof incidence.

incorporated in our single atom code. Deitz et al had concluded that a satisfactory match
could be made between experiment and theory for Cu(001) with a model that included
only two diffraction events. This implies that the pre-emergent beam is only turned back
by the surface potential barrier once. This lead them to interpret the fine structure
features as interferences rather than surface barrier resonances. Previously, these
features had been referred to as resonances. This interpretation was further supported by
Le Bosse et al [6] in a theoretical study where they were able to show that the surface
states were so short lived that they could not reasonably be described as resonances.
Read and Christopoulos [25] have also analysed a number of surfaces where they found
that the match between experiment and theory could be improved by varying the
number of diffraction events. We assume that our model was sufficiently complete to
allow the inclusion of all necessary diffraction events, and that our model of the spatial
and energy dependence of the inelastic potential was accurate enough to produce the
correct intensity for the higher order diffraction events in the barrier region.

4.2 The Oxygen Exposed ($\sqrt{2}$×2$\sqrt{2}$)R45° O Surface

The exact structure of the oxygen exposed surface has been problematic. For many years
it was thought that there were two stable structures on the surface, a c(2×2) O and a ($\sqrt{2}$
×2$\sqrt{2}$)R45° O surface forming at higher exposures to oxygen. This view has been
revised after careful measurement that showed fractional order beams appearing in the
LEED pattern even at very low exposures [16]. LEED analysis by Zeng et al [15]
indicated that the surface reconstructs in a missing row model with every fourth row in
the (001) direction absent. This was later confirmed by STM measurements [26, 27] and
found to be consistent with high resolution EELS measurements [28]. This model
reconciled all the existing evidence, though there remains some uncertainty about the
position of the atoms in the reconstruction.

   The VLEED data we had previously collected indicated a strong dependence of the
fine structure on oxygen exposure. We attempted to model this using our multi-atom
code. The ($\sqrt{2}$×2$\sqrt{2}$)R45° O surface clearly involves two domains at right angles to each
other. We used the missing row model of Zeng et al as shown in figure 3. We found that

Figure 3. The missing row model of Zeng and Mitchell [1]. The copper atoms are the
larger spheres, while the oxygen atoms are smaller.

we could achieve acceptable agreement only when we allowed the inner potential of the
surface layer (the missing row layer) to differ from the potential of the substrate by 3.5
eV. This assumption is similar to that made by Pfnur et al [2] in their analysis of
VLEED from oxygen on Ru, where they found it necessary to use a two step model of
the surface potential barrier. It is certainly clear that the potential in the surface region
differs from the clean surface as the work function is known to change with adsorption
of oxygen. We found that the barrier origin, Z_0, also moved out to some 4.5 Bohr radii,
from its value of 3.2 Bohr radii on the clean surface. The parameters of our best fit are
shown in table one.

	Clean	c(2×2)	(√2×2√2)R45
Z_0 (Atomic units)	-3.2±0.2	-4.5±0.2	-4.5±0.2
λ (Atomic units)	0.62±0.03	0.68±0.03	0.60±0.03

Table 1. This shows the structural and surface barrier parameters used to obtain the
theoretical fits.

In fitting the data to the theory, it is clear that some quantities are not well known. In
particular, the dependence of the inelastic potential on energy is not known. In our
calculation we have assumed that it increases monotonically in a power law. We
assumed that the spatial variation in damping in the barrier region decayed like:

$$i(z) = \frac{VI}{\exp[(z-z_1)^2 / \alpha)]} \qquad (2)$$

where VI was the imaginary part of the potential inside the crystal, z_1 and α were fitting
parameters. The exact shape of the barrier is also unknown, and given the degree of
reconstruction of the surface it is surprising that a one dimensional barrier is able to do
such a good job. It is also known that the inner potential varies with energy in this
regime [29].We have assumed a constant inner potential of 11.56 eV in the substrate in
these calculations.

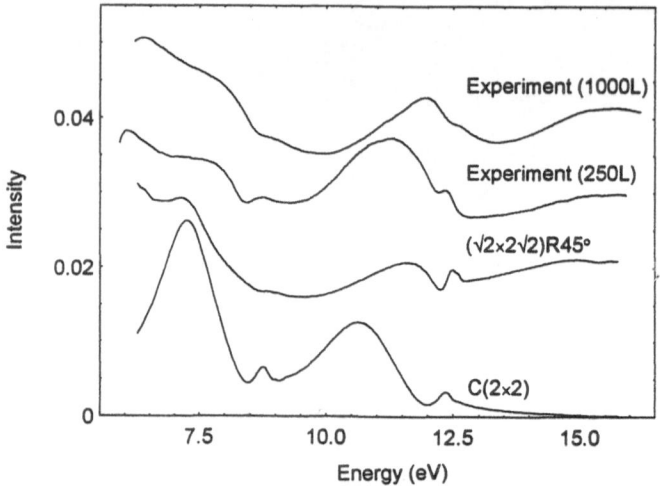

Figure 4. Comparison between the experimental curve (top two curves) for oxygen
exposures of 250L and 1000L, and the theoretical curve for the (√2×2√2)R45° O
and c(2×2) surfaces.

4.3 The c(2×2) Structure.

While the (√2×2√2)R45 structure appears to explain most of the features in the data
from surface exposed to 1000L, it does not account well for the data at 250L exposure,
as can be seen in figure 4. We attempted to fit this data with the c(2×2) model of
Lederer et al [30], which puts an oxygen atom in the four fold hollow site, 0.8 Å above
the plane of atoms. We suggest on the basis of this result that the 250L surface and the
1000L surface both have contributions from both the (√2×2√2)R45 and the c(2×2)
reconstructions.

5. Conclusions

This work illustrates that the features induced in LEED I/V curves by the surface
potential barrier can be well accounted for in a calculation that involves more than one
atom per unit cell. This should help to overcome the reluctance of some towards using
this region to make structural determinations. Indeed, we have shown that some
structural information can be derived from the shape of the image induced peaks
themselves. However, in circumstances where the shape of the surface potential barrier
is not known, the energy dependence of the inner potential is not known, and the
dependence of the imaginary potential is not known, it would be impudent to attempt to
derive too much independent structural information from the fine structure features
alone. None-the-less, the sensitivity of the technique to short range order can sometimes
be used to great benefit. Pfnur [2] et al pointed out that multiple scattering is often less
important in this region and so the VLEED can be used to examine disordered
adsorption. Indeed, we have shown in this work that the relatively disordered system of
adsorption on C(001) at room temperature without annealing can be sensibly analysed
in this way and that the VLEED indicates the existence of an intermediate c(2×2) O that
forms at exposures between the clean surface and the (√2×2√2)R45° O.

Acknowledgments

We gratefully acknowledge the generosity of Matti Lindroos for allowing us to use his programs. This project was supported by the Australian Research Council.

References

[1] H. C. Zeng and K. A. R. Mitchell, Surf. Sci. Letts, 239 (1990) L571.

[2] H. Pfnur, M. Lindroos and D. Menzel, Surf. Sci., 248 (1991) 1.

[3] M. Lindroos, H. Pfnur and D. Menzel, Phys Rev. B, 33 (1986) 6684.

[4] E. G. McRae and C. W. Caldwell, Surf. Sci., 2 (1964) 509.

[5] N. V. Smith, Phys Rev B, 32 (1985) 3549.

[6] J. C. Le Bosse, J. Lopez, C. Gaubert, Y. Gauthier and R. Baudoing, J. Phys. C: Solid State Phys., 15 (1982) 3425.

[7] C. Gaubert, R. Baudoing and Y. Gauthier, Surf. Sci., 147 (1984) 162.

[8] P. M. Echenique, F. Flores and F. Sols, Phys. Rev. Letts., 55 (1985) 2348.

[9] G. Hitchen and S. M. Thurgate, Phys. Rev. B, 44 (1991) 3939.

[10] G. Hitchen, S. Thurgate and P. Jennings, Aust. J. Phys., 43 (1990) 519.

[11] J. M. Baribeau, J. D. Carette, P. J. Jennings and R. O. Jones, Phys. Rev. B, 32 (1985) 6131.

[12] P. J. Jennings and R. O. Jones, Phys. Rev. B, 34 (1986) 6699.

[13] R. O. Jones and P. J. Jennings, Surf. Sci. Rep., 9 (1988) 165.

[14] F. Besenbacher and J. K. Norskov, Prog. in Surf. Sci., 44 (1993) 5.

[15] H. C. Zeng and K. A. R. Mitchell, Surf. Sci. Letts, 208 (1989) L7.

[16] R. Mayer, C. S. Zhang and K. G. Lynn, Phys. Rev. B, 33 (1986) 8899.

[17] S. M. Thurgate and G. Hitchen, Appl. of Surf. Sci., 24 (1985) 202.

[18] S. M. Thurgate and G. Hitchen, Surf. Sci., 197 (1988) 24.

[19] M. A. Van Hove and S. Y. Tong, Surface Crystallography by LEED. (Springer, Berlin, 1979).

[20] G. Malmstrom and J. Rundgren, Compt. Phys. Commun. 19 (1980) 263.

[21] J. Rundgren and G. Malmstrom, Phys. Rev. Letts., 38 (1977) 836.

[22] M. N. Read and P. J. Jennings, Surf. Sci., 74 (1978) 54.
P. J. Jennings, Surf. Sci., 75 (1978) L773.

[23] R. E. Deitz, E. G. McRae and C. W. Caldwell, Phys. Rev. Lett, 45 (1980) 1280.

[24] P. J. Jennings, R. O. Jones and M. Weinert, Phys. Rev. B, 37 (1988) 6113.

[25] M. N. Read and A. S. Christopoulos, Phys. Rev. B, 37 (1988) 10407.

[26] F. Jensen, F. Besenbacher and I. Stensgaard, Phys. Rev B, 42 (1990) 9206.

[27] C. Woll, R. J. Wilson, C. Chiang, H. C. Zeng and K. A. R. Mitchell, Phys. Rev. B, 42 (1990) 11926.

[28] M. Wuttig, F. Franchy and H. Ibach, Surf. Sci. Letts, 224 (1989) L979.

[29] P. J. Jennings and S. M. Thurgate, Surf. Sci., 104 (1981) L210.

[30] T. Lederer, D. Arvanitis, G. Comelli, L. Toroger and K. Baberschke, Phys Rev B, **48,** 15 390 (1993).

Surface Structure Investigation with Ion Scattering and Scanning Tunneling Microscopy at Oxygen and Nitrogen Covered Cu₃Au Surfaces

Horst Niehus[1], **Matthias Voetz**[2], **Carlos Achete**[3], **Karina Morgenstern**[2], **George Comsa**[2]

[1]Humboldt-Universität zu Berlin, FB Physik/Atomstoßprozesse,
  Invalidenstr. 110, D10115 Berlin, Germany
[2]Forschungszentrum Jülich, Institut für Grenzflächenforschung und
  Vakuumphysik, KFA Jülich, D5170 Jülich, Germany
[3]permanent address: COPPE-UFRJ Rio de Janeiro, CP68501,
  21945 Rio de Janeiro RJ, Brazil

Abstract

The investigation of clean $Cu_3Au(100)$ and (110) with 180° low energy ion scattering and detection of neutrals (NICISS) shows that the surface is terminated at room temperature with the Au rich plane. The clean $Cu_3Au(100)$ surface exhibits a LEED c(2x2) superstructure and has been imaged by scanning tunneling microscopy (STM) in ultra high vacuum with atomic resolution. The measured different grey scale heights in the unit cell can be attributed to the position of Au and Cu atoms, respectively. Adsorption of atomic nitrogen starts with the formation of islands. From the NICISS data it has been concluded that after saturation with O or N the surfaces are terminated with a Cu plane. On the basis of STM and direct recoil spectroscopy (DRS) the fourfold hollow site of oxygen or nitrogen above Cu atoms has been deduced for the $Cu_3Au(100)$ surface. $Cu_3Au(110)$ is less inert and oxygen adsorption is readily achieved by gas exposure. An added row structure similar to the structure discovered at Cu(110)-(2x1) O has been found.

1. Introduction

The Cu_3Au alloy system is well studied in the past, basically because it behaves as a classical ordering alloy /1-10/. At room temperature the $Cu_3Au(100)$ surface exhibits a c(2x2) superstructure converging into (1x1) at the phase transition temperature. By following the superstructure reflex intensities vs. temperature a transition with first order kinetics is proposed. Surface segregation and the order disorder behaviour has been also studied in a low energy ion scattering spectroscopy (ISS) experiment /11,12/. We make no attempt to add any additional information to the order-disorder discussion, instead an investigation of the surface structure on an atomic scale for ordered $Cu_3Au(100)$ and (110) at room temperature and the influence of gas adsorption (here oxygen and nitrogen) will be presented. In future investigations, also other binary alloys e.g. important in heterogene catalysis may become an example for custom design surfaces and the strategy of combined investigation shown below with NICISS (180° impact collision ion scattering spectroscopy with neutral particle detection) and STM may have some model character. At Cu(110) oxygen can be adsorbed at room temperature while nitrogen has to be offered as atomic nitrogen or to be implanted first as ions and than successive annealing of the sample can result in a smoothing and ordering of the surface. Obviously pure Au is not only passive against both, oxygen and nitrogen adsorption at room temperature, but also against considerable oxide or nitride formation. In case of the $Cu_3Au(100)$ the surface acts more like a pure Au surface, i.e. very passivated against oxygen or nitrogen adsorption. In contrast, oxygen could be adsorbed at the corresponding (110) plane. The surface structure and the influence of adsorption on both Cu_3Au surfaces will be determined in the following.

2. Experimental

The Cu_3Au alloy crystal have been cut by spark erosion from a single crystal rod with the nominal composition of Cu_3Au, oriented by X-ray diffraction and mechanically polished by standard procedures. After transfer into the NICISS ultra high vacuum system /15/ the sample was prepared by 1keV Ar^+ sputtering at room temperature followed by annealing at 800K until no detectable impurities were found with Auger spectroscopy (AES). Thereafter the sample was held for 10h at about 500K to allow bulk and surface ordering getting complete. The resulting LEED superstructure was checked in situ, in case of (100) a c(2x2) superstructure with sharp LEED reflexes has been observed.

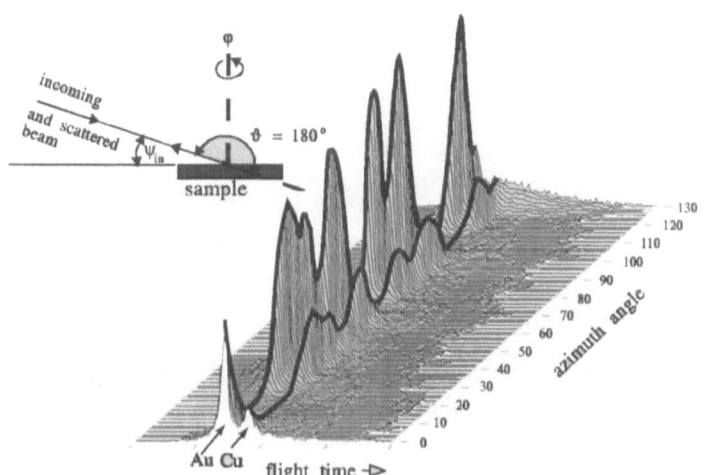

Fig.1 Experimental NICISS azimuthal scan (φ: 4° to 130°) at constant grazing incidence angle $\psi = 4$° of He. Primary energy of He^+ ions: 3KeV. Note the appearance of flight time discriminated scattering at Au and Cu atoms, respectively.

For the (110) sample in accordance with earlier investigations a (4x1) at low temperature and a (2x1) superstructure at T>425K has been monitored. Nitrogen was adsorbed after dissociation at a hot W filament or, alternatively by low energy (200eV) N^+ implantation with the help of a sputter ion gun, similar as described for the N/Cu experiments /13,14/. Also oxygen ions have been implanted in the same manner in case of the $Cu_3Au(100)$ surface, whereas pure oxygen adsorption readily occurs at $Cu_3Au(110)$. The amount of adsorbed gas was checked with AES (cylindrical mirror analyser and a primary electron energy of 3keV). In NICISS the back scattered intensity of He particles at Cu or Au atoms are a measure for the position of the corresponding surface atoms and have been used to uncover the surface structure /16,17,20/.

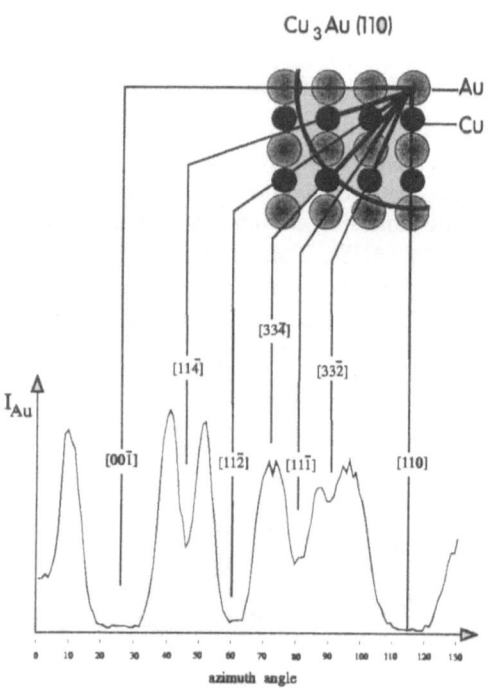

Fig.2 Ion beam surface crystallography: signal of back scattered He at Au atoms obtained at grazing angle of incidence. Same data set as in fig.1.

A set of time of flight (TOF) spectra in an azimuthal scan (φ: 4° to 130°) at constant grazing incidence angle $\psi=4°$ is shown in fig.1 for the clean $Cu_3Au(110)$ surface. He back scattering at Au or Cu atoms could be easily discriminated. From such a dataset the surface crystallography can be extracted by plotting e.g. the scattering signal from Au (fig.2). As has been discussed recently

/20/ in the NICISS azimuth scans deep intensity minima are expected to occur at low indexed surface directions due to ion shadowing and blocking effects at grazing angle of incidence.

After accomplishing the NICISS investigations, the sample has been transferred ex situ into another UHV chamber equipped with a 'beetle' type STM /18/, LEED optics, a heater stage, a sputter gun and a gas manifold. Following the same surface treatments as mentioned above, the same $Cu_3Au(100)$ or (110) crystals were imaged with atomic resolution by scanning tunneling microscopy.

3. Results and Discussion

3.1 Clean Cu_3Au

The $L1_2$ crystallographic structure of $Cu_3Au(100)$ (fig3b) results in two possible terminations of a (100) surface, either the gold rich 50%Au-50%Cu plane (A) or the pure 100% Cu plane (B). In order to come to a decision about the surface structure, i.e. the appearance of A versus B termination of the clean (100) surface we investigated the φ [001] geometry with the help of a NICISS polar scan. In the FAN model calculation /16/ a critical angle of incidence for the first appearance of a head-on collision situation, necessary for the occurrence of 180° He back scattering at Au atoms, is reached at $\psi = 13°$ in case of the (A) termination /12/.

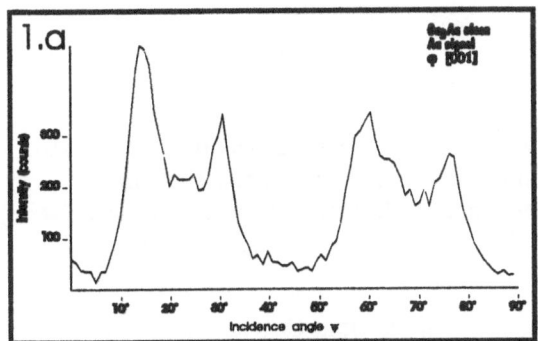 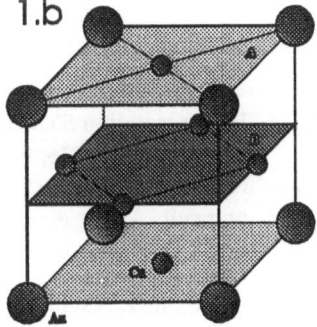

Fig.3 1.a) Experimental NICISS polar scan of He back scattered at Au atoms from clean $Cu_3Au(100)$ in the in the <001> geometry. Note the first layer Au peak around 5°. 1.b) Crystallographic structure model. (A) gold rich, (B) pure Cu plane.

In contrast, for surface termination (B) the first appearance of He back scattering at Au atoms is not reached before $\psi = 27°$. This is because in the case of a (B) termination, no first layer Au atoms are available and the projectiles have to find a head-on collision situation with second layer Au atoms. The measured polar scan for the Au signal is given in fig.3.1.a. The intensity rises sharply between 10° to 15° clearly identifying first layer Au atoms, only available in the (A) configuration. The other scattering peaks appearing at around 30°, 55° and 75° can be easily explained by scattering from third layer Au atoms, but are not of primary interest in this discussion. In fact, clean $Cu_3Au(100)$ is unequivocally terminated by the gold rich plane (A). Such a finding is also in agreement with other ISS analysis data /11,12/.

Similar to the (100) plane also a Au rich or pure Cu termination may occur. On the basis of the NICISS experiments shown in fg.2 it has been concluded that as for $Cu_3Au(100)$ also the $Cu_3Au(110)$ surface appears to be terminated by the Au/Cu plane /19/.

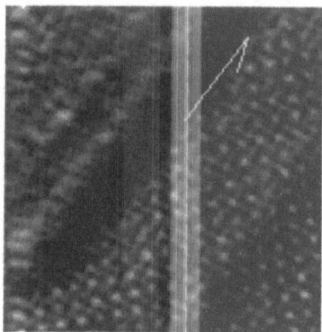

Fig.4 High resolution STM grey scale image of clean $Cu_3Au(100)$ surface. Two terraces are visible, on the lower one in the lower right part atomically resolved Cu and Au positions appear as different grey level heights in the unit cell. Arrow along <011>. U_{Tip} = 20mV, I_{Tunnel} = 10nA.

In a high resolution STM image of the clean $Cu_3Au(100)$ surface we can see the surface topography on an atomic scale (fig.4. Two terraces can be seen, parallel to the indicated <011> direction a step edge appears in the image. On the lower terrace it is possible to distinguish between the position of Au and Cu atoms by different grey levels in the unit cell at very low tip to sample voltages of U_T = 20mV. At this stage it is not possible to identify e.g. the brighter spots either with Au or Cu. However, at least it may be plausible to connect the bright spots with the position of the Au atoms also because of a theoretically proposed first layer buckling of about -7% (contraction) for Cu and +0.23% (expansion) for Au atoms /10/.

3.2 Oxygen and Nitrogen on Cu₃Au

In the adsorption experiments we used the implantation technique recently also successfully applied for N^+ implantation at Cu(110) /13,14/. NICISS patterns of both, nitrogen and oxygen saturated surface layers have been performed. The relevant structure data for Cu and Au at the surface are obtained by the polar and azimuth scans and are very similar for both cases. The polar scan in fig.5 was obtained in the φ [001] geometry for the oxygen adsorption with the same experimental parameters as for the measurement for the clean surface shown in fig. 3. The comparison of the two NICISS patterns shows immediately the lack of the first layer Au peak in case of oxygen adsorption, hence a surface geometry with no Au atoms in the topmost layer has appeared.

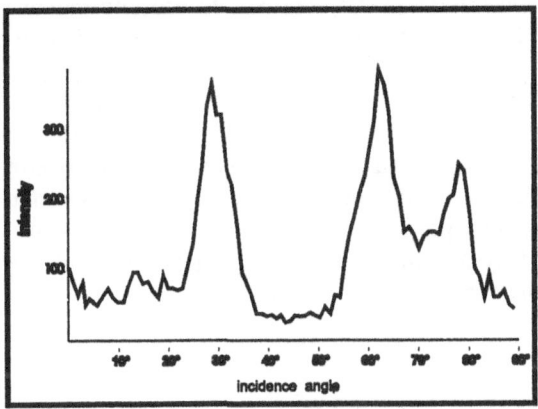

Fig.5 Experimental NICISS polar scan of He back scattered at Au atoms from oxygen covered Cu₃Au(100) in the in the $<001>$ geometry. Note the first appearance of the Au peak around 27° showing that the surface is no longer terminated with the Au rich (A) but by a Cu surface.

The same effect has been found for the nitrogen covered surface. Vanishing of the first layer Au peak in connection with the appearance of an increase in the NICISS Au signal at $\psi = 27°$ incidence angle is direct evidence for a Cu termination with surface (B) with the location of Au atoms in the second layer.

By looking in detail on the terraces, the atomic structure on both surfaces could be imaged with STM. A high resolution image of the oxygen saturated Cu₃Au(100) surface an area of approximately 60Å x 60Å is shown in fig.6a. Rows of bright protrusions can be seen, running in the $<001>$ directions with a inter-protrusion distance of about 3.7Å. Two basic differences as compared with the clean surface (fig.4) shall be pointed out: firstly, the measured corrugation for the O/Cu₃Au surface is about five times greater and largely insensitive to variation of U_T; secondly, it was not possible by variation of the STM imaging parameters to obtain any protrusion structure in the centre of one of the square sized unit cells.

Even at extremely small tip to sample voltages of a few mV, the image of the unit cells never changed into a centred structure comparable with the ones shown in fig.4. We attribute the protrusions in the STM image in fig.6a to the location of Cu atoms. Similar as for oxygen on plain copper surfaces, the measured corrugation enhancement would be a result of chemical bonding of copper to oxygen atoms. Oxygen itself is also not visible in the STM image in fig.6a. Because of the almost equal grey level height visible for all Cu atoms shown in fig.6a (and DRS measurements /12/) we infer as a possible oxygen position tentatively the fourfold hollow site in the centre of the (100) unit cell of Cu atoms. Such a site would result in a homogeneous partition of corresponding height modulation measured in the STM images.

The implantation technique inherently leads to rough surfaces which have to be annealed for smoothing and restructuring. In order to monitor the first stage of adsorption induced changes we also exposed the surface directly with 'reactive' species, i.e. with dissociated nitrogen as obtained from N_2 impinging at a hot tungsten filament. The effect of the N adsorption was monitored with STM. After submonolayer coverage, small dark clusters start to appear on the terraces (no preferred decoration of step edges can be seen). The area around one of such a cluster is presented in fig.6b in a high resolution STM image. A cluster of four by four protrusions (showing high grey level corrugation is visible in the upper right of fig.6b) is surrounded by an area of clean $Cu_3Au(100)$ indicated by the less well expressed height variations.

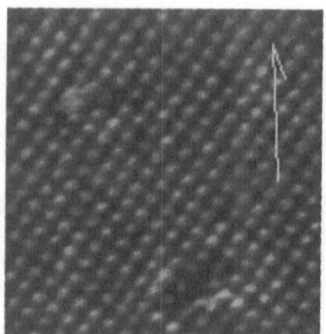 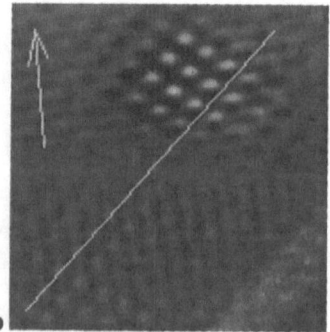

Fig6a. Fig.6b

a) High resolution STM grey scale image of the oxygen saturated $Cu_3Au(100)$ surface. $U_{Tip} = 20mV$, $I_{Tunnel} = 5nA$; arrow along $<011>$.
b) of a partially nitrogen covered surface. The Cu-N cluster is shown at the upper right. Note the phase shift along the line through the unit cells.

In the area of the clean surface, although less well atomically resolved as in fig.4, still the centred unit mesh can be recognised with corresponding atom positions of Au and Cu atoms. As already mentioned above, the chemical influence of oxygen on the local electronic distribution at the surface prevents us to measure directly the height difference between the $Cu_3Au(100)$ terrace and the N-Cu cluster, the non centred unit cell in this area is visible in the upper right area. Similar to measurements at O/Cu(110) and N/Cu(110), the adsorbate covered surface area reflects not the expected topographic height, but looks darker ('deeper') than the surrounding terrace. Probably also in the case of $Cu_3Au(100)$, the N-Cu cluster is located on top and not below the surrounding clean surface. In any case, the inspection of the STM image in fig.6b shows directly the appearance of a mono-step (up or down) between clean and adsorbate covered surfaces. Indeed, the extension of a straight line drawn across the top of protrusions at the clean area falls in the middle between protrusions of the N-Cu cluster. From the crystallographic model we can conclude that atoms from both areas cannot be located in the same plane, i.e. with (A) termination, but instead in two planes (A) and (B). This finding is in accordance with the NICISS data showing the change from (A) type termination (fig.3) for the clean surface into (B) type Cu termination (fig.5) after adsorption of nitrogen or oxygen.

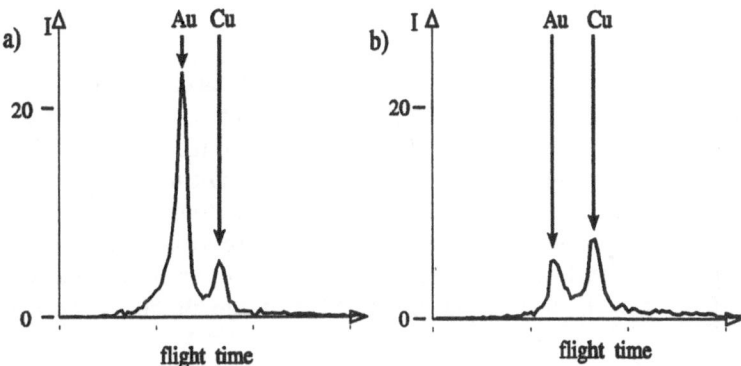

Fig.7 a) Au and Cu NICISS peak for the clean b) for the oxygen covered surface. Note the increase of the Cu peak upon oxygen induced Cu segregation and the corresponding decrease of the Au signal. $\psi = 27°$, φ [001]

Also in case of the (110) plane we found a similar segregation effect of Cu to the surface of the former Au/Cu terminated surface which is documented in the NICISS data shown in fig. 7. In contrary to the (100) plane, here oxygen has been adsorbed at room temperature followed by subsequent annealing at 800K.

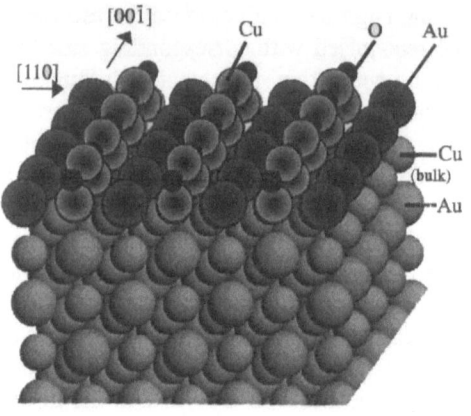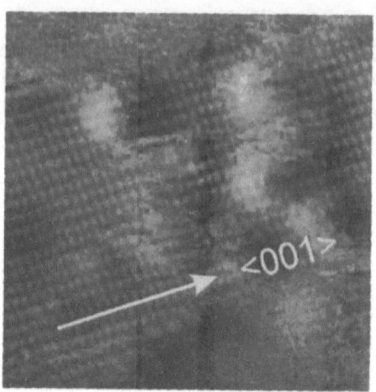

Fig.8 a) surface reconstruction model of added -O-Cu- rows. b) High resolution
STM grey scale image of the oxygen covered $Cu_3Au(110)$ surface. The
added rows along $<001>$ are visible. Area: 130Å x130Å.
U_{Tip} = -400mV, I_{Tunnel} = 1nA.

The LEED pattern changes from a faint (4x1) and (2x1) for the clean
surface to an intense (2x1) superstructure with bright half order spots in case of the
oxygen covered $Cu_3Au(110)$ surface. On the basis of the complete NICISS data set
the following structure model has been proposed: -O-Cu-O-Cu- added rows along
the $<001>$ directions are formed by Cu segregation to the surface. These rows are
similar to the structure at Cu(110)-(2x1) O and here located above the Cu rows of
the former clean surface. The added row structure shown in fig.8a as deduced from
the NICISS data is also visible in the STM image fig.8b of the corresponding
surface.

4. Summary

On the basis of a combined lateral averaging and microscopic research
with NICISS and STM, we found the following properties of clean and adsorbate
covered $Cu_3Au(100)$: In accordance with recent investigations, both methods demonstrate for the clean sample the Au rich surface termination. For the first time we
could image the surface with atomic resolution showing Cu and Au atoms with different grey level heights. The surface is highly inert; direct adsorption of oxygen or
nitrogen at room temperature from the molecular species does virtually not occur.
Implantation of adsorbates followed by low temperature annealing restores the surface and both, nitrogen and oxygen covered $Cu_3Au(100)$ is terminated by the Cu

surface with no Au atoms in the first layer. Similar termination is found after adsorption of dissociated nitrogen. Atomic resolution is also achieved for such a surface, exhibiting in the STM data considerably larger height corrugation as compared with the clean surface. The positions of the adsorbates are not directly visible in STM, but from the grey scale images it has been proposed that nitrogen and oxygen is located in one type of the fourfold hollow sites, probably the hollow sites above Cu atoms. $Cu_3Au(110)$ is less inert and oxygen adsorption is readily achieved by gas exposure. An added row structure similar to the structure found at $Cu(110)$-$(2x1)$ O has been deduced. Both clean surfaces, $Cu_3Au(100)$ and $Cu_3Au(110)$ are terminated by the Au rich surface.

References:

1.   V.S.Sundaram, R.S.Alben, W.D.Robertson, Surf.Sci.**46** (1974) 653
2.   H.C.Potter, J.M.Blakely, Journ.Vac.Sci.Technol.**12** (1975) 635
3.   S.F.Alverado, M.Campagna, A.Fattah, W.Uelhoff, Z.Phys.**B66** (1987) 103
4.   H.Dosch, L.Mailänder, H.Reichelt, J.Peisl, R.L.Johnson, Phys.Rev.**B43** (1991) 13172
5.   B.E.Warren, X-ray Diffraction, Addison -Wesley, Reading, MA, 1969, Chap. 12
6.   S.E.Nagler, R.F.Shannon, C.R.Harkless, M.A.Singh, R.M.Nicklow, Phys.Rev.Lett.**61** (1988) 718
7.   K.F.Ludwig, G.B.Stephenson, J.L.Jordan-Sweet, J.Mainville, Y.S.Yang, M.Sutton, Phys.Rev.Lett.**61** (1988) 1859
8.   A.Stuck, J.Osterwalder, L.Schlapbach, H.C.Poon, Surf.Sci.**251/252** (1991) 670
9.   B.Gans, P.A.Knipp, D.D.Koleske, S.J.Sibener, Surf.Sci.**264** (1992) 81
10.   W.E.Wallace, G.J.Ackland, Surf.Sci.Lett.**275** (1992) L685
11.   T.M.Buck, G.H.Wheatley, L.Marchut, Phys.Rev.Lett.**51** (1983) 43
      T.M.Buck, G.H.Wheatley, D.P.Jackson, Nucl.Instr.Phys.**218** (1983) 257
12.   H.Niehus, C. Achete, Surf.Sci.**289** (1993) 19
13.   H.Niehus, R.Spitzl, K.Besocke, G.Comsa, Phys.Rev.**B43** (1991) 43
14.   R.Spitzl, H.Niehus, G.Comsa, Surf.Sci.Lett.**250** (1991) L355
15.   The UHV system has been described in detail /14/ and includes NICISS, AES and LEED facilities.
16.   H.Niehus, R.Spitzl, Surf.Interface Anal.**17** (1991) 287
17.   H.Niehus, Journ.Vac.Sci.Technol.**A5** (1987) 751
18.   K.Besocke, Surf.Sci.**181** (1987) 145
19.   K.Morgenstern Diplom work, 1992, KFA Jülich
20.   H.Niehus, W.Heiland, E.Taglauer, Surf.Sci.Report **17** 4/5 (1993)

II.

Surface Electronic Properties

Surface States on Metals

A. Goldmann and R. Matzdorf

Universität GH Kassel, Fachbereich Physik,
Heinrich-Plett-Straße 40, D-34132 Kassel, Germany

Abstract. First we report on the experimental identification of electronic surface states on metals. Then we discuss a few selected examples on how such states may be exploited to study the surface potential in its dependence on surface orientation, sample temperature and effects of adsorbate atoms incorporated into the outermost layer. Finally we shortly address the following topics: linewidths of surface state photoemission peaks, identification of surface resonances, and search for electronic states at step atoms.

1. Introduction

At the surface of solids the changed periodicity of the crystal lattice may lead to the development of electronic states not present in the bulk. In the discussion of these surface states it has become customary to distinguish "Tamm" states from "Shockley" states and "true" surface states from surface "resonances" /1-4/. True surface states are located energetically within gaps of the bulk bands projected onto the surface Brillouin zone (SBZ). Resonances are energetically degenerate with bulk bands and therefore it is generally not trivial to identify them experimentally. The appearance of Tamm states may be visualized as a band bending effect over the distance of one atomic layer: d-like surface states on transition metals, for example, which have atomic-like localized wavefunctions, can be split-off energetically from bulk bands by the surface potential /1, 3, 4/. Shockley states arise primarily from the special boundary conditions introduced by the surface. These states occur in energy gaps caused by the hybridization of crossed bands, e.g. in s,p gaps for which the lower band has odd parity at the zone boundary, while the upper band is even there /1, 2, 4/. Such gaps exist often in the energy range near the Fermi level E_F. Surface states are generally investigated best by the techniques of angle-resolved photoemission /1/ and inverse photoemission /1, 5-7/. Since results from inverse photoemission experiments are treated in a separate article of this book, we will in what follows concentrate exclusively on photoemission studies of occupied surface states. Due to space limitations we will discuss only very few examples, but in sufficient detail to point out the basic ideas and strategies.

2. Experimental identification

Surface atoms experience a different local environment relative to the bulk atoms which is reflected in a changed local electronic structure. However, screening lengths of metals are characteristically so small that the charge distributions are more or less bulk like already for the second and deeper layers. This is clearly evident from Fig. 1 which reproduces contours of equal charge density calculated for a Cu(100) surface. Surface states are thus located essentially in the topmost one or two layers. Since the typical sampling depths of angle-resolved electron spectroscopies are 0.5 - 1.5 nm, these techniques give information about bulk and surface electronic states simultaneously. We will therefore discuss first, how surface properties can be safely identified in experimental spectra. As an example we have chosen the well-known Tamm-state observed by several groups around the $\overline{M}$ point of the surface Brillouin zone (SBZ) on Cu(100).

Inspection of the Cu bulk bands (reproduced in Fig. 2) shows that a large gap exists between about -2 eV and + 1.8 eV along the XW-line of the bulk Brillouin zone (BBZ). At -2 eV a d-band occurs between X and W with almost no dispersion. There is thus a high density of d-states available, and if the potential within the outermost atomic layer differs from that of the bulk a Tamm state may be split-off. How can we observe it?

The connection of the ΓXWK plane of the BBZ and the $\overline{\Gamma M}$ direction of the SBZ is visualized in Fig. 3: the XW-line projects directly onto the $\overline{M}$ point. Since the

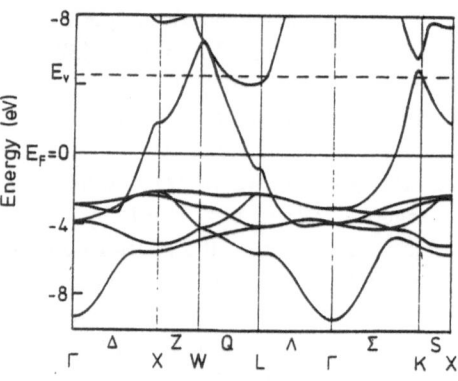

Fig. 1: Contours of equal charge density calculated /8/ for Cu(100). Contours differ by √2. The crystal is cut perpendicular to the surface at the top of the figure. The lattice constant of fcc-Cu is a = 0,36 nm.

Fig. 2: Energy bands of bulk Cu along selected symmetry lines of the bulk Brillouin zone. Nonrelativistic calculation with neglected spin-orbit coupling.

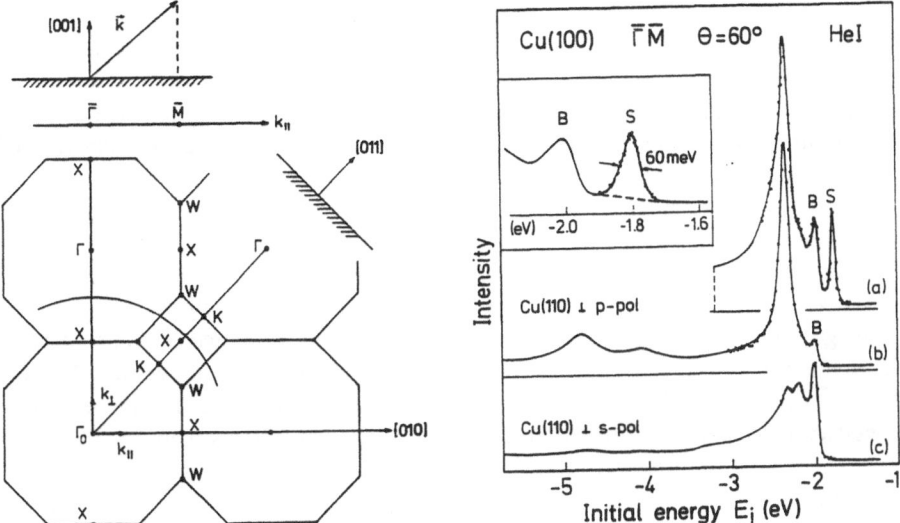

Fig. 3: Correspondence between the ΓXWK mirror plane of the fcc bulk Brillouin zone and the Γ̄M̄-direction of the surface Brillouin zone of Cu(100). Shaded: relative orientation of the (001) and (011) surfaces, respectively. The circle indicates the position of final states for bulk direct transitions at the upper d-band edge (- 2 eV) with photons of energy ℏω = 21.2 eV from Cu(001) and Cu(011).

Fig. 4: Electron energy distribution curves observed in photoemission from Cu(100) at θ = 60° off-normal (trace a) and in normal-emission from Cu(110) (b, c). Photon energy ℏω = 21.2 eV. Unpolarized light for curve a. Linearly polarized light was incident with the $\vec{E}$-vector oriented perpendicular (s-pol) and parallel (p-pol) to the ΓXWK plane when curves b, c were measured. Insert: magnified part of curve a.

photoelectron wave-vector component parallel to the surface

$$k_\parallel = \sin \theta \cdot [(2m/\hbar^2) \cdot E_{kin}]^{1/2}$$

is conserved /1/, (θ = electron emission angle with respect to the surface normal, Ekin = kinetic energy of photoelectron in vacuum, m = electron rest mass), a surface state split-off the X-W bulk band can be searched for by adjusting the kinematical parameters to M̄ /9/. At a photon energy ℏω = 21.2 eV and for initial state energies around E_i = - 2 eV, the M̄ point of Cu(100) is observed at θ ≈ 60°. The corresponding electron energy distribution curve is reproduced in Fig. 4 (a), where B indicates the upper edge of the bulk 3d-bands (XW) and S labels the Tamm state split-off B. That S really corresponds to the (100) surface, but not to the bulk states, can be proven by spectra taken along the surface normal of Cu(110), see Fig. 4 (b, c): the bulk band edge B is in "energy-coincidence" /1, 9/ with the one in Fig. 4(a). This was expected from Fig. 3: both experiments probe the immediate vicinity of the bulk X-point. However, the state S observed in Fig. 4(a) neither corresponds to bulk bands nor to the (110) surface. Therefore it is missing in traces (b, c), but present in Fig. 4a. The absence of S in (b, c) cannot be due to selection rules: both polarization directions give the identical message. Polarization dependent spectra have also been measured from Cu(100) at $\overline{M}$. They clearly show /10/ that the symmetry behaviour of both S and B is identical, in support of the model that S is split off the bulk band by the surface potential.

Another piece of evidence for the surface character of S is its sensitivity to surface contamination. Fig. 5 shows again the energy interval of the inset in Fig. 4, but now at considerably higher resolution /11/: besides S_1 (S in Fig. 4) and B_1 (B) a new peak S_2 is resolved, about 65 meV below B_1. The sensitivity of S_1 and S_2 to oxygen adsorption (and other contaminants) gives evidence of their surface character. Moreover, the experimentally determined energy dispersions E_i ($k_{\parallel}$) along $\overline{\Gamma M}$ (reproduced in Fig. 6) and $\overline{\Gamma X}$ /11/ are independent of $k_{\perp}$, the electron wavevector perpendicular to the surface. The latter fact follows from the observation that E_i ($\vec{k}_{\parallel}$) determined at several photon energies $\hbar\omega$ /9 - 11/ does not depend on $\hbar\omega$.

Inspection of Fig. 6 shows the gap for $E_i > - 2$ eV, as expected from the non-relativistic band structure reproduced in Fig. 2. From Fig. 2 an additional gap ($E_i = -2.1$ eV at $\overline{M}$) was not expected. In fact it is due to a spin-orbit splitting of the bulk bands calculated relativistically along XW, and this gap supports the surface band S_2. For a detailed analysis we refer to /11/. We mention in passing that the FWHM for S_1 at $\overline{M}$ is 28 meV, the smallest band width reported to date. Taking the analyzer resolution (16 meV) into account, the intrinsic width of S_1 is probably about 23 meV only /11/.

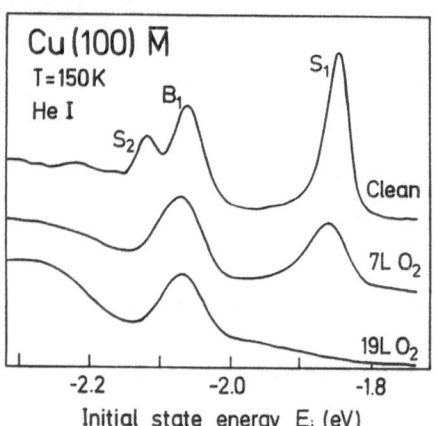 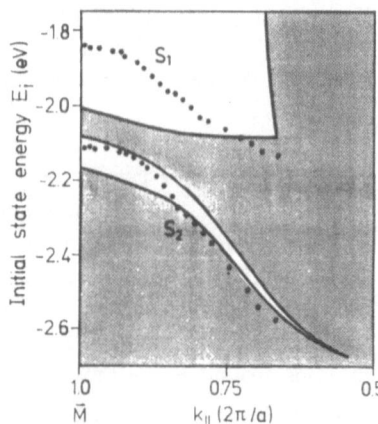

Fig. 5: Angle-resolved photoelectron spectra taken at the $\overline{M}$ point of Cu(100) as a function of oxygen exposure (1 L = 1.3 · 10^{-6} mbar · s). Photon energy $\hbar\omega = 21.2$ eV (unpolarized), sample kept at T = 150 K.

Fig. 6: Experimental /11/ dispersion relations E_i ($k_{\parallel}$) for S_1 and S_2 (compare Fig. 5), along with the projected (shaded) relativistic bulk band structure /11, 12/.

3. Interim statements

In section 2 we have discussed surface states around $\overline{M}$ on Cu(100) to give a transparent example for the experimental identification. Over the past decade many metal surfaces have been studied using angle-resolved photoelectron techniques /1/ and surface states have been identified on almost all the low index faces of the metallic elements in the periodic table. For several illustrative examples and more complete references we refer to chapters 4 and 10 of /1/. A particular type of surface states is supported by the potential well created in front of a metal surface by the attractive force acting between an electron and its image charge in the metal /2, 5-7/. The corresponding "image states" form a series of states converging towards the vacuum energy E_v and can be measured with high precision by two-photon photoelectron spectroscopy /13, 14/. Their experimental binding energies and effective masses may be used, in combination with the same parameters for empty as well as occupied Shockley surface states, to model the shape of the potential barrier outside the metal surface. The barrier potentials thus obtained allow then for a consistent description of energies and effective masses of all known surface states for the faces considered. Such an analysis has been performed recently for the low-index faces of Cu /15/ and the result is reproduced in Fig. 7. We summarize from an experimentalist's point of view that much systematic knowledge is available, and the physical origin as well as the identification of true surface states are rather well understood. Therefore these surface states may now be exploited as analytical tools to probe the surface potential in its dependence on adsorbate coverage and/or sample temperature. Moreover, surface states generally exhibit rather small linewidths and therefore can give us lower limits for the lifetime of the photohole. Both applications will be discussed further below. Much less is known about surface resonances and electronic states located at surface defects like steps, kinks, and so on. We will come back to that later on.

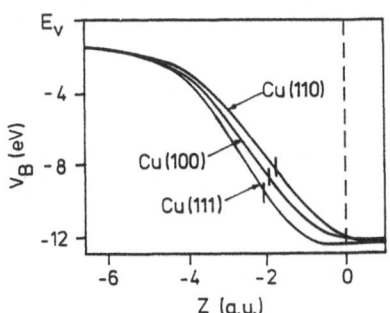

Fig. 7: Barrier potentials for the low-index faces of Cu from a simultaneous analysis /15/ of photoemission, inverse photomission and two-photon photoemission data. The center of the outermost atomic plane is at z = 0. Short solid lines mark the different image planes. Inside crystal $z \geq 0$.

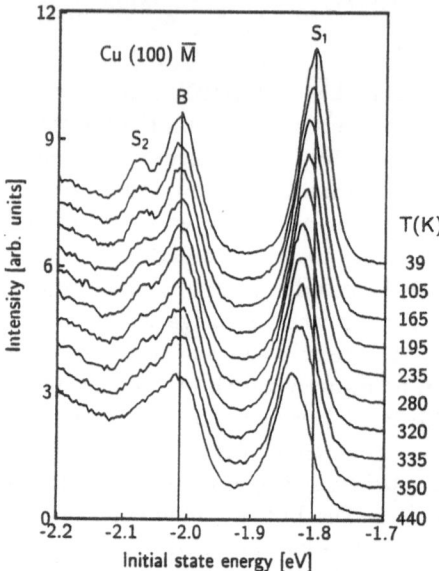

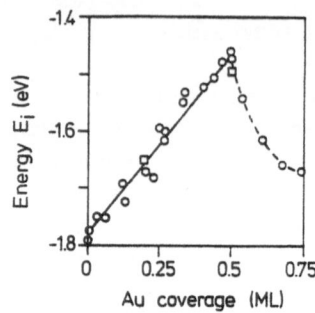

Fig. 9: Initial state energy of the $\overline{\text{M}}$ Tamm state S_1 on Cu(100) versus Au coverage at room temperature /19/.

Fig. 8: Temperature dependence of the d-like Tamm state S_1 at $\overline{\text{M}}$ on Cu(100). Photon energy $\hbar\omega = 21.2$ eV /16/. The spectra are shifted against each other along the ordinate direction.

4. Probing the surface potential

Fig. 8 shows how the photoelectron spectra of the $\overline{\text{M}}$ surface state S_1 on Cu(100) depend on the sample temperature T. Besides a temperature-induced loss of intensity at increasing T and a concomitant increase of the linewidth, in particular an energy shift with respect to the position of the bulk peak B is clearly revealed. A similar decrease of the energetic distance between B and S_1 is observed from samples after gentle argon ion bombardment and insufficient annealing. Thus the observed T-induced shift might be identified with the effect of thermal disorder and/or the thermal lattice expansion. However, both experimental and theoretical studies indicate /17/ that the X-point bulk energy does not shift with T, in agreement with the observed independence of peak B on T. We therefore believe that the T-dependent energy distance between B and S_1 reflects a change of the potential in the outermost layer of Cu(100). We have started a theoretical investigation to explore this. If our interpretation is correct, the study of Tamm states represents a sensitive tool to monitor the T-depence of the surface potential.

Surface-localized Tamm states may also be used as a specific probe of the chemical environment in the surface plane /18, 19/. For deposition of 0.5 monolayers (ML) or less of Au on Cu(100), Au is incorporated into the outermost layer forming an ordered surface alloy, with a c(2x2) LEED pattern at 0.5 ML. Correspondingly, the energy of S_1 at $\overline{M}$ shifts linearly with the concentration of Au to lower binding energies. This result of /19/ is reproduced in Fig. 9. At coverages > 0.5 ML, S_1 shifts back towards the energy of the clean surface. This is interpreted by a decreasing Au content of the outermost Cu layer, in support of a de-alloying model /20/. The example shown here demonstrates nicely how Tamm states may provide information on the substrate surface composition. This high sensitivity to the in-layer environment is easily explained: the S_1 state is composed of d_{xy} derived orbitals, which are localised in directions parallel to the (100) surface.

The Cu(100)c(2x2) Au surface alloy may be described structurally /21/ as a checkerboard pattern of Au and Cu atoms in the surface plane, with a pure Cu(100) layer beneath. This makes the two top layers almost identical to the outermost two layers of $Cu_3Au(100)$, compare Fig. 10. In fact, angle-resolved photoelectron spectra demonstrate /18, 22, 23/ that a d-like Tamm state is observed on $Cu_3Au(100)$ at almost the same initial state energy as on Cu(100)c(2x2) Au. Our results /23/ are reproduced in Fig. 11: Besides S_1, as second surface state S_2 is observed, in close analogy to the case of Cu(100). The corresponding dispersion curves $E_i(k_\parallel)$ along the [100] direction of the SBZ are shown in Fig. 12. We have also studied the T-dependence of the energy of S_1 and the results /23/ are very similar to those reported above for S_1 on Cu(100). Calculations are in progress to model the T-dependent potential within the surface layer of $Cu_3Au(100)$.

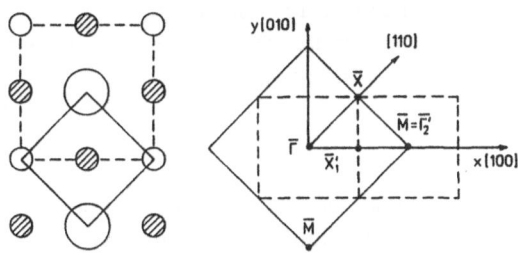

Fig. 10: Schematic top view (left) of the $Cu_3Au(100)$ surface: topmost layer with Au and Cu atoms (large and small empty circles) and second layer with Cu atoms (hatched circles) only. The 2D unit cells for Cu(100) and $Cu_3Au(100)$ are indicated by the solid and dashed lines, respectively. Right: corresponding 2D surface Brillouin zone.

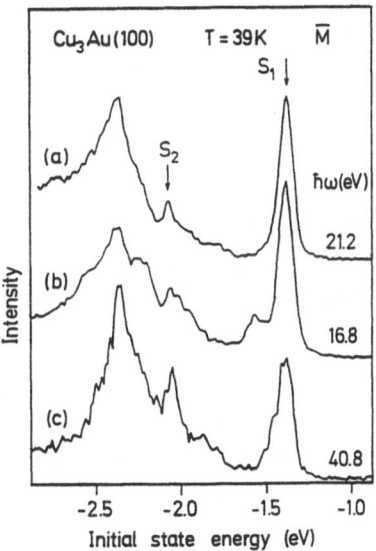

Fig. 11: Photoelectron spectra taken at the $\overline{M}$ point of the SBZ (compare Fig. 10) from $Cu_3Au(100)$ at different photon energies $\hbar\omega$.

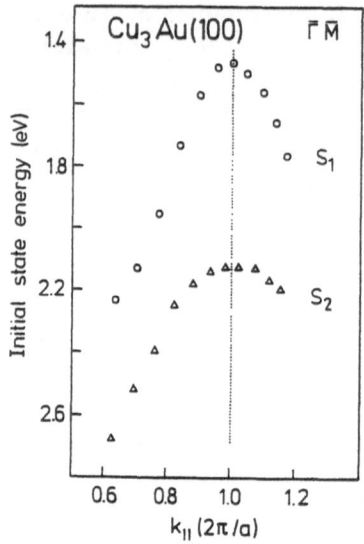
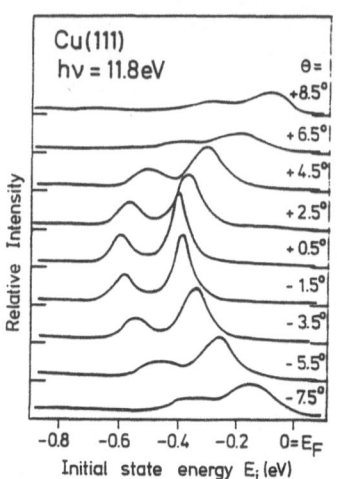

Fig. 12: Dispersion E_i ($k_\parallel$) of the surface bands S_1 and S_2 on $Cu_3Au(100)$ along the [100] direction of the SBZ. Sample temperature T = 39 K.

Fig. 13: Experimental photoelectron spectra of the s,p-like Shockley state for several emission angles θ around $\bar{\Gamma}$ on Cu(111) /25/. Since the ArI doublet was used to excite photoemission, two emission lines are actually observed.

5. Linewidths

Peak widths in photoemission are related to the electron and hole lifetimes. For the case of normal emission out of a surface state this relationship may be approximately expressed as /1, 24/ $\Gamma_m = \Gamma_h + (v_h/v_e)\Gamma_e$ where Γ_h, Γ_e are the final state hole and electron inverse lifetimes, and v_h, v_e are the corresponding band velocities v = δE/δk taken normal to the surface. Γ_m is the experimental linewidth. For a 2D state $v_h = 0$ and $\Gamma_m = \Gamma_h$. Also 3D transitions may occur at k-space points where $v_h = 0$. Thus inverse hole lifetimes, which give a direct experimental access to the imaginary part of the self-energy (describing the fully interacting many-electron system and therefore going beyond the usual one-electron band structure picture) may be measured in principle. Simple physical arguments lead to the expectation that $\Gamma_h \to 0$ for $E_i \to E_F$: the final-state hole is filled by radiationless processes, for which the rate goes to zero since the relevant phase space becomes vanishingly small at E_F. High-resolution experiments /25/ on the Shockley surface state near $\bar{\Gamma}$ ($k_\parallel = 0$) on Cu(111), which are reported in Fig. 13, demonstrate the opposite, however: already visual inspection shows an increase of Γ_m when $E_i \to E_F$. This surprising result was explained /25/ by a dominant

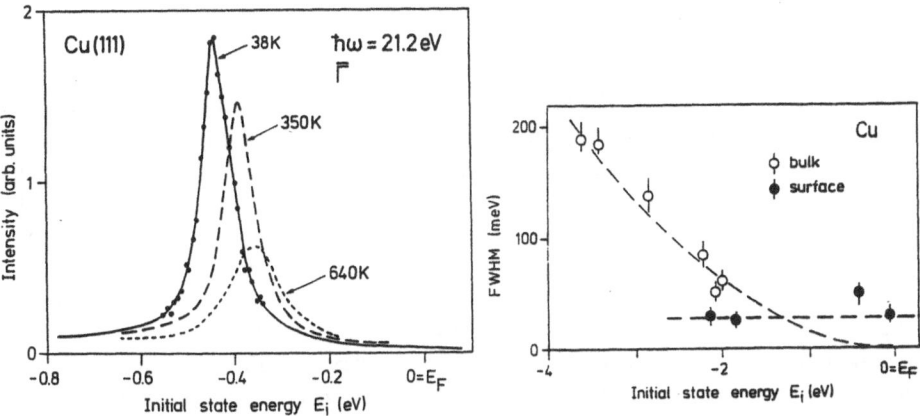

<u>Fig. 14:</u> Temperature-dependent photoemission from the Shockley state at $\bar{\Gamma}$ on Cu(111). Data from /16/.

<u>Fig. 15:</u> Experimental photoemission linewidths observed /26/ for transitions at several initial state energies on Cu(100). The dashed parabola starting at E_F is given by $\Gamma = 1.5 \cdot 10^{-2}$ eV^{-1} $(E_i - E_F)^2$. The dashed horizontal line is constant at 28 meV.

(large as compared to Γ_h) effect on Γ_m of elastic scattering from a low concentration of surface impurities or defects. If this is correct, it is not trivial to extract the correct Γ_h from Γ_m. As a further check we have studied the T-dependence of this surface state at $\bar{\Gamma}$ /16/, see the results shown in Fig. 14: Phonon-electron interactions scatter electrons out of the detector solid angle, and therefore the intensity goes down drastically with increasing T. Also a shift in E_i is observed, correlated with the T-dependence of the potential supporting this state. However, no change with T of the linewidth Γ_m is observed below 350 K, in contrast to the expectation. This result strongly supports the idea that Γ_m is dominated by an extrinsic scattering mechanism as proposed in /25/.

Nevertheless, experimental Γ_m values may supply an upper limit for the inverse hole lifetime. This is demonstrated in Fig. 15, where we have collected several results /26/ from bulk and surface state transitions on Cu(100) and Cu(111). The Landau theory of Fermi liquids (as is copper) predicts a quadratic dependence of Γ_h on the energy distance from E_F. We have fitted such a parabolic dependence to the bulk data, see Fig. 15. This curve, representing an upper limit to Γ_h, is much below the observed surface state widths Γ_m on approaching $E_i \rightarrow 0$. We conclude that present day experiments to explore Γ_h near E_F are probably still dominated by limitations like sample imperfection and other "purely" experimental facts. Nevertheless several groups around the world are presently working on the problem of approaching Γ_h, the ultimate resolution obtainable in photoemission.

60

6. Surface resonances

A surface resonance is an electronic state located at the surface, but energetically overlapping bulk states. Therefore its experimental investigation is generally more difficult as compared to true surface states. One of the few cases where a unique identification was possible is summarized in Figs. 16 and 17. As reported elsewhere /28/ the peaks labeled 2-6 are due to direct bulk transitions. In the following we discuss evidence /27/ that the emission line D2 in Fig. 16, which is particularly sensitive to T, results from states which are spatially localized within the topmost layer of the reconstructed Au(111)(22 x √3) surface. By gently sputtering the sample kept at T = 50 K with 200 eV Ne$^+$ ions the corrugated surface layer can be removed, and the sample exhibits a very clear (1x1) LEED pattern, characteristic of an unreconstructed fcc(111) surface. The corresponding photoelectron spectrum measured at 72 K is reproduced at the bottom of Fig. 17. As is evident all bulk emission peaks are clearly resolved while D2 is missing. By warming up the thermodynamically unstable (1x1) surface, the (22 x √3) reconstruction is reestablished, monitored by the corresponding LEED pattern. Simultaneously, the

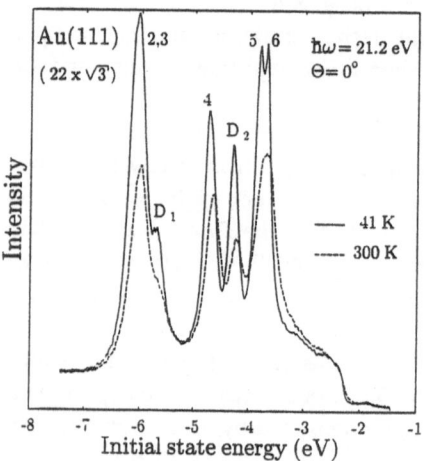

Fig. 16: Normal emission photo-electron spectra from the recon-structed Au(111) surface taken at two different sample temperatures /27/.

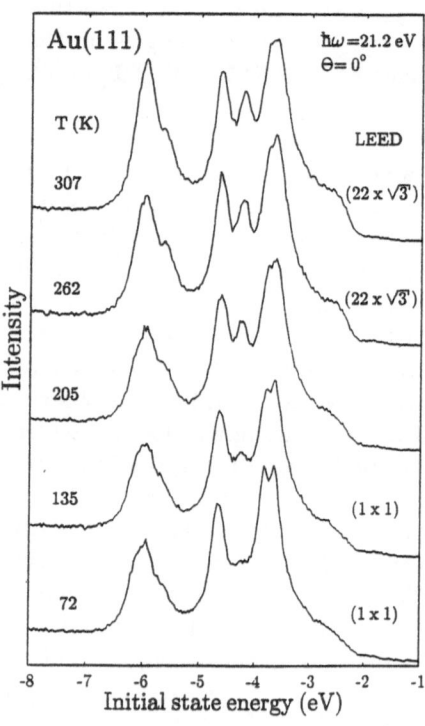

Fig. 17: Normal emission photoelectron spectra taken at different temperatures T from the unreconstructed (bottom) and fully reconstructed (topmost) Au(111) surface /27/.

emission line D2 reappears. Further experimental evidence is discussed in /27/. In summary, these data present clear experimental evidence that D2 corresponds to a d-like (Tamm-type) surface resonance localized spatially in the corrugated surface layer.

7. Localization at steps

One should also expect the formation of electronic states (surface resonances) localized at step edges, and which exhibit more or less 1D-character. Only very recently, the observation of such a step resonance was reported /29/. The main message is summarized in Fig. 18. The topmost curve shows a normal emission spectrum taken from a clean Ni(7 9 11) stepped surface. After adsorption of 0.38 ML of Na some features are predominantly attenuated, thereby exhibiting their surface character. These are shown more pronounced in the difference curve plotted at the bottom of Fig. 18. By comparison to analogous data taken from the flat Ni(111) surface, and checking for the influence of other adsorbates as well as for the effects of different light polarization, the authors present experimental evidence /29/ that the peak labeled S_t is an already well-known Shockley-type surface state associated with the flat (111) terraces. In contrast, peak S_s can only be explained by an electronic state associated with the monoatomic steps oriented along the [121] bulk direction. This example demonstrates /29/ that the investigation of the electronic properties of steps is not at all trivial and that almost all necessary work remains to be done in the future, may be with different techniques like e.g. scanning tunneling spectroscopy.

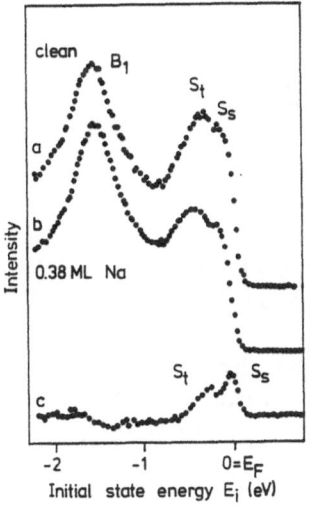

Fig. 18: Normal emission photoelectron spectra taken at $\hbar\omega = 10$ eV from a stepped Ni (7 9 11) surface (a) clean and (b) after adsorption of 0.38 ML of Na. (c) difference curve /29/.

62

8. Outlook

In this short article only very few examples for the investigation of surface states and surface resonances on metals could be discussed. We hope, however, they make clear how photoelectron spectroscopy contributes to their experimental identification and, in particular, how they may be exploited to investigate specific properties of the surface potential. The occurence of true surface states is basically well understood. The quantitative understanding of surface resonances and electronic properties connected to steps, defects or adatoms, which may be crucially important for the interaction of the surface with a gaseous, liquid or solid environment, is still in its infancy.

Acknowledgement: Our work is continuously supported by the Deutsche Forschungsgemeinschaft (DFG).

References

/1/ "Angle-resolved Photoemission" (S.D. Kevan, editor), Vol. 74 of "Studies in Surface Science and Catalysis", Elsevier, Amsterdam, 1992, and many references therein

/2/ N.V. Smith, Phys. Rev. B32 (1985) 3549

/3/ S.G. Davison and M. Steslicka, "Basic Theory of Surface States", Clarendon Press, Oxford 1992

/4/ F. Forstmann, Progr. Surface Sci. 42 (1993) 21

/5/ N.V. Smith, Appl. Surface Sci. 22/23 (1985) 349

/6/ V. Dose, Surface Sci. Rept. 5 (1985) 337

/7/ G. Borstel and G. Thörner, Surface Sci. Rept. 8 (1988) 1

/8/ F.J. Arlinghaus, J.G. Gay and J.R. Smith, Phys. Rev. B21 (1980) 2055; B23 (1981) 5152

/9/ P. Heimann, J. Hermanson, H. Miosga and H. Neddermeyer, Phys. Rev. B20 (1979) 3059

/10/ D. Westphal and A. Goldmann, Surface Sci. 95 (1980) L249

/11/ P.L. Wincott, D.S.L. Law, N.B. Brookes, B. Pearce and G. Thornton, Surface Sci. 178 (1986) 300

/12/ H. Eckardt, L. Fritsche and J. Noffke, J. Phys. F. (Met. Phys.) 14 (1984) 97

/13/ W. Steinmann, Appl. Phys. A49 (1989) 365; Progr. Surface Sci. 42 (1993) 89

/14/ W. Steinmann and Th. Fauster, in "Laser Spectroscopy and Photochemistry on Metal Surfaces" (H.L. Dai and W. Ho, editors), World Scientific, Singapore, 1994)

/15/ M. Graß, J. Braun, G. Borstel, R. Schneider, H. Dürr, Th. Fauster and V. Dose,
      J. Phys. Condens. Matter 5 (1993) 599

/16/ R. Matzdorf, G. Meister and A. Goldmann, Surface Sci. 286 (1993) 56

/17/ J.A. Knapp, F.J. Himpsel, A.R. Williams and D.E. Eastman,
      Phys. Rev. B19 (1979) 2844

/18/ G.W. Graham, Surface Sci. 184 (1987) 137

/19/ J.C. Hansen, M.K. Wagner and J.G. Tobin, Solid State Commun. 72 (1989)
      319

/20/ J.C. Hansen and J.G. Tobin, J. Vac. Sci. Technol. A7 (1989) 2475

/21/ Z.Q. Wang, Y.S. Li, C.K.C. Lok, J. Quinn and F. Jona,
      Solid State Commun. 62 (1987) 181

/22/ S. Löbus, M. Lau, R. Courths and S. Halilov, Surface Sci. 287/288 (1993) 568

/23/ R. Paniago, R. Matzdorf, A. Goldmann and R. Courths,
      Surface Sci., to be published

/24/ N.V. Smith, P. Thiry and Y. Petroff, Phys. Rev. B47 (1993) 15476

/25/ J. Tersoff and S.D. Kevan, Phys. Rev. B28 (1983) 4267

/26/ R. Matzdorf, R. Paniago, G. Meister and A. Goldmann, to be published

/27/ R. Paniago, R. Matzdorf and A. Goldmann, Europhys. Letters, in press

/28/ R. Courths, H.-G. Zimmer, A. Goldmann and H. Saalfeld,
      Phys. Rev. B34 (1986) 3577

/29/ H. Namba, N. Nakanishi, T. Yamaguchi and H. Kuroda,
      Phys. Rev. Letters 71 (1993) 4027

Low-Dimensional States on Metal Surfaces

N. Memmel and V. Dose

MPI für Plasmaphysik, Euratom Association, D-85748 Garching, Germany

Abstract

Two-dimensional electronic states on metal surfaces have been studied quite extensively in the past twenty years. If these surface states are confined not only perpendicular but also parallel to the surface, their dimension is further reduced - analogous to the "quantum wires" and "quantum dots" well known in semiconductor physics. Examples of two-, one- and zero-dimensional states are given.

List of Abbreviations

E_F	Fermi energy	PBS	projected bulk band structure
E_{vac}	vacuum energy	RT	room temperature
IPE	inverse photoemission	SBZ	surface Brillouin zone
LEED	low-energy electron diffraction	STM	scanning tunnelling microscopy
ML	monolayer	2PPE	two-photon photoemission
MR	missing row		

1. Introduction

Only shortly after the discovery of Bloch's theorem *[Bloch 1928]* for electronic states in systems with translational symmetry, it was realized, that in the case of a three dimensional solid, whose translational symmetry is broken perpendicular to the surface, new bound states of two-dimensional character occur. As the wave functions of these states are localized near the surface they were properly named "surface states"*[Tamm 1932, Maue 1935, Shockley 1938, Goodwin 1939]*. Surface states can exist, if their energy lies within a band gap of the projected bulk band structure and below the "vacuum escape threshold" $E_{vac} + \hbar^2/2m^* k_{//}^2$ ($k_{//}$ denotes the component of the electron wave vector parallel to the surface), so the electron can neither penetrate into the crystal nor escape into the vacuum. The situation is analogous to that of a potential well, with walls confining the motion perpendicular to the surface, but infinitely wide (with a periodic potential) parallel to the surface.

If these surface states are confined also along one or even both directions parallel to the surface their dimension will be reduced from two ("quantum well") to one ("quantum wire") to zero ("quantum dot"). As the dimensionality is a universal quantity determining the physical behaviour of a system (range of interactions, scaling laws, critical exponents etc.), the existence of two-, one- or zero dimensional surface states may have important consequences for the physics on solid surfaces.

In this paper we show examples of two-, one- and zero-dimensional states, occurring on clean and modified transition metal substrates. These metals exhibit large Shockley inverted bulk band gaps, which are almost completely located above the Fermi energy. Inverse Photoemission *[Dose 1985]* and Two-Photon Photoemission *[Steinmann 1994]* are used to probe the unoccupied electronic states in these gaps. Whereas 2PPE with its superior energy resolution is particularly useful to study sharp, long-living states near the center of the SBZ, IPE allows also the investigation of surface states in band gaps near the SBZ boundary, which are not accessible to 2PPE.

2. Two-Dimensional States: Ni(110) and 1 ML Na/Ni(110)

In this chapter we focus on two-dimensional surface states and the concepts that can be used to understand their dispersion behaviour. We illustrate these ideas for a Ni(110) surface, both clean and covered with a quasihexagonal monolayer of sodium *[Memmel 1993]*. Structure models and SBZ's of both surfaces are shown in fig.1. This chapter is divided into three parts: In the first part we discuss the surface states of the clean surface, then we proceed to the two-dimensional band structure of a hypothetical unsupported Na monolayer in vacuum and finally we end with the band structure of a monolayer of Na on Ni(110).

Fig.2a (right) displays the E vs. $k_\parallel$ diagram for clean Ni(110) along the $\overline{\Gamma Y}$ azimuth of the SBZ. Experimental data points as determined by IPE are shown as open circles. An IPE spectrum of clean Ni(110) near the $\overline{Y}$ point is shown in fig.3a. The PBS (grey area in fig.2) exhibits

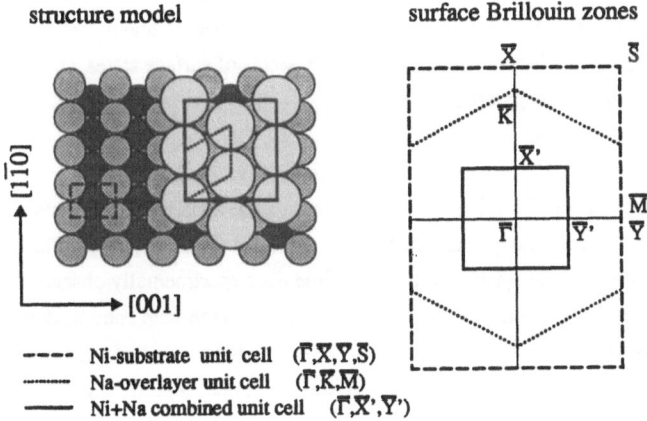

Fig.1 Structure and unit cells for Na/Ni(110) near monolayer coverage in real (left) and reciprocal space (right).

a large gap (white area), centered around the SBZ boundary at $\overline{Y}$. Inside this band gap two bands with opposing dispersion behaviour are clearly visible which are attributed to surface states of the clean Ni(110) surface. They are of two-dimensional character for the following reasons: They occur inside a gap of the projected bulk-band structure and therefore cannot be three-dimensional bulk states. Furthermore they exhibit a strong dispersion along $\overline{\Gamma Y}$ and therefore they are delocalized along this direction. In order to prove that these states are also delocalized in the perpendicular direction, dispersion data along $\overline{YS}$ are needed. Unfortunately no such data exist. However, for the analogous Cu(110) band gap, dispersion of the lower surface state band S_1 was found both along $\overline{\Gamma Y}$ and $\overline{YS}$ [Su 1994], thereby proving the two-dimensional character of this state. The second surface state band S_2 was not observed in this study, probably because it only occurs as a surface resonance outside the bulk band gap.

To understand the dispersion behaviour of these two-dimensional surface states, we performed surface state calculations [Smith 1985, Chen 1987] based on the simple potential diagram depicted in the left part of fig.2a. The semi-infinite periodic potential is described within a nearly-free electron two-band approximation, which takes into account only the first coefficients $V_0+2V_Gcos(Gr)$ of the Fourier expansion of the periodic crystal potential giving rise to the gap in the bulk band structure of width $2V_G$. The values of V_0 and V_G were chosen such as to reproduce position and width of the projected bulk band gap. However, to compensate for the deficiency of the two-band model that - independent of $k_{||}$ - the band gap width always amounts to $2V_G$, the value of V_G has to be adjusted for each $k_{||}$ to reproduce the actual gap width of the PBS. The reciprocal lattice vector $G=2\pi/a\,(111)$ associated with the band gap at $\overline{Y}$ possesses a component normal as well as parallel to the surface. Therefore the pseudopotential $V_0+2V_Gcos(Gr)$ inside the crystal varies not only perpendicular but also parallel to the surface.

The immediate surface region is described by a flat potential of depth V_0 and width d. The value of d was set to 1.2Å in order to fit the experimentally determined surface state energies at $\overline{Y}$. In the vacuum region we used an image potential barrier, decaying as $1/4z$ towards the vacuum level E_{vac}. This gives rise to a Rydberg-like series of surface states, converging towards the vacuum escape threshold parabolas

$$E_{vac} + \hbar^2/2m^*k_{||}^2 \quad \text{or} \quad E_{vac} + \hbar^2/2m^*(k_{||}-G_{||})^2 \,, \tag{1}$$

respectively, which are shown as dashed lines in the E vs. $k_{||}$ diagram.

As a result of the calculations the lowest two surface state bands (a crystal induced state and the first member of the image state series) are drawn in fig.2a (right) as thick solid lines. Considering the simplicity of the model they describe the experimentally observed dispersion behaviour quite well. Note that the dispersion of the upper surface state band S_2 deviates slightly, that of the lower surface state band S_1 markedly from a free-electron like behaviour. This is solely a consequence of the "leakage" of these states into the crystal region, as in the calculation the potential in the near surface and the vacuum region was assumed to be constant parallel to the surface, which would give a free-electron dispersion.

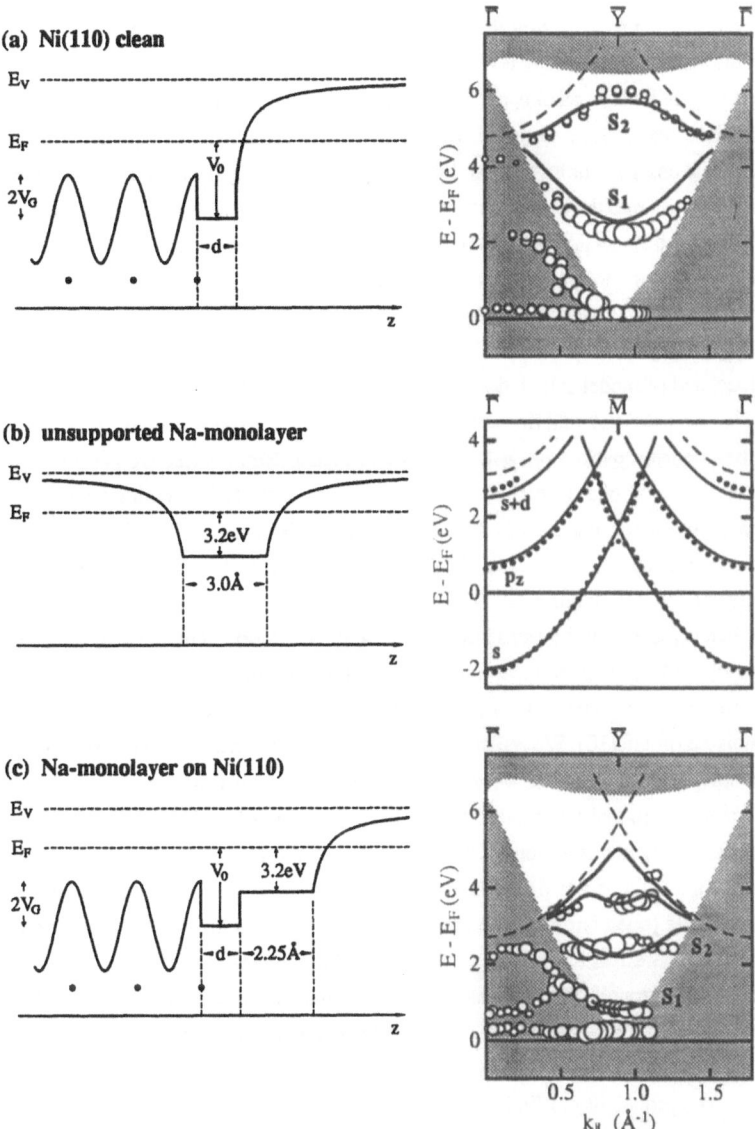

Fig.2   Model potentials (left) used for the surface state calculations and $E(k_{||})$-dispersions (right) for (a) the clean Ni(110) surface, (b) an unsupported Na monolayer in vacuum, and (c) a Na monolayer on Ni(110). Open circles denote experimental data points, solid lines the results from the surface state calculations. Dotted lines in (b) are from FLAPW calculations *[Wimmer 1983]*. The size of experimental symbols indicates the observed spectral intensity. (From *[Memmel 1993]*).

The resulting band dispersion therefore is the consequence of two opposing effects [Chen 1987]: If the surface state energy is close to the "escape threshold parabolas", the electron wave function is mainly located outside the crystal in a region with no (or only weak) potential corrugation parallel to the surface. Consequently we obtain a free-electron like behaviour. However, if the surface state energy is close to a band gap boundary, the electron wave function extends to an appreciable amount into the crystal with its strong potential corrugation and the surface state dispersion follows the dispersion of the bulk band which defines the respective band gap boundary.

Before we proceed to the Na/Ni(110) system, let us briefly consider the band structure of a truly two-dimensional, although only hypothetical system: an unsupported sodium monolayer in vacuum. Fig.2b (right) shows the results of self-consistent FLAPW calculations [Wimmer 1983] (dotted lines) and of model calculations [Lindgren 1989] (solid lines), where the monolayer was modelled by the potential well depicted in fig.2b, left part: A flat potential 3Å wide (=distance between two densely packed Na bulk layers) and 3.2eV deep (=effective Na potential in free electron theory), confined on both sides by image-potential barriers with a vacuum energy of 3.1eV above E_F (=work function of a free Na monolayer [Wimmer 1983]). Parallel to the surface the potential is assumed to be constant. FLAPW and model calculation agree very well with each other. Only at the SBZ boundary the FLAPW calculation exhibits a small gap of ≈0.5eV, which of course is not present in the "structureless" model calculation. The close agreement demonstrates that such a free Na monolayer represents an almost perfect two-dimensional electron gas.

In the final step we combine the results of the previous parts to calculate the band structure of a Na monolayer on Ni(110). We replace the image potential barrier at the left side of the unsupported monolayer by the potential chosen to represent the Ni(110) substrate- and near surface region. The resulting model potential is shown in fig.2c, left. The width of the Na potential was reduced to 2.25Å, which is about the distance between the Na overlayer and the outermost Ni layer. In the right hand part of fig.2c the calculated bands (solid lines) are compared with the experimental results (open circles). The agreement is fairly good, except for band S_2, which according to the experiment has an almost flat dispersion, whereas the calculation yields a slight upward dispersion around $\overline{Y}$. However, with ≈0.5eV the absolute deviation is about the same as in the preceding steps. The main result of both experiment and calculation is, that the lowest three states in the gap show a rather flat dispersion. This is a consequence of the hybrid character of these states: The probability to find the electron in the periodic crystal or in the corrugation free potential region outside, respectively, is roughly equal. Whereas the periodic crystal favours an upward dispersion around $\overline{Y}$, the corrugation-free region outside favours a free-electron like dispersion, i.e. a downward dispersion around $\overline{Y}$.

In the above calculations the Na monolayer was always considered to be structureless parallel to the surface. Therefore the calculated bands for the Na-on-Ni system exhibit the periodicity of the Ni substrate. From LEED investigations it is known that 1 ML Na/Ni(110) forms a superlattice as shown in fig.1, with double periodicity along [001], i.e. along $\overline{\Gamma Y}$ in reciprocal space

[Gerlach 1969]. Since the Na monolayer and the Ni substrate alone both have only single period-icity along [001] and since the pseudopotential corrugation of Na is weak anyway (as indicated by the small band gaps in fig.2b), we can to a first approximation include the doubled periodicity in the calculation by backfolding the bands into the smaller Brillouin zone, without considering further interactions of the backfolded bands with already existing bulk or surface bands. The ex-perimental data (fig.2c) also indicate this doubling of the periodicity. The band S_1 "originally" dispersing around $\overline{Y}$ seems to be present also around $\overline{\Gamma}$.

Depending on the point of view the electronic states of the Na/Ni(110) system can either be considered as the surface states of the clean surface, modified by the presence of the alkali over-layer (indeed the evolution of the Na/Ni(110) states out of the clean surface states has been fol-lowed experimentally *[Memmel 1991]*) or as the states of the free alkali monolayer disturbed by the underlying substrate. Adopting the latter point of view the free-electron like band dispersion of the unsupported alkali-metal film may be severely distorted on substrates like Ni(110) with strongly varying pseudopotentials (i.e. with large gaps in the bulk band structure), whereas on jellium-like substrates with their small pseudopotential corrugation the dispersion (not the en-ergy!) of the alkali-monolayer bands will change only slightly, in agreement with experimental observations *[Heskett 1987]* and ab-initio calculations *[Ishida 1989]*.

3. One-Dimensional States: H/Ni(110)

In this chapter we will show that one-dimensional states do not only exist in truly one-di-mensional systems but may also occur on two-dimensional surfaces, if the pseudopotential varia-tion along one direction is sufficiently strong to cause a localization of electronic states along this direction. Such a strong potential variation may exist due to the presence of steps *[Himpsel 1994]* or individual rows of atoms on the surface *[Bischler 1993]*, as they occur on MR-recon-structed fcc(110) surfaces.

Such a MR reconstruction is induced on Ni(110) by adsorption of hydrogen at room tem-perature. From scanning tunnelling microscopy *[Nielsen 1991]* it is known that already at low hydrogen coverages a missing/added row reconstruction occurs, where some of the densely packed Ni rows are removed from the surface layer and form additional rows along the $[1\overline{1}0]$ di-rection on top of the original terraces. The missing and added rows occur rather irregularly on the surface. At higher coverages the average distance between the rows amounts to twice the lattice constant, but different distances occur as well. Even at saturation coverage of 1.5 ML (produced by cooling the sample in hydrogen atmosphere from 400K down to 200K) the periodicity is strongly disturbed. LEED shows a (1x2) pattern with streaky half-order spots, indicating a one-dimensional disordered surface.

If hydrogen is adsorbed at low temperatures (110K) the formation of the one-dimensional disordered MR reconstruction is kinetically hindered. LEED shows ordered lattice gas structures up to 1 ML and a well-ordered (1x2) pattern at saturation coverage (1.5 ML), due to the formation of a pairing-row reconstruction *[Penka 1984]*.

In the following we will study the influence of this one-dimensional disordering onto two different crystal-induced surface states: The surface state $S_{\bar{Y}}$ (=S_1 in the notation of chapter 2) with $k_\parallel$ along $\overline{\Gamma Y}$ and the surface state $S_{\bar{X}}$ with $k_\parallel$ pointing along $\overline{\Gamma X}$ (the gap in the PBS at $\overline{X}$ extends from ≈2.1 to 9.5 eV above E_F). $S_{\bar{Y}}$ is a state propagating along [001], i.e. along the direction in which the one-dimensional disordering takes place in the MR-reconstructed phase, whereas $S_{\bar{X}}$ propagates along [1$\overline{1}$0], i.e. along the close-packed rows of the Ni(110) surface.

IPE spectra of both states are depicted in fig. 3a ($k_\parallel ≈ \overline{Y}$) and fig. 3b ($k_\parallel ≈ \overline{X}$) *[Bischler 1993]*. In the low-temperature experiment both states exhibit a rather similar behaviour: upon hydrogen adsorption they shift to lower energies towards the Fermi energy. However, the important point to note here is, that both states persist on the hydrogen covered surface. The situation is quite different for the one-dimensional disordered MR reconstructed surface. In this case the surface state $S_{\bar{Y}}$ is completely quenched already at low coverages, whereas $S_{\bar{X}}$ persists at all coverages without any noticeable intensity loss.

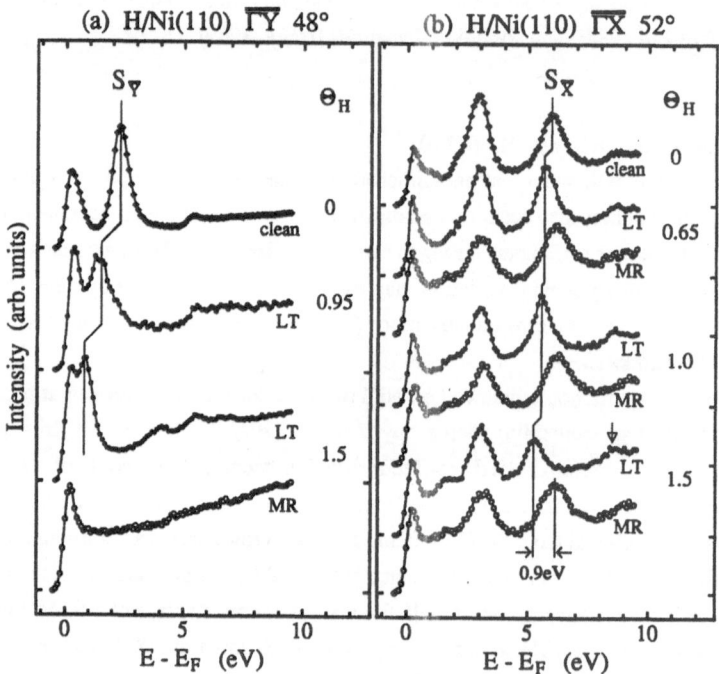

Fig.3 IPE spectra of H/Ni(110) in the low-temperature (LT) and missing-row (MR) reconstructed phase. Spectra are taken near the $\overline{Y}$ (left panel) and $\overline{X}$ (right panel) of the SBZ. Note that for the one-dimensionally disordered MR-reconstructed phase the surface state $S_{\bar{Y}}$ is quenched completely whereas the surface state $S_{\bar{X}}$ persists at all coverages. (From *[Bischler 1993]*).

The insensitivity of $S_{\bar{X}}$ towards the one-dimensional disorder indicates, that this state is localized with respect to the [001] direction. Therefore $S_{\bar{X}}$ is a one-dimensional state extending along the densely packed rows of the Ni(110) surface. $S_{\bar{X}}$ is an almost dispersionless state along $\overline{\Gamma X}$ both on the clean and the H-covered surface. Therefore we cannot exclude that $S_{\bar{X}}$ is also localized along the [1$\bar{1}$0] direction. However, the absence of dispersion is only a necessary, but not a sufficient criterion for a localized surface state *[Bischler 1993]*.

Finally we would like to comment on the upward shift of $S_{\bar{X}}$ upon proceeding from the low-temperature H-covered surface to the MR-reconstructed phase. A similar shift of $S_{\bar{Y}}$ was observed upon the alkali-induced MR-reconstruction of fcc(110) surfaces *[Memmel 1991]*, *[Tang 1993]*, where the MR-reconstruction can also be suppressed at low temperatures. In this work *[Memmel 1991]* the upward shift could be quantitatively explained by a lateral averaged surface potential model, which took account of the reduced substrate atom density of the MR-reconstructed surface.

A different explanation of the upward shift is the one-dimensional localization of the surface state due to the MR-reconstruction. If upon formation of the MR reconstruction $S_{\bar{X}}$ becomes a one-dimensional state localized to a single atomic row, but is a delocalized state on the clean and low-temperature H-covered surface, this localization process will result in an energetic upward shift of $S_{\bar{X}}$. The expected upward shift for the lateral confinement to a single row for the surface covered by 1.5ML of hydrogen can be roughly estimated in a simple one-dimensional particle-in-a-box model. An obvious choice for the width of the box is the lattice constant of Ni which is 3.52Å. The parameters for the bottom and top of the potential well are deduced in the following way: For the completely delocalized state (d→∞) the energy of the lowest eigenstate of the potential well (i.e. $S_{\bar{X}}$) approaches the bottom of the well, which therefore is set to the low-temperature value of $S_{\bar{X}}$, which is 5.3eV. Reducing the width of the well, which simulates the removal of Ni-rows from the surface layer due to the MR reconstruction, results in an upward shift of the energy eigenvalues. For d→0, which is equivalent to the removal of the complete surface layer, the energy approaches the top of the potential well. However, after removal of a complete Ni layer from the semi-infinite substrate the original energy spectrum has to be reproduced, only the labeling by the quantum numbers is different. All surface states appear upshifted by one quantum number *[Memmel 1991]*. Therefore the top of the potential well is chosen to be 8.4eV, i.e the energy of the next higher surface state of the low-temperature phase. This state is indicated by an arrow in fig.3b. Using this simple model we obtain an upward shift of $S_{\bar{X}}$ by 1.1eV which compares surprisingly well with the experimental value of 0.9eV.

In both explanations discussed so far - the laterally averaging and the confinement model - the upward shift has the same origin, namely the MR-reconstruction of the surface A further - but in our opinion rather unlikely - explanation of the observed energy shift may be different adsorption sites of H in the MR reconstructed and the low-temperature phase.

4. Zero-Dimensional States: Ag/Pd(111)

The final chapter of this paper is devoted to electronic states that are localized in all three dimensions of space, i.e. zero-dimensional electronic surface states. We illustrate this for a special class of surface states - the image states. As already discussed before, these states are mainly localized in front of the surface and form a Rydberg-like series close to the vacuum escape threshold. For $k_\parallel=0$ the energies of these states are given in good approximation by [Echenique 1990]

$$E_n = E_{vac} - \frac{0.85eV}{(n+a)^2} \qquad (2)$$

with the quantum defect a depending on the properties of the substrate. On inhomogeneous surfaces the question arises, whether the vacuum energy, which these states are pinned to, is determined by the macroscopic laterally averaged work function or by the local work function of the different patches existing on the surface - a question which is intimately connected with the lateral extension of the electronic states.

This problem was studied by Fischer et al. [Fischer 1993a, 1993b] for ultrathin films of Ag grown on a Pd(111) substrate. At room temperature these films grow in a layer-by-layer mode [Eisenhut 1993]. Upon deposition of Ag the macroscopic work function decreases from 5.44eV by ≈0.9 eV with a linear variation between the completed monolayers. Both Pd(111) and Ag(111) exhibit a gap in the PBS near E_{vac} thus meeting the requirements for the existence of image states.

Fig. 4 shows 2PPE spectra at $k_\parallel=0$ for various room temperature deposited Ag films on Pd(111) [Fischer 1993a]. The clean surface spectrum exhibits a prominent peak at 4.9eV which is due to the n=1 image state of the clean surface. The weak feature observed at lower energies results from a crystal induced surface state and will not be discussed in the following. Upon Ag deposition the image potential state loses intensity and is no longer visible for 1 ML Ag/Pd(111). Simultaneously a new state appears at 4.1eV above E_F reaching maximum intensity at 1 ML. At higher coverages this state loses in-

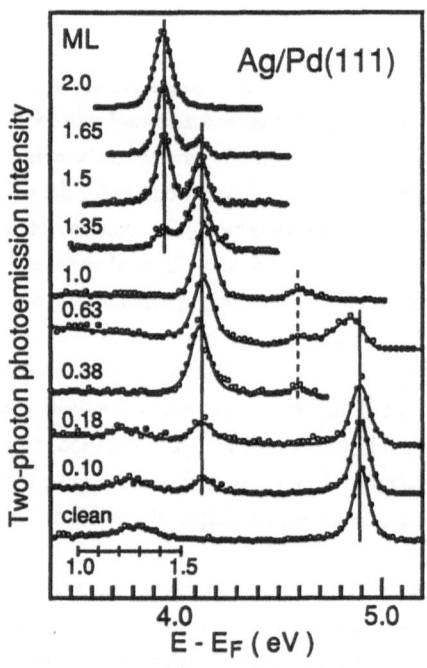

Fig.4 2PPE spectra at $k_\parallel=0$ for various amounts of Ag deposited onto Pd(111) at RT. The inserted energy scale is valid only for the crystal-induced surface state (seen in the lowest three spectra) due a different excitation process. (From [Fischer 1993a]).

tensity and disappears at 2 ML. Above 1 ML a second peak evolves at an energy of 3.95eV. The new states appearing upon Ag deposition are attributed to the n=1 image states of 1 and 2 ML Ag/Pd(111). Note that for fractional coverages two n=1 states exist. Independent of coverage the image states are observed at fixed energies, although the macroscopic work function varies continuously with coverage. From this Fischer et al. concluded that the observed image states "feel" the local work function of the different patches existing on the surface at intermediate coverages. Using equation (2) the local work function (and quantum defect a) can be calculated from the measured energies E_1 and E_2 of the n=1 and n=2 image states (the energy of the n=2 image state is indicated in fig.4 by the dashed line). As E_1 and E_2 remain constant until the next monolayer is completed, the calculated local work function also stays constant. Whenever a full layer is completed it equals the macroscopic work function within 20meV *[Fischer 1993a]* . The fact that the energies of the image states are determined by the local work function also implies that an electron residing in such a state is localized laterally to the region of this particular patch on the surface.

At room temperature the individual patches present on the surface have diameters larger than 100Å *[Eisenhut 1992]*. This is much larger than the typical extension of n=1 image states perpendicular to the surface, which is of the order of several Å. Therefore the lateral confinement to these large patches does not result in a measurable energy shift or broadening of the 2PPE peaks. The situation is different for films grown at 90K, where much smaller islands are formed due to a reduced atom mobility. Fig.5 shows 2PPE spectra obtained for various submonolayer coverages of Ag deposited at 90K. Compared to the room temperature n=1 image state whose position is indicated by the dashed line, the image state is asymmetrically broadened and shifted towards higher energies. At a coverage of 0.25 ML the energy shift amounts to 200meV. Increasing the coverage and/or heating reduces the energy position. The observed energy shift is ascribed to the lateral confinement of the image state to rather small Ag islands on the Pd(111) surface. The smaller the islands are, the larger the observed energy shift. The asymmetric broadening results from the simultaneous existence of islands with different sizes and reflects the island size distribution.

To quantify these ideas, the spectra were analyzed using the following assumptions: 1) The wave function can be separated into a lateral and perpendicular part. The latter is independent of coverage and given by that of the surface homogeneously covered by 1 ML of Ag. 2) The lateral confinement to the Ag islands can be described by a two-dimensional circular potential well of diameter d. The height of the potential well is given by the energy difference of the n=1 states of clean Pd(111) and 1 ML Ag/Pd(111). The energy shift relative to the 1 ML RT spectrum then approximately scales as $\sim 1/d^2$. The experimental spectra at various coverages were now fitted by a weighted superposition of shifted 1 ML RT spectra using a least square procedure. The energy shift is related to the island size according to the second assumption, whereas the weighting factor gives the total area covered by islands of that particular size. The resulting island size distri-

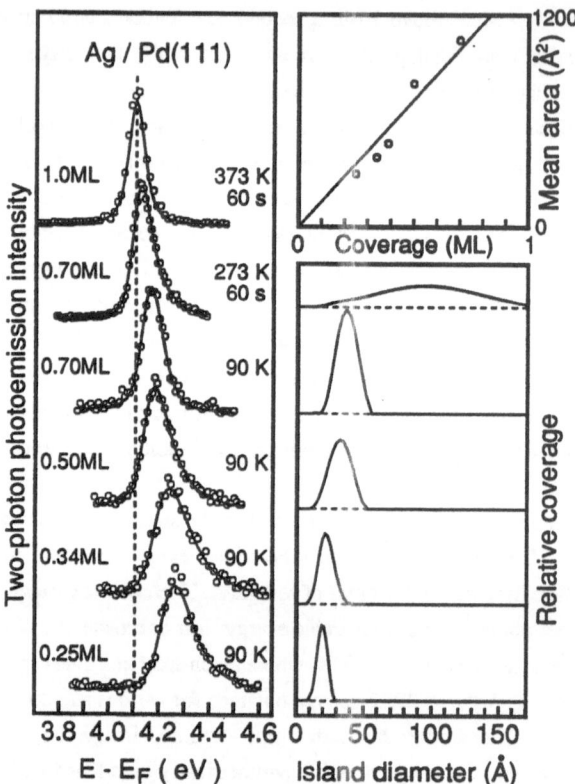

Fig.5 Left panel: 2PPE spectra for various amounts of Ag deposited onto Pd(111) at 90 K. The energy of the n=1 image state on the surface homogeneously covered with 1 ML of Ag is shown as the dashed line. Right panel: Island size distribution (bottom) and mean island area (top) derived from the experimental data. (From [Fischer 1993b]).

butions show the expected behaviour (see fig.5 bottom right): At small coverages the island size amounts to only ≈20Å, in agreement with STM observations for several metal/metal systems [Hwang 1992, Becker 1993]. Larger coverages result in larger islands sizes. After annealing the 0.7 ML film to 273K a broad distribution of island sizes with a mean value of 100Å is obtained. At 90 K the mean area of the islands increases linearly with coverage (see fig.5 top right), indicating a constant island density of $\approx 7 \cdot 10^{12} cm^{-2}$ for coverages up to 0.7 ML [Fischer 1993b]. Above this coverage coalescence of the individual islands may occur.

5. Outlook

Examples of two-, one- and zero-dimensional surface states were presented. In all of these examples the surface states are completely unoccupied and therefore do not contribute to total energy of these systems. It remains to explore the consequences of the reduction in dimensionality from two-dimensional surface states down to one- or zero-dimensional states in cases where a surface state is partially occupied and therefore plays an active role in the physical processes occurring on the surface.

Acknowledgements

It is a pleasure to thank E. Bertel for numerous stimulating discussions and R. Fischer and Th. Fauster for discussions about their results on Ag/Pd(111). Support from SFB 338 is gratefully acknowledged.

References

A.F. Becker, Th. Michely and G. Comsa, Surf. Sci. **272**, 161 (1992).

F. Bloch, Z. Physik **52**, 555 (1928).

U. Bischler and E. Bertel, Phs. Rev. Lett. **71**, 2296 (1993).

C.T. Chen and N.V. Smith, Phys. Rev. B **35**, 5407 (1987).

V. Dose, Surf. Sci. Reports **5**, 339 (1985).

P. Echenique and J.B. Penrdy, Prog. Surf. Sci. **32**, 111 (1990).

B Eisenhut, PhD thesis, University of Munich, unpublished (1992).

B. Eisenhut, J. Stober, G. Rangelov and Th. Fauster, Phys. Rev B **47**, 12980 (1993).

R. Fischer, S. Schuppler, N. Fischer, Th. Fauster and W. Steinmann, Phs. Rev. Lett. **70**, 654 (1993a).

R. Fischer, Th. Fauster and W. Steinmann, Phs. Rev. B **48**, 15496 (1993b).

R.L. Gerlach and T.N. Rhodin SS **17**, 32 (1969).

E.T. Goodwin, Proc. Camb. Phil. Soc. **35**, 205, 221, 232 (1939).

D. Heskett, K.H. Frank, E.E. Koch and H.J. Freund, Phys. Rev. B **36** , 1276 (1987).

F. Himpsel et al., submitted for publication (1994).

R.Q. Hwang, C. Günther, J. Schröder, S. Günther, E.Kopatzki and R.J. Behm, J. Vac. Sci. Technol. A **10**, 1970 (1992).

H. Ishida, Phys. Rev. B **40**, 1341 (1989).

S.Å. Lindgren and L Wallden in *Physics and Chemistry of Alkali Metal Adsorption*, edited by H.P. Bonzel, A.M. Bradshaw and G. Ertl, Elsevier, Amsterdam (1989).

A.W. Maue, Z. Physik **94**, 717 (1935).

N. Memmel, G. Rangelov, E. Bertel and V. Dose, Surf. Sci. **251/252**, 503(1991).

N. Memmel, G. Rangelov and E. Bertel, Surf. Sci. **285** (1993) 109.

L.P. Nielsen, F. Besenbacher, E. Laegsgaard and I. Stensgaard, Phs. Rev. B **44**, 13156 (1991).

V. Penka, K. Christmann and R. Ertl, Surf. Sci. **136**, 307 (1984).

W. Shockley, Phys. Rev **56**, 317 (1939).

N.V. Smith, Phys. Rev. B **32**, 3549 (1985).

W. Steinmann and Th. Fauster, in *Laser Spectroscopy and Photochemistry on Metal Surfaces*, edited by H.L. Dai and W. Ho., World Scientific (1994).

C. Su, D. Tang, D. Heskett, submitted for publication (1994).

I. Tamm, Physik. Z. Sowjetunion **1**, 733 (1932).

S. Tang, C. Su and D. Heskett, Surf. Sci. **295**, 427 (1993).

E. Wimmer, J. Phys. F **13**, 2313 (1983).

Electronic Properties of Semiconductor Surfaces and Fermi Surface Studies Using Photoelectron Spectroscopy

J. D. Riley, R. Leckey, Y. Cai, X. Zhang and J. Con Foo
School of Physics, La Trobe University, Bundoora Vic. 3083, Australia

1. Introduction

This paper discusses measurement of the band structure of a semiconductor, GaAs, and the Fermi surface of Copper determined from angle resolved photoemission measurements. The dissimilarity in the materials and the information derived is not as divergent as may first appear. The measurements techniques are identical, the methods of interpretation are similar, both provide information of bulk electronic states and anyway, the Fermi surface is way of describing aspects band structure.

Angle resolved photoelectron spectroscopy using synchrotron radiation as a light source is the established technique for the determination of the electronic energy band structure of conducting solids. Many experimental band structures have published for a large variety of metals and semiconductors. A summary of some of these can be found in the Landolt-Börnstein series [1].

The measured transitions occur from band states of the crystal to an excited state which permits the electron to emerge into the vacuum. It is not possible to determine all quantum numbers describing the initial state from a single measurement, as the perpendicular component of momentum is not conserved across the crystal surface. The triangulation process [2] using two measurements of the same transition observed from different surfaces, does offer an absolute determination of the momentum vector but it has only been applied in a limited way to a few metals for which the transitions are relatively sharp.

This lack of knowledge of the perpendicular component of k tends to hide the fact that in order to describe a transition between two states both are required before the transition can be said to be explained. The lack of knowledge means that both states are not known and modelling is required. It is possible to assume knowledge of one or other of the states in which the perpendicular component of k can be determined. In the usual experiments it is not possible to determine the band structure independent of a model of the photoemission process.

In the published results the experimental results often confirm the theoretical model chosen by the experimentalist for the purposes of comparison. This is not altogether surprising, for in the case of simple materials such as the III-V semiconductors, even the "ab-initio" calculations have parameters which permit the models to be adjusted to agree with experimentally determined critical points. The result is that all the models provide transitions which can be assigned to observed transitions within an acceptable accuracy. It is argued here that is not possible to determine the best model using photoemission, as the differences between models are smaller than the precision of the measurements. In this paper the validity of the claim that photoemission determines the bandstructure of materials and the models used to interpret photoemission experiments will be discussed.

The problem of determining the perpendicular component of momentum is resolved by applying a model to supply the missing parameters. The usual approach is to supply intermediate states between the initial state of the electron in the crystal and the final state of the electron in the detector known as the three step model. The steps are (i) the initial excitation from the ground state to an excited state, (ii) the translation of the electron to the surface and (iii) the negotiation of the surface barrier to a state which is observed in the detector. These intermediate states are considered physical and are characterised by (i) a final state determined by an optical transition from the initial state (ii) a mean free path relating to the distance the electron travels through the crystal and (iii) the loss of energy and momentum from the transition through the barrier. The final step in resolving the problem requires a model of one of the unknown states the most usual being the intermediate final state.

The most popular choice is the free electron model in which the final state is taken to be a free electron state but still inside the crystal. The life time of the state is rather short which increases the uncertainty in k and limits the effect of the crystal potential. This model has one adjustable parameter, the inner potential and the model permits the calculation of the missing momentum components if the work function of the material is known. There are subtleties associated with the selection of a reciprocal lattice vector to be involved in the transition which will be illustrated briefly below. Such analysis gives results which show reasonable agreements between calculated initial states and experiment. One problem is that the quality of the fit is not particularly sensitive to the value of the inner potential. Another problem is that this modelling is equivalent to fitting a parabolic approximation to the final state bands which is a good approximation over a limited range of energies. The range of validity for this approximation is smaller at lower kinetic energies than at higher kinetic energies supporting the usual view that the model applies at kinetic energies greater than ~16 eV.

Another possibility is to use the calculated final states of the crystal as the final states of the photon excitation. This has been successful for low photon energies where the calculations are considered reliable but the rather limited range of energies over which this it true severely limits the usefulness of this process.

The other alternative is to assume that the initial state calculations are accurate and to use them to determine the intermediate final state. Such analyses have been published to demonstrate the inadequacy of the calculated excited states of the materials examined.

2. Interpretation of Photoelectron Spectra.

Photoelectron Spectroscopic data is usually obtained in either of two ways: the intensity of emission normal to the surface as a function of kinetic energy at a particular photon energy but for a range of photon energies or the intensity of emission as a function of polar angle in a plane perpendicular to the surface and kinetic energy for a fixed photon energy. The first is the most usual method of data collection which has the advantage that the electrons are emitted from states which have momentum parallel to the normal to the surface which is usually a principle direction in the Brillouin Zone.

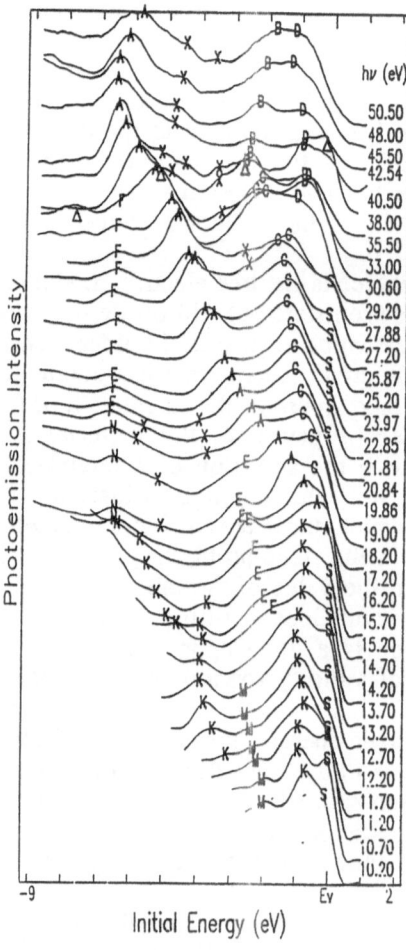

Figure 1. Normal emission spectra obtained from a GaAs (001) surface for a range of photon energies. Binding energies are referred to the valence band maximum.

A data set is shown in Figure 1 for emission normal to the the GaAs (100) surface [3]. The peaks in the spectra are the transitions from initial states in the bandstructure and are plotted on a diagram described as a structure plot shown in Figure 2 whose axes are binding energy with respect to the valence band maxima and the photon energy. This plot has the virtue that experimental data is plotted without interpretation except for the choice of the valence band maxima which in this plot is chosen to be the lowest binding energy of any observed transition. The process of optical excitation involves a reciprocal lattice vector and transitions can occur using different G vectors. Thus transitions from a particular initial state can occur on the diagram at different points and the G vector involved must be identified to permit interpretation.

(A ---- X) : EXPERIMENTAL DATA
(CURVES) : THEORETICAL TRANSITIONS (FEFS)

GaAs(001)

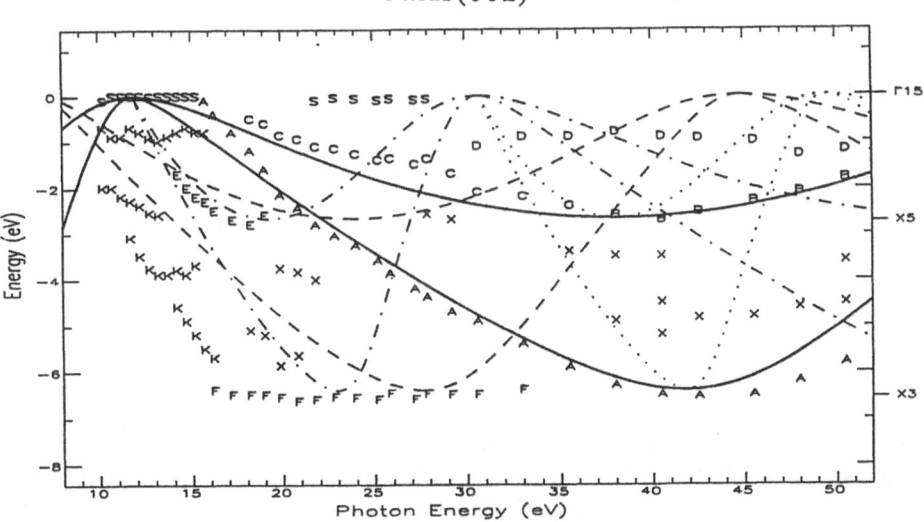

Figure 2. Normal emission structure plot showing experimentally determined transitions plotted as a function of binding energy and photon energy. The solid lines are primary cone transitions predicted using an LMTO calculation as described in the text and free electron final states. Secondary cone transitions are shown as dashed lines.

Predicted transitions can be plotted onto this diagram using a calculated initial state band structure and different choices of G vector permitting the identification of the initial states from which they arise. For the structure plot shown, the initial states have been calculated using an LMTO density functional calculation [5] and transitions are then predicted using these initial states and free electron final states. The value of the inner potential, the base value for the free electron final states with respect to the valence band maximum, is chosen so that the predicted transitions at the X point coincide with the observed transitions. Usually the transitions involving the G vector perpendicular to the surface are strongest and conventionally referred to as the primary cone transitions while transitions using other G vectors are known as secondary cones.

Examination of the structure plot shows that there is a reasonable agreement between the observed primary cone transitions labelled A and the calculated transitions while there are some disagreements at low photon energies. These transitions come from a Δ_1 band along the $\Gamma\Delta X$ direction. Agreement is also observed between the secondary cone transitions labelled E and the calculated values. These arise from the $\Delta_{4,5}$ band along the $\Gamma\Delta X$ direction. A detailed discussion had been given of the agreement and divergences between theory and experiment for this data by Cai et al.[3] and the conclusion drawn is that this modelling will account for a majority of the transitions observed.

3. Off normal data.

The alternative method of data acquisition and presentation is to determine the photoemission intensity as a function of kinetic energy over all polar angles in a plane perpendicular to the surface for a fixed photon energy, referred to as "off-normal" data. The momentum of the states observed does not in general lie along high symmetry directions of the Brillouin Zone so the interpretation and comparison with theory is more difficult.

A plot of the peaks observed from GaAs (001) surface for a detection plane containing the [100] azimuth, the (010) plane is shown in figure (3). Calculated transitions are superimposed on the data using the LMTO calculation referred to above and free electron final states. In this case the bandstructure is required for the whole detection plane and transitions are determined by plotting constant energy difference contours. These are plotted onto the experimental binding energy - emission angle axes. Circles with an included cross show the calculated surface state dispersion from the calculation of Pollman and Pantelides [4] for the dangling bond state. Primary cone transitions are shown in continuous lines and secondary cone transitions in hatched lines.

The agreement achieved is subjective with agreement occurring between the points labelled C for the primary cone. Those points labelled E show a variation which follows the calculated secondary cone transitions though not an exact fit to the energies. The points labelled F which are from regions of the Brillouin Zone near the X point show a small variation in energy which mimics the calculated secondary cone transitions suggesting that these transitions are direct and have a greater intensity than other secondary cone transitions due to the high density of states in the region.

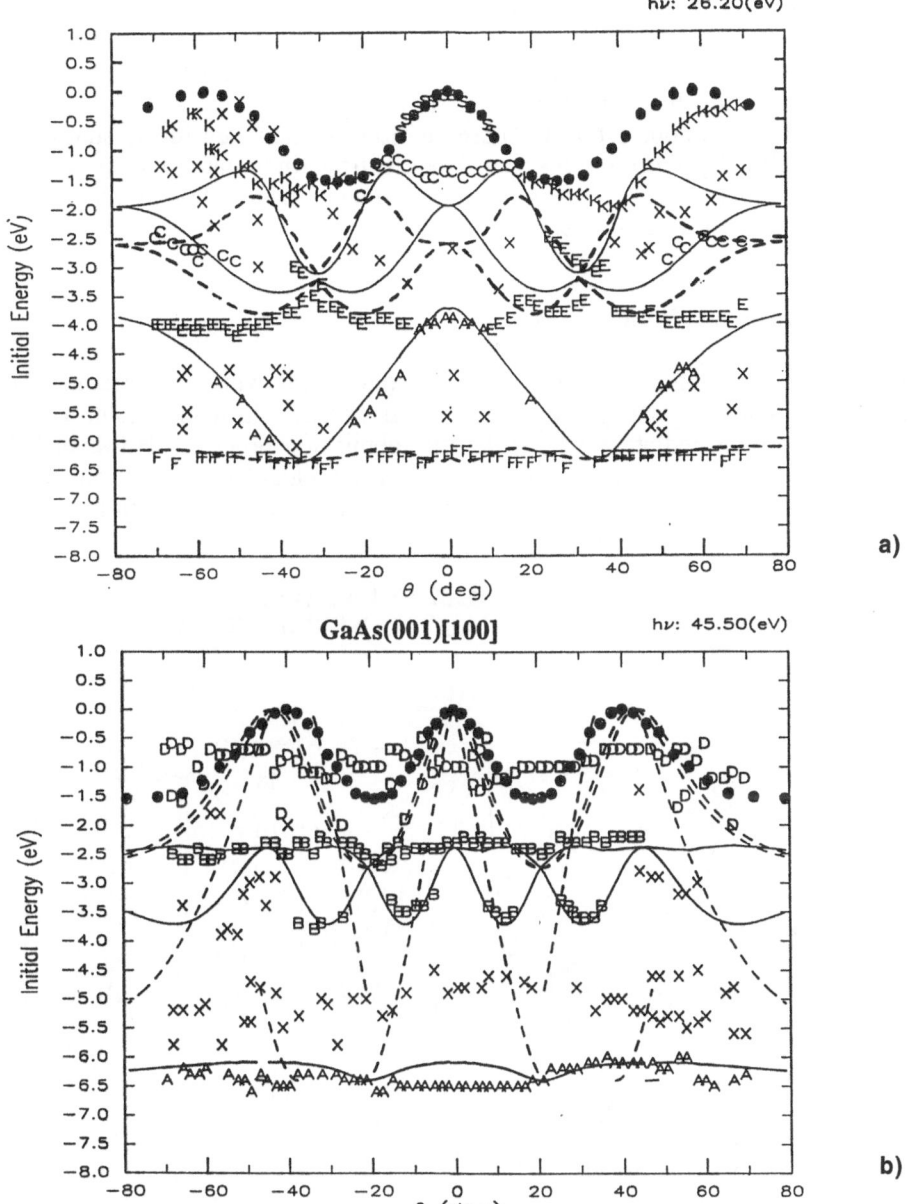

Figure 3.   Off normal emission structure plot showing experimentally determined transitions as a function of binding energy and emission angle at (a) 26.2 eV and (b) 45.5 eV. The solid lines are primary cone transitions predicted using LMTO intial states and free electron final while the secondary cone transitions are shown as dashed lines.

Large numbers of such data sets have been examined and the agreement between the calculation and the experimental data is better for the higher photon energies as seen in figure (3). In each case the agreement between the primary cone transitions and the calculations is good.

Thus it would appear that the combination of calculated initial states with matched critical point energies and free electron final states will give some agreement at lower photon energies and good agreement at higher energies for transitions in semi-conductors.

4. Experimental final states.

Band structure calculations show excited state bands which are ground state bands and not necessarily the states for an excited electron - hole, a more appropriate intermediate state for the photoemission process. However it is instructive to compare the experimental data and calculated transitions using both calculated initial and final state bands. The result is shown in figure (4) for the normal emission date used above.

(A ---- X) : EXPERIMENTAL DATA
(CURVES) : THEORETICAL TRANSITIONS (LMTO)

GaAs(001)

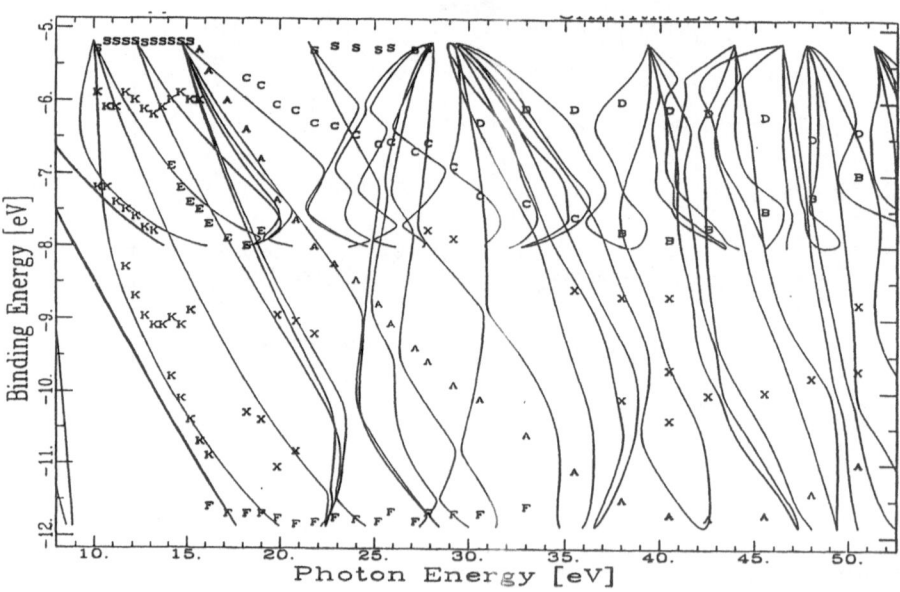

Figure 4. Normal emission structure plot with the experimental data as for figure 2 with the predicted transitions determined using LMTO initial and final states.

The agreement is reasonable for transitions at lower photon energies. Transitions labelled E are close to calculated transitions as are the low photon energy transitions A. There appear to be few calculated transitions which correspond in any systematic way with the observed transitions at higher photon energies. This is consistent with the expectation that the accuracy and adequacy of the calculation will deteriorate at higher photon energies due to the limited basis set used in the calculations.

Calculated initial states and lower energy final states do give agreement with observed transitions suggesting that the discrepancies observed in the the predictions arises largely from problems with the final states.

5. Determination of the final state bands.

Ab-initio bandstructure calculations are either able to or organised to produce bands whose critical points agree with observed values. They can also match the curvature near the Γ point so and would only be judged as good band structure calculations if the shape of the initial state bands agreed largely with simpler calculations or data from experiments. Variations between acceptable calculations of initial states will therefore be small while variations for the conduction bands are considerable. It seems reasonable then in the face of an unknown in the measurements, the perpendicular component of the momentum k, to choose the initial states as accurate and to use the measurements to determine the intermediate final states.

This has been done for normal emission from the two surfaces of GaAs (001) and (110). In each case the wavevector k is determined from the initial state band and the observed binding energy of the initial state. The final state energy and momentum is then plotted onto the LMTO calculated final states [5]. In those regions where the initial state band are reasonably flat a significant uncertainty is introduced into the value of k. Also there are regions of the initial state bands where two or more bands have the same binding energies at different k values. Various structure plots assist in determining the particular initial state from which the transition occurs and this process has been discussed in detail in Zhang et al [6].

Figures (5) show the experimentally determined final states superimposed on the LMTO calculation. The band for the (110) surface show a reasonable agreement with the calculation shown here with states falling along the apparent free electron band. Agreement with the calculation at the Γ point is also quite good.

The data for the (001) surface is more limited with the agreement being reasonable except for the transitions labelled B which show significant deviation from the calculated curves and arise from transitions which deviated significantly from the predicted transitions using the free electron final states. Data from several experiments confirms this behaviour but the predicted final state dispersion appears too large.

The determination of final state bands and their comparison with theory provides a spur for improved calculations of the conduction bands of semiconductors.

GaAs BAND STRUCTURE:

——— : LMTO Calculated Bands

Symbols : **Experimental Data (Zhang et al., 1993)**

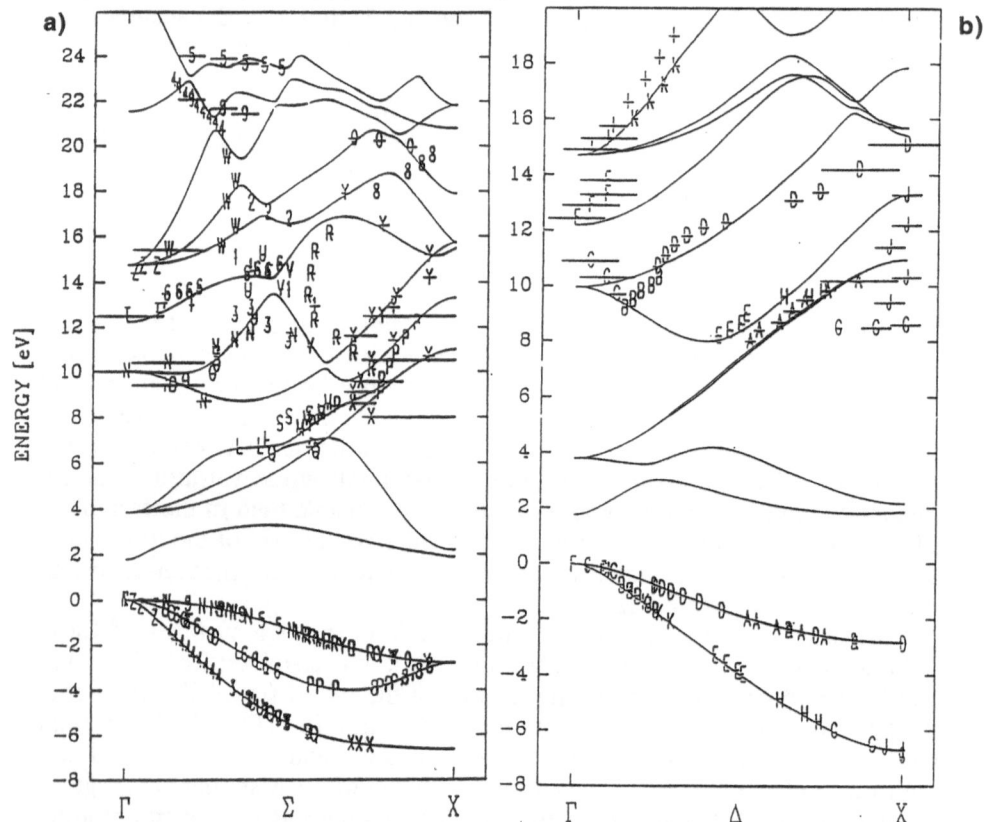

Figure 5. Experimentally determined transitions at normal emission from (a) GaAs (001) and (b) GaAs (110) surfaces plotted onto the LMTO initial states so as to predict the final state for the transition. The conduction band states are LMTO [5] calculated final states.

6. Fermi Surface of Copper.

The measurement of the Fermi surface can be regarded as a measurement of one aspect of the band structure of the material, for if the band structure is known the Fermi surface can be extracted. De Haas Van Alphen measurements have been the usual method for determining the Fermi surface of materials but recently photoemission has been used to determine the Fermi surface of surface states for several metals, the Surface Fermi surface. The bulk Fermi surface can also be obtained from photoemission by determining the k for bands which cross the Fermi level. This suggests that if the emission intensity is measured for initial states with binding energy at or near the Fermi energy as a function of azimuthal angle, regions of high intensity will map transitions from the Fermi surface.

The Fermi surface is a 3 dimensional surface in k space. In photoemission measurements the component of k perpendicular to the surface depends largely on the photon energy. Thus to plot the 3 dimensional shape of the Fermi surface, measurements at many different photon energies are required. At any photon energy transitions may not be observed from all available points on the surface since for transitions to be observed a final state is required. Thus the observed transitions may only show parts of the Fermi surface. A further complication occurs as different G vectors may be involved in the transition though it is expected that transitions using the primary cone G vector will dominate the results.

7. Measurements.

Extensive measurements of transitions from the Fermi surface of Copper have been made on 3 surfaces using a toroidal display analyser [7] enabling all transitions to be observed from the sample surface for a range of photon energies from 10 eV to 90 eV. Similar measurements have been reported by Aebi et al [8] using 21.22 eV. Examples of the results are shown below.

As a preliminary validation of the measurements the Copper Fermi surface as parametrised in Halse [9] has been used to describe the initial states for the transition and a free electron as the final states. Transitions can then be predicted and compared with the measurements.

Figure (6) shows intensity variations as grey scale images for photon energies of 25 eV and 55 eV from the GaAs (001) surface. The predicted transitions have been calculated and are shown below the experimental results. The agreement between prediction and experiment is good and indicates that the observed transitions show the expected behaviour in k space.

Figures (7) and (8) show the intensity variations from the GaAs (111) surface at 21.2 eV and 55 eV and from the GaAs (110) at 24 eV. In each case the predicted transitions reproduce the experimental results showing that they do indeed originate from the Fermi surface of the material.

To determine whether photoemission will measure the Fermi level with a significant accuracy, it will be necessary to calculate the k value for the transitions and map them onto a three dimensional surface.

Cu Fermi Surface.

Crystal Surface (001) Photon Energy 25.0 eV.   Crystal Surface (001) Photon Energy 55.0 eV.

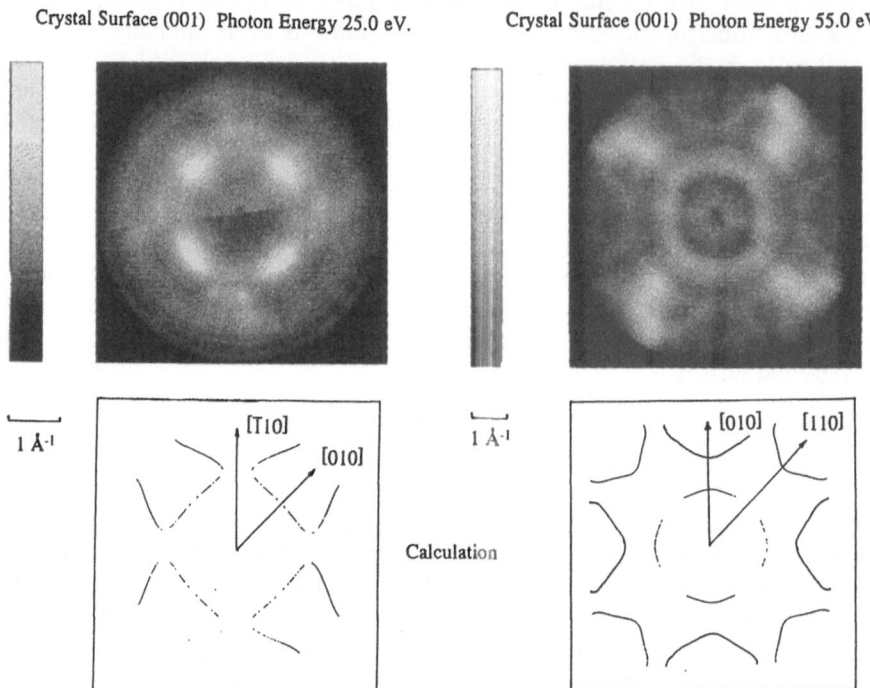

1 Å⁻¹   1 Å⁻¹   Calculation

Figure 6. Intensity variation of emission at the Fermi energy from a Cu (001) shown as a grey scale image at photon energies of 25 eV and 55 eV as a function of polar angle and azimuthal angle. Predicted transitions are shown based on the known Fermi surface [9] and free electron final states.

Cu Fermi Surface.
Crystal Surface (111) Photon Energy 21.21 eV.
Crystal Surface (111) Photon Energy 55.0 eV.
1 Å-1
[1̄1̄2]
[1̄10]
1 Å-1
Calculation
[1̄1̄2]
[1̄10]

Figure 7. Intensity variation of emission at the Fermi energy from a Cu (111) shown as a grey scale image at photon energies of 21.2 eV and 55 eV as a function of polar angle and azimuthal angle.

Cu Fermi Surface.
Crystal Surface (110) Photon Energy 24.0 eV.
Calculation
1 Å-1
[001]
[1̄11]
[1̄10]

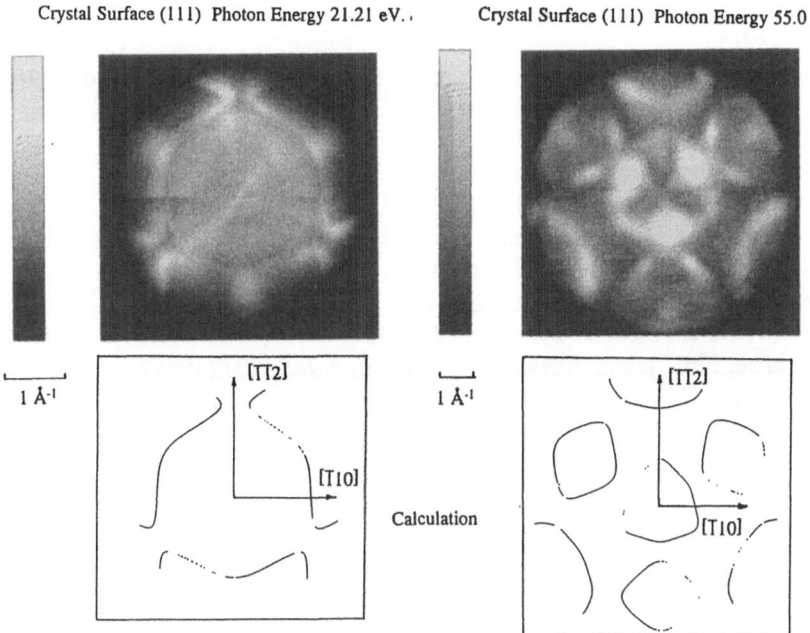

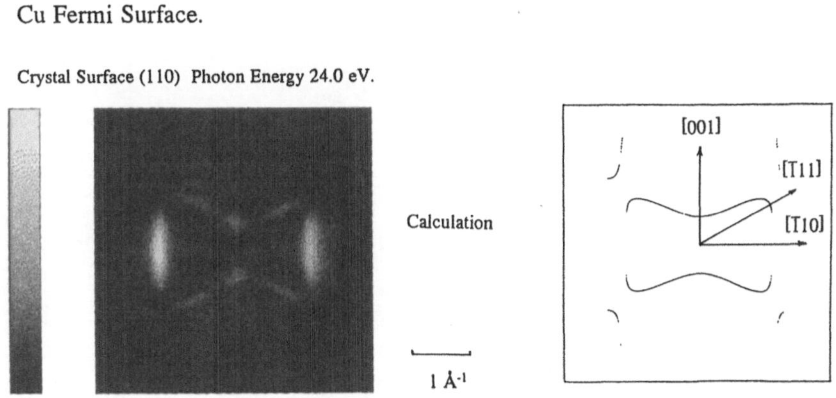

Figure 8. Intensity variation of emission at the Fermi energy from a Cu (110) shown as a grey scale image at photon energies of 24 eV as a function of polar angle and azimuthal angle.

References

[1]   Landolt-Börnstein, „Zahlenwerte u. Funktionen aus Naturwiss. u. Technik",
       Bd. 23 (1989)
[2]   P. Heimann, P. Miosga and H. Neddermeyer, Solid State Comm. **29**, 463 (1979)
[3]   Y.Q. Cai, A.P.J. Stampfl, J.D. Riley, R.C.G. Leckey, B. Usher and L. Ley,
       Phys. Rev. B46, 6891-6901 (1992)
[4]   Pollman and Pantelides, Phys. Rev. B18, 5524 (1978)
[5]   N.E. Christiansen, Phys. Rev. B30, 5753 (1984)
[6]   X.D. Zhang, J.D. Riley, R. Leckey and L. Ley, Phys. Rev. B48, 17077 (1993)
[7]   R.C.G. Leckey, R. Fasel, D. Naumovic, L. Schlapbach,
       Surface Science and J.D. Riley, Appl. Surf. Sci. **22/23**, 196 (1985)
[8]   P. Aebi, J. Osterwalder **307-309**, 917 (1994)
[9]   M.R. Halse, Phil. Trans. Royal Soc. (London) A **265**, 507 (1969)

Recent Progress in Investigation of Luminescent and Electrical Phenomena from Semiconductor Cleavage

D. Haneman, D.G.Li, N.S.McAlpine and C.J.Kaalund
School of Physics, University of New South Wales, P.O.B.1, Kensington, Australia 2033

Abstract. Measurements of luminescence from cleavage of a range of semiconductors in vacuum and in air, have shown that some overall generalisations are possible. In the cases of Si, Ge, Ge_xSi_{1-x} of three different compositions, and GaAs and InP, a short duration light emission (20 microsec. or less) was found to occur on cleaving. This luminescence took place whether the cleavage was in air or high vacuum. It has been interpreted as originating from defect regions on the cleavage surfaces. Another emission which only occurs upon cleavage in vacuum, is associated with bulk band recombination of electrons excited by the cleavage process. This has a relatively long duration in the cases of the elemental semiconductors, attributed to their indirect band gaps. Its spectral distribution reveals that hot electrons are present. A surface state signal has also been detected from Si. Preliminary work on cleavage in ambients other than air appears to indicate a signal-shortening effect. Studies of electrical conduction across the crack during cleavage of Si show that voltage pulses up to about 400 mV are produced at cleavage, with currents up to about 5 mA being detected. This is explained in part as due to generation of a dipole moment on cleavage. Such a dipole can only occur if scission takes place on the three-bond plane and not on the single-bond plane. This has important consequences for surface structure models.

1. Introduction

We have already reported the generation of light when Si, Ge, GaAs, InP and 3 different compositions of Ge_xSi_{1-x} are cleaved in vacuum or in air.[1-5] This luminescence is caused by the de-excitation of electrons that have gained energy from the processes of bond rupture. The precise details of this process are not yet understood. In all cases, except possibly GaAs where the emission was rather weak,[4] there can occur at least two luminescence signals of different characteristics, and in Si three were observed. A schematic diagram is shown in Fig.1(a). A method that we frequently use to carry out the experiment is shown in Fig.1(b). Analysis of these emissions has led to new insights into the nature of cleaved surfaces, and into the nature of the cleavage process.

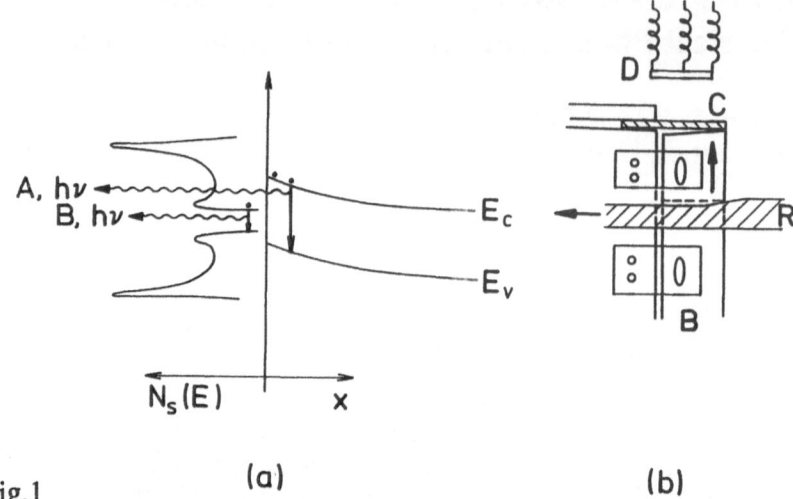

Fig.1 (a) (b)

(a) Schematic diagram to illustrate the three kinds of luminescence observed upon cleaving Si. The bulk and defectt recombinations have been observed on all seven types of semiconductor tested.

(b) Diagram of one type of frequently used cleavage device, featuring gradual increase of bending stress. C is the semiconductor wafer, D the detector on an anti-vibration mounting, B is a block with a bevelled top and R a rod mounted on a vacuum-sealed linear motion feedthrough that can force B up slowly.

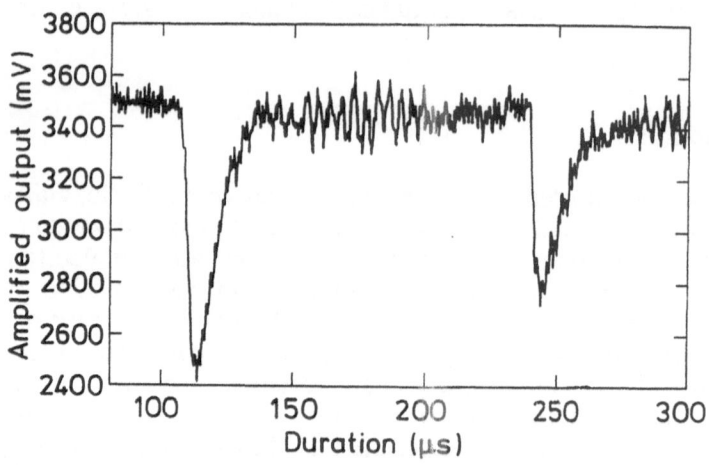

Fig.2

Typical short duration luminescence signals, called C. The figure shows emission from two sequential cracks in Ge cleaved in air.

2. Emission from Surface Defects

In all the seven kinds of semiconductor studied, there can occur a relatively short duration signal (20 microsec. or less), that appears whether the material is cleaved in vacuum or in air. A typical signal from Ge, cleaved in air, is shown in Fig.2. The vacuum used for these experiments is usually only 10^{-4} to 10^{-5} Torr, since no significant contamination can occur at these pressures in the experiment duration of less than 1 millisec. As a time comparison, the time of progression of cleavage is usually a few microseconds[2] for the thin wafers that were mostly used. It is possible that this ubiquitous signal, called C, has an even shorter intrinsic emission time, with the measured emission time being a convolution of the intrinsic duration and the duration of crack progression, regarding the crack as a continuously developing source.

The fact that the signal C is observed from so many different semiconductors, and is also observed upon cleavage in air, when appreciable contamination of an exposed surface may be supposed to occur, has led us to propose that its origin is recombination of carriers at surface defects.[5] These occur on all kinds of cleavage surfaces. Furthermore, defects exist which are known to be affected by gas exposure at a rate several orders of magnitude less than on a free surface.[6,7] Such defects occur in narrow crevices, which are readily found in the rougher parts of cleaved surfaces, and also near the origin regions where the initial flat zone meets the typical terrace regions, see Fig.1. Hence recombination centres at these sites would not necessarily be affected by gases in the first 20 or so microseconds after cleavage, even if carried out in air.

An additional possibility is that the defects occur as vacancies on flat surfaces, such that contamination by an adsorbed molecule does not greatly affect the recombination characteristics of the vacancy.[5]

3. Emission from Excited States

With regard to the other widely observed signal called A, and seen only from vacuum cleavages, the minimum energy in its spectral range, where recorded, is that of the bulk band gap. Therefore this luminescence is ascribed to bulk band gap recombination. If air contamination is present, many intermediate states can be created at the surface across the energy gap range, so that recombination can occur by a number of small steps, producing long-wavelength photons which are undetectable, hence accounting for the lack of observed signal.

A very important finding in the above work, from spectral measurements of the A radiation, is that the electrons excited by the cleavage in the elemental semiconductors can have quite high energies, up to 2.7 eV in the case of Si, and 1.46 eV in the case of Ge.[5] A histogram summarising the results from many Si cleavages in vacuum is shown in

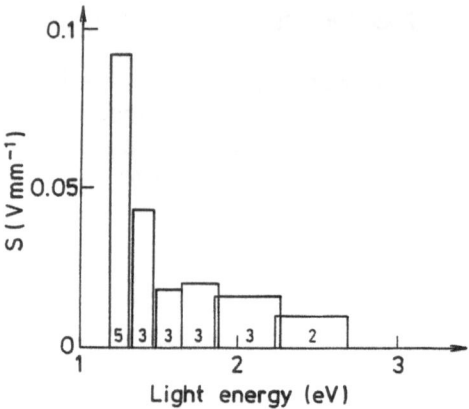

Fig.3

Histogram of Si cleavage luminescence intensity per unit length of crack, S, measured in vacuum, as function of energy ranges of light transmitted by optical filters. The numbers inside the blocks refer to the numbers of experiments that were averaged.

Fig.3. The data was obtained by a laborious set of experiments using a different optical filter in front of the detector for each cleavage. In the case of the III-V compounds, the radiation is confined to the band gap energy. This is explained in terms of the band structure. Since the recombination in InP and GaAs is direct, electrons excited to high energies can rapidly fall to the bottom of the conduction band, E_C, since states there are emptied immediately by direct transitions to the valence band. On the other hand, in Si and Ge electrons at Ec are delayed by the necessity for phonon participation in conserving momentum while transiting to the valence band. Therefore electrons excited by the cleavage to states high in the conduction band cannot immediately drop to lower states since these are still occupied. Therefore there is the possibility of phonon-assisted transitions occurring from occupied excited states to the top of the valence band. This phenomenon is not generally observed because excited electrons would flow to another part of the specimen, but we have already inferred the presence of a small potential trap at the freshly cleaved surface,[4] and this prevents the excited electrons from escaping. One notes that the upper value of 2.7 eV for Si exceeds the bond-breaking energy of about 2.3 eV. This shows that the energy release on bond rupture, about which little is known, can be of a cooperative nature so that some electrons receive more than the maximum energy from rupture of a single bond.

In the case of Si, it was possible to carry out more exhaustive searches for other radiation, and a 3rd signal was found, called B, corresponding to surface state recombination.[1] The situation is shown schematically in Fig.1. A summary of signal types from all the materials studied is given in reference 5.

As expected, this luminescence B did not appear in air cleavages, doubtless due to contamination affecting surface states. The energy of this radiation was quite sharply defined as 0.26 eV,[3] giving an experimental value for the mimimum surface state gap. Due to the relatively long duration of this emission, about 200 microsec., it was deduced that the minimum surface state gap was probably indirect. For example, the direct gap in InP caused signal durations of less than 10 microsec.[4] It is however a possibility that the long duration can be explained by the participation of intraband trap states. It is of interest that the surface state band structure calculated for both the Pandey[8] and the Three bond Scission (TBS)[9] models of Si(111)2x1 show almost parallel bands along the minimum surface state gap region of J-K, indicating that the gap can be indirect, the degree depending on the details of the structural parameters adopted for the computations.

The intensities[5] of all the above radiations are relatively weak. The C signals for all the materials correspond to an emission of about 10^8 photons per cm^2, while the A signals are about 10^{10} cm^{-2} for Ge and Si, 10^8 for InP and about 10^6 for GaAs. The B (surface state) signal for Si corresponded to about 10^{12} photons cm^{-2}. The low density for the C signal is consistent with its origin being a relatively low density of defects, and the highest value for the surface state signal is consistent with its origin being on the surface and least subject to absorption losses.

4. Ambient Effects

Some studies have been made of the effect of cleaving in different gases, mostly with Si due to its greater availability. Studies were carried out in water vapour, hydrogen, helium, neon, carbon dioxide, oxygen and dried air, at various pressures. The defect signal C was widely observed, but in no case was the A signal detected. However, in the cases of atmospheric pressure of hydrogen, helium and neon, a signal of around 80-90 microsec. duration was detected, which was too large a duration for the C signal but insufficient to be identified with the A signal. A similar duration signal was also observed in 10 Torr pressure of water vapour. There is a suggestion that the A signal is truncated by water vapour, possibly because it consists of 2 components. Alternatively, new surface trap states have been introduced. More work needs to be carried out on these aspects.

In the case of InP, the A signal at 1.25 eV was studied under cleavage in different ambients.[5] Although not observed under cleavage in air, it was detected under cleavage in He, both at 2 and 20 Torr, in nitrogen at 20 Torr and in hydrogen at 760 Torr. It was also observed under cleavage in oxygen at 20 Torr and weakly even at 760 Torr. From further experiments in water vapour it was concluded that the presence of this gas was mainly effective in suppressing the signal, being much more

94

effective than oxygen. Due to the known low adsorption rate of the latter on cleaved InP, the result is understandable. Presumably the water vapour adsorbs to produce intermediate states that allow a series of undetectable low-energy carrier recombinations.

5. Cleavage Surfaces

There are a large number of studies of cleavage surfaces by a variety of techniques. Unfortunately, such surfaces are very far from ideal flat ones, except on some microscopic patches. Studies by optical and scanning electron microscopy (SEM) invariably show quite complex topography on microscopic as well as macroscopic scales.[10-14] It is possible to find ideal flat regions of dimensions up to hundreds of nm, but such areas are too small to study with techniques other than those using microprobe techniques, such as scanning tunneling microscopy (STM). Even here it is necessary to be aware of tip-surface interaction effects. In the case of optical and LEED studies, these average over surfaces of dimensions of hundreds of microns, and effects ascribed to surface non-uniformities are often large. The problem is that whereas gross effects of the latter can be identified by comparison of results from many experiments, small effects can not. Therefore subtle variations in derived band structure for example, that are sometimes but not uniformly observed, may or may not be due to surface irregularities, and it is very difficult to know what credence should be attached to small features.

Fig.4

Appearance under low power optical microscopy of cleaved Si surface. The specimen width is 0.52 mm. Note tear marks radiating from cleavage origin at corner.

Fig.5

      Scanning electron microscope picture of portion of cleaved surface showing undercut regions.

Fig.6

      SEM picture of typical origin region of cleavage, showing small flat region bounded by rough edges.

In Fig.4 we show an optical micrograph of the cleavage surface of a wafer, showing the typical tear marks that radiate from the origin region of the cleavage. Often the origin occurs at a corner site, with tear marks radiating out along the flatter portion of that part of the surface that was under tensile stress. There is a clear boundary along the long dimension of the wafer cross section, to the much rougher portion that was under compressive stress prior to cleavage. When these surfaces are examined by high power SEM, they reveal a wealth of detail, with flat regions only extending very short distances as mentioned above. Fig.5 shows a typical region that gives evidence of undercuts, invoked to account for the cleavage luminescence signal C being observed in air. In Fig. 6 we show the frequently observed small flat region around the origin, with the precipitous transition to a very rough region due to the advancing crack diverting onto other planes. These diversions are due to interaction with various stresses and with elastic waves reflected from the specimen boundaries. Even a region that looks very flat at magnifications of several thousand, reveals striations as shown in Fig.7 when observed under higher magnification.

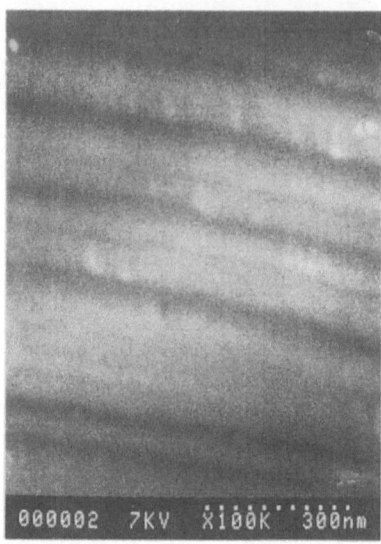

Fig.7

High resolution SEM picture showing stepped regions on portions that appear flat on lower magnification.

An example of small effects is the dispersion along the gamma to J direction in the surface state band structure of Si(111)2x1. A small dip in the otherwise uniformly sloping dispersion was observed by one laboratory,[15,16] and theoretical attempts were made to match it, but there is no knowledge as to whether this is a genuine effect or not. Our recent *ab initio* Hartree-Fock computations of the band structure for both the Pandey and TBS models find no such dip.[9] Of course all such calculations are for ideal, flat surfaces.

6. Electrical Effects

We now describe another very interesting phenomenon associated with cleavage. We have found that voltages are generated upon cleaving silicon, with magnitudes up to about 400 mV. The technique consists of attaching ohmic electrical contacts to the ends of a strip of Si wafer, in a manner used previously,[2] and detecting the voltage across the contacts, or between the contact and ground, as a function of time before and after cleavage using fast recording electronics.

We initially connected a battery to the circuit so as to detect the time when the circuit was broken by the occurrence of full cleavage. These results have been reported previously,[2] and gave a method of measuring the time for the crack to pass through the specimen. However, certain aspects of the data led us to suspect that there were complexities, including voltage generation by the sample. It had already been reported that current bursts occurred on cleaving Si.[17] However, those measurements were made by detecting current passing through the specimen parallel to the crack. These bursts lasted for only a few microseconds.

With the technique used by us, the voltage and current were measured across the cleavage. A number of surprising phenomena were found.

(a) There is a voltage generated between the contacts. Its magnitude is reduced if the impedance is low so that higher currents can flow. If the circuit impedance is about 100 Kohm or higher, the full voltage appears. It is generally a few hundred mV. (The highest value detected just after cleavage so far is 390 mV.)

(b) If the circuit impedance is reduced to essentially that of the specimen, about 100 Ohm, then currents up to 5 mA have been detected, with voltages of a few tens of mV.

(c) By connecting each side of the specimen through an independent meter to ground, we found that the currents generated by the cleavage persisted well after the cleavage had completed.

The above facts show that the structure was acting like a charged capacitor. The two plates, i.e. the separated halves of the specimen, independently discharged to ground over periods that depended on the respective impedances.

One may ask why a voltage should be generated. If cleavage of Si or Ge occurs across the single bond (111) plane, the so-called shuffle plane, then there is symmetry on either side and no dipole moment is expected between the two newly created sides.. We have confirmed this by a calculation[18] using a commercial molecular orbital computation package, DMOL. However, if cleavage takes place across the alternative (111) plane, the three-bond glide plane as it is called, then there is no symmetry at the time of bond rupture and a dipole moment could be expected. Calculation with the above-mentioned method confirms that a dipole moment will occur. Thus a voltage will appear between the separating surfaces. Note that the Pandey chain model[19] of the 2x1 structure on cleaved surfaces features shuffle plane cleavage, whereas the Three Bond Scission model[20,21] is based on the assumption of cleavage along the glide planes. If the observed voltage is due to this dipole moment, then glide plane cleavage is strongly indicated. This matter is currently being explored in detail.

In conclusion, the new technique of measuring light emission, and recently electrical effects, on cleaving semiconductors in various ambients has yielded a variety of novel information which is adding to the understanding of semiconductor cleavage surfaces and the nature of the bond rupture process.

This work was supported by the Australian Research Council.

References

[1]   D. Haneman and N. McAlpine, Phys. Rev. Lett. 66, 758 (1991)
[2]   D.G.Li, N.S.McAlpine, and D.Haneman, Appl.Surface Sci. 65/66, 553 (1993)
[3]   D.G.Li, N.S.McAlpine, and D.Haneman, Surface Sci. Lett. 289, L609 (1993)
[4]   D.G.Li, N.S.McAlpine, and D.Haneman, Surface Sci. Lett. 281, L315 (1993)
[5]   D.G.Li, N.S.McAlpine, and D.Haneman, Surface Sci. Nov./Dec. (1993)
[6]   B.P.Lemke and D.Haneman, Phys. Rev. B17, 1893 (1978)
[7]   M.F.Chung and D.Haneman, J.Vac.Sci. Tech. A2, 1475 (1984)
[8]   J. E. Northrup, M. S. Hybertsen, and S. G. Louie, Phys. Rev.Lett. 66, 500 (1991)
[9]   B.Chen and D.Haneman, submitted for publication
[10] D.R.Clarke, Ch.2 in Semiconductors and Semimetals, vol.37, Academic Press, (New York) (1992)
[11] Y.Mera,T.Hashizume, K.Maeda and T.Sakurai, Ultramicroscopy 42-44, 915 (1992).
[12] H.Tokumoto, S.Wakiyama, K.Miki, S.Murakami, S.Okayama, and K.Kajimura, J.Vac.Sci.Tech. B9, 695 (1991)
[13] J.C.McLaughlin and A.F.W.Willoughby, J.Crystal Growth 85, 83, (1987)
[14] P.A.Bennett, H. Ou, C.Elibol, and J.M.Cowley, J. Vac. Sci. Tech. A3, 1634 (1985)
[15] R.I.G.Uhrberg, G.V.Hansson, J.M.Nicholls, and S.A.Flodstrom, Phys. Rev. Lett. 48, 1032 (1982)
[16] P.Martensson, A.Cricenti, and G.V.Hansson, Phys.Rev.B32, 6959 (1985)
[17] S.C.Langford, D.L.Doering, and J.T.Dickinson, Phys. Rev. Lett. 59, 2795 (1987)
[18] B.Chen and D.Haneman, unpublished
[19] K. C. Pandey, Phys. Rev. Lett. 47, 1910 (1981)
[20] D.Haneman, Rep.Prog.Phys.50, 1045 (1987)
[21] D. Haneman and M. G. Lagally, J. Vac. Sci. Technol. B6, 1451 (1988)

III.

Surface Composition Analysis

Problems of Quantitative Surface Composition Analysis Using LEIS

R.J. MacDonald, D.J O'Connor, B.V. King, Li Ting and Zhu Li
Department of Physics, The University of Newcastle, Callaghan, NSW, 2308, Australia

Abstract. Low Energy Ion Scattering (LEIS) is used to determine composition and structure of single crystal surfaces. The scattering event and the atomic exchange processes with the surface are complex events, however. A number of processes may affect the charged state of scattered particles, but in some cases where high surface layer specificity is desirable, the analysis would relate mainly to the ionised component of the scattered particle yield. In this paper we look at the factors important here. We present some detailed studies of inelastic loss processes which will affect quantitative analysis of surfaces.

1. Introduction

There has been no lessening of the interest in accurate compositional analysis as a function of depth in a surface. So many of the properties of a material which are of technical importance are associated with the composition and the structure of the outermost layers of the surface. Corrosion will initiate from a surface whether it be the absolute termination of the lattice or surfaces generated by cracks or similar mechanical defects. In a similar way the level of catalytic activity sustainable at the surface is often related to the energetics of the surface layers and hence to their composition and structure.

Many of the modern techniques of surface analysis, whether related to composition or structure, provide an average over a number of layers at the surface. Depending on the probe characteristics, the number of layers may vary from one to many. One of the attractions of Low Energy Ion Surface Analysis (LEIS) has always been its extreme sensitivity to the outermost one or at most two surface layers. This is particularly true when one uses inert gas ions as the analysing probe. Other ions such as alkali metals allow for differentiation of surface and below surface contributions but there is a merging of the respective signals from the surface and below it.

LEIS has been used very effectively over the last few years for the precise location of the atoms in the surface layer. Most of the applications have been to elemental targets, with a significant amount of work on the location of the adsorption sites at various sub-monolayer levels of coverage for electronegative adsorbates such as oxygen. Some of the pioneering work of Aono et el (1) was on TiC but only recently has it been applied to other binary alloys or to compounds. Over the last few years, however, there has been a significant amount of interest in binary alloys. Some of these have been ordered while others have not.

With this interest in the structure of the surface has come an interest in the composition of the succeeding layers at the surface. Some attention has therefore focused on the practicality of surface composition analysis using LEIS.

2. The Basis for Compositional Analysis Using LEIS

All applications of LEIS use a low energy ion beam (typically 1keV) impinging on a surface. Scattered particles are energy analysed at a particular scattering angle. The particular particles which are most sensitive to the surface are those which are ionised in the scattered state. Time-of-flight (TOF) techniques are being used to analyse the neutral component of the scattered flux but these are in general of lower energy resolution and the differentiation of particles scattered from the outermost one or two layers is harder than in the case of the ionised component.

If we consider the ionised component alone we can always relate the signal strength measured to the surface composition as follows:

$$I \, dEd\Theta \; = \; I\sigma(E,\Theta)P(E,\Theta)dEd\Theta \tag{1}$$

where $I \, dEd\Theta$ is the number of scattered ions detected in the energy range dE at E and the angle range $d\Theta$ at Θ,

I_0 is the number of ions incident per unit area of the surface,

$\sigma(E,\Theta)$ is the cross-section for scattering of an ion of incident energy E_0 into the range $dEd\Theta$ at E,Θ respectively

$P(E,\Theta)$ is the probability that an incident ion will be scattered into the detector and survive the trajectory as an ion.

If the scattering event preserves energy and momentum the relationship between the incident energy E_0, the scattered energy of the incident ion E_1, the scattered energy of any recoil particle E_2 and the scattering angle Θ uniquely defines the mass of the scattering particle M_2 and hence offers a method of determining the elements present on the surface. If one knows the parameters which characterise the interaction; i.e. the cross-section and the charge exchange parameter P and the equipment is itself suitably characterised, one can also determine the number of atoms of type M_2 on the surface.

Using inert gas ions experimental observation tells that we only detect charged particle scattered from the outermost surface layer. Charge exchange to neutral inert gas atoms occurs for all ions which penetrate the surface itself to be scattered from atoms below the surface layers. Alkali metal ions on the other hand emerge from within the target material in an ionised state. Similar results apply to the scattering of hydrogen ions. Recently, however, we have analysed the scattering of hydrogen in respect of the depth from which ions in the "surface peak" originate. Only the top two perhaps three layers contribute .

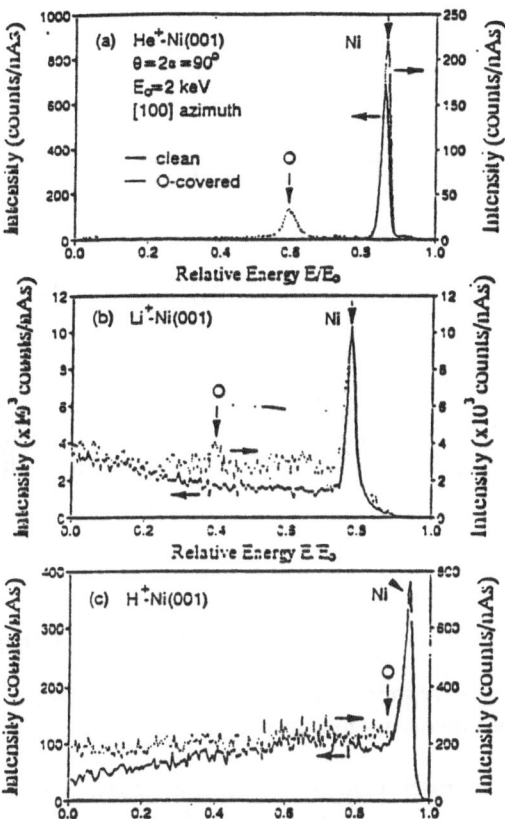

Figure 1 - Typical Spectra obtained for H⁺, He⁺ and Li⁺ scattering from
clean and oxygen-covered Ni(001). The scattering conditions
represent double alignment.

The differing forms of typical energy spectra for these types of scattering are
illustrated in Fig. 1.

The maximum in the compositional resolution with respect to depth will occur
using inert gas ions.

We will concentrate on a discussion of the use of LEIS with inert gas ions
incident and the ionised component of the scattered flux detected.

3. Neutralisation processes in low energy ion scattering

The probability of charge exchange leading to neutralisation has been a topic of
great interest to those using LEIS for surface analysis. The usual approach is to
use the model of Hagstrum (2) which was developed to explain the properties of
secondary electrons emitted in the case of ions of energy 10eV or so interacting
with a solid surface. Such ions will not in general penetrate the surface and their

104

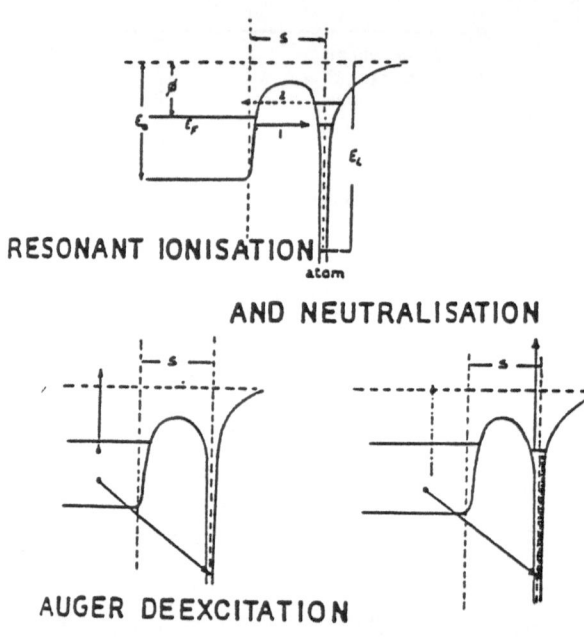

RESONANT IONISATION

AND NEUTRALISATION

AUGER DEEXCITATION

AND NEUTRALISATION

Figure 2 - A schematic representation of the charge exchange
processes involving ions near surfaces.

distance of closest approach is large in an atomic sense. At the energies typical
of LEIS (1keV) the ions will penetrate the surface and the distance of closest
approach in the binary collision responsible for the major deflection of the ion
will be of order of the orbital radii. The suggested mechanisms for the charge
exchange involve resonant or Auger processes as indicated schematically in Fig. 2.

Under this model the probability of the exchange is expressed in terms of the
distance of the ion from the surface so the exchange is not identified with the
localised electrons on the ion cores but rather with the distributed band electrons
in the surface region. The probability of the exchange is then determined by the
time spent in the surface region and the transition rates between states in the
surface and on the ion.

An alternate proposal due to Woodruff et al (3) suggests the exchange is
related to the distance between atoms involved in the binary collision rather than
the undifferentiated surface. This has the potential to take account of
enhancements in exchanges due to orbital overlap but the simplest form of the
model uses a probability function which is exponential in separation distance.
The probability function for the survival of the ion in an ionised state under both
models is similar. One would expect, however, that the survival probability as a
function of ion trajectory would be different for the two cases. Under the
Hagstrum model, as the angle of the trajectory to the surface normal increases

towards glancing exit, the survival rate will decrease because the ion is spending more time within the exchange distance with the surface. For an ion with a speed v after scattering, the probability of the survival will be

$$P = \exp - \frac{v_c}{v_\perp} \qquad (2)$$

where $v_\perp$ is the perpendicular component of ion velocity with respect to the surface. This will vary as vcosO.

Under the Woodruff model, the probability of survival should be independent of the angle to the surface until such time as the ion trajectory passes close enough to the nearest neighbour surface atom to have a probability for exchange large enough to distort the distribution from the original scattering. The distances characteristic of the exchange are about 2A, which is close to the interatomic spacing in the surface region. One would then expect the probability distribution will only begin to feel the effects of the second atom when the angle of the trajectory to the surface is 50 or more.

MacDonald, O'Connor and others have done a lot of work to characterise this exchange (4). Using the geometry in Fig. 3 they have determined the neutralisation parameters for a range of ion-target combinations. A typical experimental result is shown in Fig. 4. This is the ion yield as a function of exit angle for a scattering angle of 90°. In Fig. 4b this result is analysed according to the Hagstrum model under which the ln (I/I_0) should be linear in $1/v_\perp$. The fit is quite good and certainly supports the Hagstrum model but the result in Fig. 4a also shows a distribution which could be consistent with the Woodruff model.

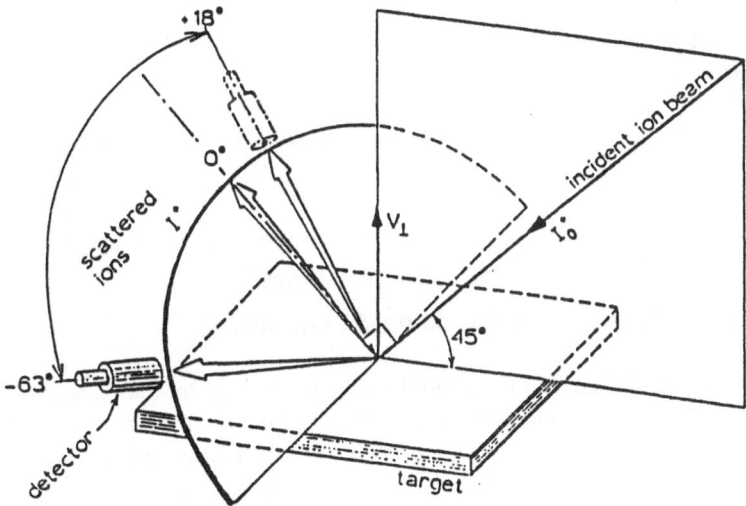

Figure 3 - A schematic diagram of the method used to determine charge exchange parameters in LEIS.

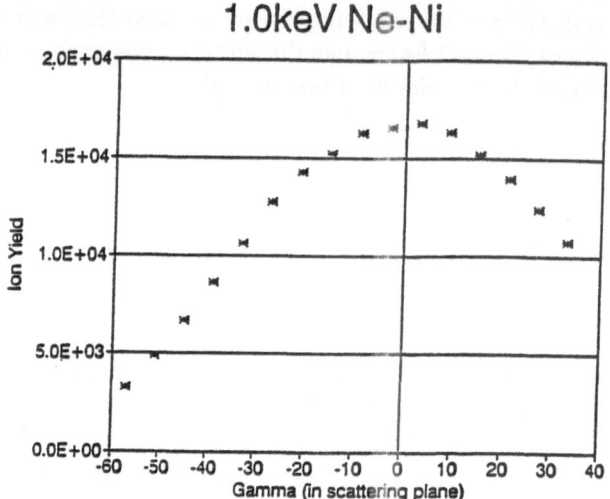

Figure 4a - A typical result for the measurement of the ion yield as a function of angle to the surface using the method shown in Fig. 3.

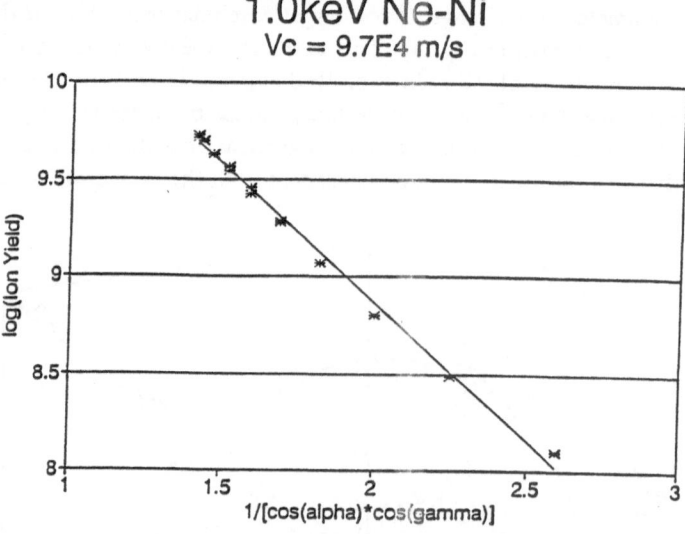

Figure 4b - The results of Fig. 4a fitted to equation 2.

It is well established that the neutralisation is effected by the distance of closest approach. The oscillations in the ion yield as a function of energy for selected ion-target combinations (e.g. He$^+$ on Pb) have been interpreted in terms of a quasi-resonant exchange which arises when the distance of closest approach mixes orbitals appropriately. The oscillations arise from quantum mechanical effects in that mixing but the onset of oscillations requires a threshold in the distance of closest approach. Measurements of the neutralisation parameters for ions scattering at energies corresponding to maxima and minima in the quasi-resonant exchange rate reveal no difference in these parameters. This suggests the final neutralisation probability may be the sum of probabilities determined by a number of individual processes.

The neutralisation process is in itself remarkable. Figure 1 was an example of the energy spectrum of ions scattered from a clean surface. There is a narrow peak corresponding to the energy resulting from an elastic collision between the incident ion and the surface atom. The ion yield at energies below this is very nearly zero indicating that incident ions which penetrate the surface and suffer their major scattering event below the surface are all neutralised. Using the methods indicated above, the neutralisation parameters can be evaluated for ions in the surface peak. There is no way that rate can be used to describe the neutralisation of the ions which have penetrated through the surface layer. Again a different neutralisation process must be operative for those ions. This may involve capture of an electron by the ion travelling in the solid but this is unlikely given the size of the then neutral atom. It is more likely that the incident ion travels within the solid as a charged particle and that the almost complete neutralisation of the ions occurs in transiting the surface layer. This raises the interesting question of the way in which this transit will differ for ions scattered from the outside of the top layer and ions which transit through the top layer from inside the metal.

4. Elastic and Inelastic Effects in the Scattering

In equation (1) we used the cross section for elastic scattering to calculate the number of particles scattered into the detection system. With our knowledge of the interaction potential and particularly as a result of the rapid developments in the use of computer simulation to compare with experimental results, we are able to calculate the elastic scattering cross-section with very reasonable accuracy. One would then expect that most of the problems we have with quantification would rest with the uncertainty surrounding the neutralisation process.

Recent work on the scattering event itself however would give a contrary view. It has been recognised for some time that inelastic effects would occur in the scattering. A major source of energy loss was thought to be the loss to electron excitation as the particle traverses the surface region. It is an interesting area of work because it is not readily described by current theory.

More recently, however, it is becoming evident that there are a number of inelastic effects which can occur in the binary scattering event itself. These show up as shifts in the energy of the elastic scattering peak as the scattering geometry is changed. Variations in the distance of closest approach (or the impact parameter) represent differences in the orbital overlap between the colliding partners.

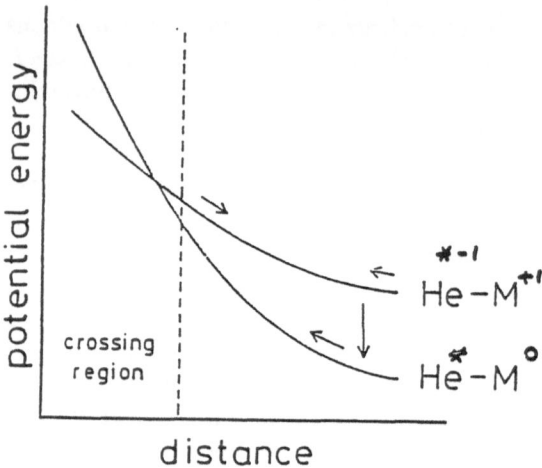

Figure 5 - A schematic representation of inelastic energy loss processes in a binary collision.

A schematic representation of the possible excitation and hence the source of the inelastic loss is shown in Figure 5. This is a common representation of the mixing and subsequent promotion of electrons in a given level used in collision physics. The incident particle, in whatever state, and the target constitute a given potential energy surface which the system will follow on its incident trajectory. As a result of interaction with the surface the potential energy surface will change, usually increasing as the interaction with the surface causes the incident particle to slow down. In some region of distance of closest approach, this potential energy surface may mix or cross another surface representing a different state of the system of incident particle and target. There may then be a form of resonance exchange of electrons between levels so the system follows a different potential energy surface on the exit trajectory with the result that there is an energy difference in the state of particle plus surface when the two are well separated. This potential energy difference will show up as an inelastic energy loss in the incident particle and possibly as a change in the charge state as well.

The experiments of O'Connor et al (5) on He$^+$-Pb and He$^+$-Sn were specifically targeted to measure the mixing distance ie the distance at which the orbital overlap and the consequent exchange occurs, in the quasi-resonant neutralisation of these systems.

An excellent example of these processes is found in the work on re-ionisation by Souda, Aono et al (6). This has been mainly done for the case of He$^+$ scattering from a variety of targets. The work stems from the observation that the He$^+$ energy spectrum sometimes showed a double peak structure with one peak at an energy close to the elastic scattering energy and a second peak approximately 20eV lower in energy. The latter peak is the one identified with the inelastic loss associated with reionisation of neutral He. One point to note is

Table 1

1	2	3	4	5	6	7	8	9	10	11	12	13	14	15	16	17	18
H $1s$																	He $1s^2$
Li $2s$	Be $2s^2$											B $2p$ $2s^2$	C $2p^2$ $2s^2$ (≤200)	N $2p^3$ $2s^2$	O $2p^4$ $2s^2$	F $2p^5$ $2s^2$ (——)	Ne $2p^6$ $2s^2$
Na $3s$ (≤200)	Mg $3s^2$ (≤200)											Al $3p$ $3s^2$ (300)	Si $3p^2$ $3s^2$ (300)	P $3p^3$ $3s^2$	S $3p^4$ $3s^2$	Cl $3p^5$ $3s^2$	Ar $3p^6$ $3s^2$ (——)
K $4s$ (≤200)	Ca $4s^2$ (≤200)	Sc $3d$ $4s^2$	Ti $3d^2$ $4s^2$ (≤200)	V $3d^3$ $4s^2$	Cr $3d^4$ $4s^2$	Mn $3d^5$ $4s^2$ (500)	Fe $3d^6$ $4s^2$ (——)	Co $3d^7$ $4s^2$ (——)	Ni $3d^8$ $4s^2$ (——)	Cu $3d^{10}$ $4s^2$ (——)	Zn $3d^{10}$ $4s^2$ (——)	Ga $4p$ $4s^2$ (——)	Ge $4p^2$ $4s^2$	As $4p^3$ $4s^2$	Se $4p^4$ $4s^2$	Br $4p^5$ $4s^2$	Kr $4p^6$ $4s^2$
Rb $5s$	Sr $5s^2$ (≤200)	Y $4d$ $5s^2$ (300)	Zr $4d^2$ $5s^2$ (300)	Nb $4d^4$ $5s$	Mo $4d^5$ $5s$ (400)	Tc $4d^6$ $5s$	Ru $4d^7$ $5s$	Rh $4d^8$ $5s$	Pd $4d^{10}$ —	Ag $4d^{10}$ $5s$ (——)	Cd $4d^{10}$ $5s^2$ (——)	In $5p$ $5s^2$ (600)	Sn $5p^2$ $5s^2$ (1100)	Sb $5p^3$ $5s^2$ (1200)	Te $5p^4$ $5s^2$ (——)	I $5p^5$ $5s^2$	Xe $5p^6$ $5s^2$
Cs $6s$	Ba $6s^2$	La $5d$ $6s^2$	Hf $5d^2$ $6s^2$	Ta $5d^3$ $6s^2$ (300)	W $5d^4$ $6s^2$	Re $5d^5$ $6s^2$	Os $5d^6$ $6s^2$	Ir $5d^7$ $6s^2$	Pt $5d^9$ $6s$ (——)	Au $5d^{10}$ $6s$	Hg $5d^{10}$ $6s^2$	Tl $6p$ $6s^2$ (——)	Pb $6p^2$ $6s^2$	Bi $6p^3$ $6s^2$	Po $6p^4$ $6s^2$	At $6p^5$ $6s^2$	Rn $6p^6$ $6s^2$

Legend:
- ELEMENT
- ELECTRON CONFIGURATION
- THRESHOLD ENERGY FOR REIONIZATION (eV)

Table 1
Threshold values of the primary He^+ energy E_0 for the reionization of neutralized He^0 measured for 29 target elements C, F, Na, Mg, Al, Si, Cl, K, Ca, Ti, Mn, Co, Ni, Cu, Zn, Ge, Sr, Y, Zr, Mo, Ag, Cd, In, Sn, Sb, Te, I, Ta. " ≤ 200" means equal to or smaller than 200 eV, and "——" represents larger than 2000 eV.

that with increasing energy the inelastic process dominates the energy spectrum, implying that an increasing proportion of the incident ions are neutralised on the incoming trajectory. This is not what one would expect from the conventional models of neutralisation outlined above.

Souda and Aono et al (6) have looked at the reionisation process in a number of ion-target combinations. Table 1, reproduced from their work, summarises the result. The observation of the reionisation event is very dependent on the combination and shows considerable variation in energy. If the results follow those for Ta referred to above, then equally one would expect to see a shift in the yield of scattered ions from the elastic to the inelastic peak. Attempts to then relate the ion yield to an elastic cross-section modified by a neutralisation process will clearly be in error.

Other examples of inelastic loss occur. Fauster et al (7) have recently published some results on the LEIS determination of the surface structure of copper in which they show that some form of neutralisation process blurs the shadow cone envelope scattering at lower energies and that one must use energies of Ne^+ of the order of 5keV or so to obtain clear definition of the onset of scattering from the near neighbour atoms. In an attempt to more clearly delineate the neutralisation process, we have commenced a detailed study of the scattering of Ne^+ from Cu and Cu- containing alloys.

In the experiments we have done, we have compared the scattering of He^+ and

110

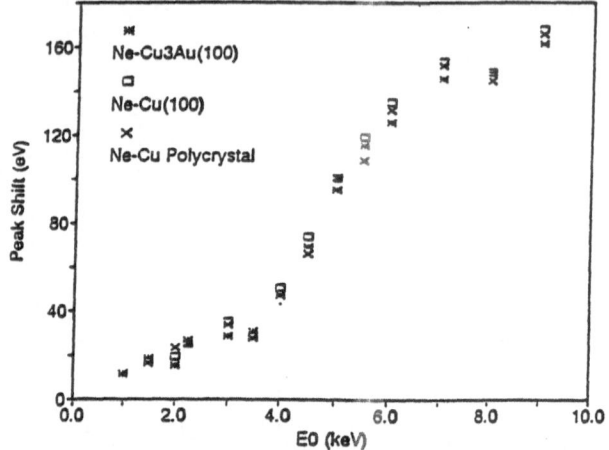

Figure 6 - Energy shift of the scattering peak from the binary scattering energy for Ne$^+$ scattered from various Cu targets.

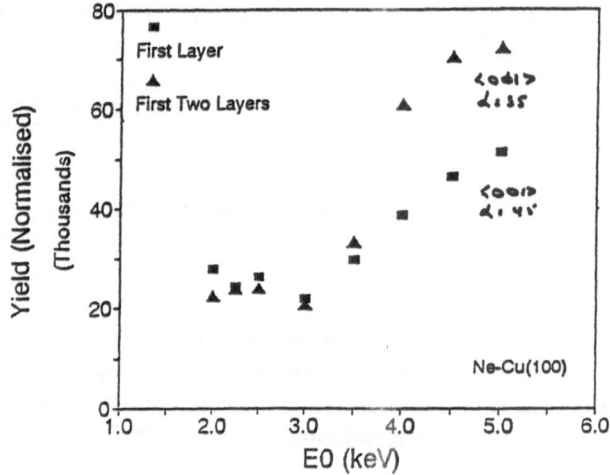

Figure 7 - The yield of Nc$^+$ ions scattered from first layer only and from (first and second) layers of a Cu(001) surface.

Ne$^+$ from polycrystalline Cu and from single crystal Cu and Cu$_3$Au along the (100) and (110) directions. There is no difference in the behaviour of the main peaks between these samples apart from the yield of scattered ions. In Fig. 6 the yield as a function of Ne$^+$ energy up to 5keV incident is shown. In all cases there is an increase in the yield with increasing energy but with an onset in the yield increase above about 3keV. At any given energy the yield is related to the Cu atoms likely to be visible, as shown in Fig 7. As a test of the apparatus the separation between the peaks corresponding to scattering from the isotopes of Cu is compared with the theoretical difference (Fig. 8) for energies up to 5keV.

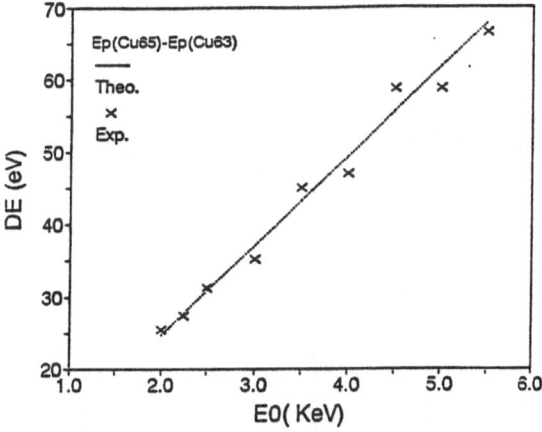

Figure 8 - Scattering from the isotopes of Cu, corrected for the inelastic loss.

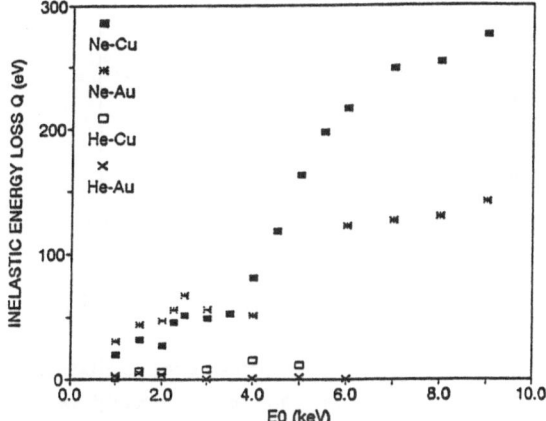

Figure 9 - The inelastic energy loss for He$^+$ and Ne$^+$ scattered
from Cu and Au in Cu$_3$Au.

There is a strong inelastic shift to the scattering of the Ne$^+$ ions. The Cu
results are the same whether the Cu is a single crystal or a polycrystal or a
component in the alloy. It is important to note that in these experiments the
scattering peak changes from an essentially elastic peak to an inelastic peak over
a range in energy, with a threshold at about the 3.5keV level. The scattering of
the He however, shows no appreciable energy loss.

The shift in the peak energy from the elastic scattering value can be used to calculate an inelastic excitation energy, using

$$Q = 2\frac{M_1}{M_2}(E_0E_1)^{\frac{1}{2}}\cos\theta + \left[1-\frac{M_1}{M_2}\right]E_0 - \left[1+\frac{M_1}{M_2}\right]E_1 \qquad (3)$$

The result of this is shown in Fig. 9. Inelastic losses up to 300eV are observed, with no evidence of any elastic scattering at all at the higher energies. He on the other hand is scattered from the same targets with no significant energy loss.

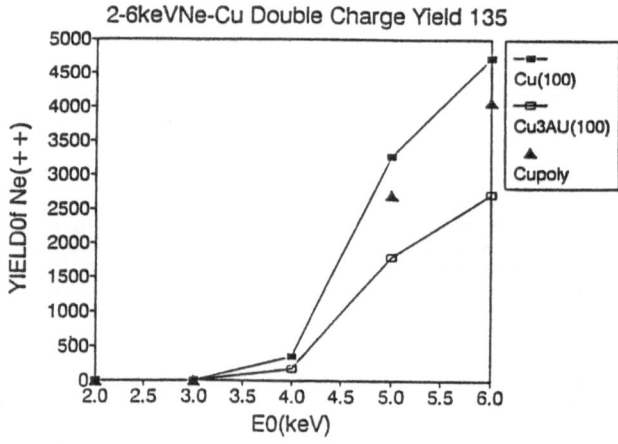

Figure 10 - The yield of Ne^{++} for Ne^+ incident on Cu, as a function of energy.

These energy losses are quite large, representing major excitation in the system. At the threshold where one begins to see the change in yield and the onset of the larger mode of inelastic loss, Fig.10 also shows the onset of a yield of doubly charged ions, Ne^{++}. At the higher energies this yield can be quite large. The yield also increases with decreasing scattering angle.

It is interesting to note that there is a similarly behaving inelastic loss for Ne^+ scattered from the Au atoms but that the magnitude of the loss is not as great. The loss in the Ne^+-Au case at the lower energy end is larger, however.

For the case of Ne^+ scattered from Cu, the threshold at which the major inelastic loss begins is a function of the experimental geometry and hence of the collision parameters such as distance of closest approach. These experiments are continuing with the view to clarifying the mechanism and the relation to the formation of the Ne^+.

5. Quantitative Analysis Using LEIS

What then is the picture for quantitative compositional analysis using LEIS? It is noteworthy perhaps that some of the reports of quantitative analysis essentially ignore the plethora of atomic contributions to the yield of the scattered ions. For example, Martin et al (12) some time ago did some analysis work on the alloy NbV. This alloy was chosen because it is miscible across a wide composition range so with care, segregation or precipitation will not be a problem. Also the sputtering yield for Nb and V are very similar so that preferential sputtering is unlikely to modify the surface composition. The alloys were analysed over a range of bulk concentrations by comparing the ion scattering yield from pure Nb and pure V with that from the alloy samples.

Shen (9) has studied the surface composition of Ni_3Al single crystals using a combination of He^+ and Li^+ ions. Orientations of the incident beam relative to the lattice are chosen which either shield the second layer or expose it. It was assumed that He^+ scatters only from the surface layer while Li^+ scatters from the visible atoms in the outermost layers. The results obtained were in accord with other determinations and with the assumptions made; i.e. the yield of the scattered He^+ was unchanged by changes in the target orientation to shield or expose the second layer.

6. Conclusion

On the basis of the results presented here one can point out that the low energy scattering event is very complex and the outcomes will vary markedly with the ion-target combination chosen. The observations reported here will not effect the applicability of LEIS to atom location on the surface of single crystals nor studies of the location of adsorbates. These are the applications of LEIS which have attracted the most attention in the last few years. However, this review of applications to surface compositional analysis and the effects which are likely to affect the outcome of those applications suggest that it is unlikely that LEIS will ever provide a simple technique for accurate surface compositional analysis.

REFERENCES

1.   M. Aono, C. Oshima, S. Zaima, S. Otani and Y. Ishizawa, *Jap.J.Appl.Phys.* <u>20</u> (1981) L829.

2.   H.D. Hagstrum, *Phys.Rev.* <u>96</u> (1984) 336.

3.   D.P. Woodruff, *Surf.Sci.* <u>116</u> (1982) L219.

4.   R.J. MacDonald, D.J. O'Connor, J.M. Wilson and Y.G. Shen, *Nucl.Instrum.&Methods* <u>B33</u> (1988) 446.

5.   D.J. O'Connor and R. Beardwood, *Nucl.Instrum.&Methods* <u>B48</u> (1990) 358.

    D.J. O'Connor, A. Kahn and R.J. MacDonald, *Surf.Sci.* <u>2</u> (1992) L83.

6.   R. Souda and M. Aono, *Nucl.Instrum.&Methods* <u>B15</u> (1986) 114.

7.   Th. Fauster and M.H. Metzner, *Surf.Sci.* <u>166</u> (1986) 29.

8.   P.J. Martin, C.M. Loxton, R.F. Garrett, R.J. MacDonald and W.O. Hofer, *Nucl.Instrum.&Methods* <u>191</u> (1981) 275.

9.   Y.G. Shen, PhD Thesis, The University of Newcastle, 1991.

Studies of Alloy Surfaces by Low-Energy Ion Scattering

Y.G. Shen[1], D.J. O'Connor[1], R.J. MacDonald[1] and K. Wandelt[2]

[1]Department of Physics, University of Newcastle, New South Wales 2308, Australia
[2]Institut für, Physikalische und Theoretische Chemie der Universität Bonn,
Wegelerstrasse 12, D-5300 Bonn 1, Germany

Abstract. Low energy ion scattering (LEIS) techniques have been used to study the initial interaction of oxygen with $Ni_3Al(100)$ and $Ni_3Al(110)$, and the structure of Cu/Ru(0001). Dosing $Ni_3Al(100)$ and $Ni_3Al(110)$ surfaces with oxygen at a low pressure of 2×10^{-7} mB at 700°C resulted in the formation of an aluminium oxide. This oxide AlO_x film was examined by measuring the azimuthal angle dependence of the scattered ion yield at fixed scattering angle under different coverages of oxygen. The results indicated that oxygen adsorption was accompanied by significant changes in the surface composition and that oxide growth proceeded by the formation of island, with the oxide layer disordered with respect to the underlying single crystal substrate. Using 1 keV Li^+ ion scattering on Cu/Ru(0001), it was found that Cu grew layer-by-layer for the first two layers. Experimental results obtained by analysing the incident angle dependence associated with shadowing of Ru by Cu confirmed that the first overlayer Cu was in a normal registry position, i.e. there was a continuation of the hcp stacking sequence. The second layer Cu atoms have also been determined to be at the fcc sites by analysing the shadowing critical angles for the second layer Cu focusing onto the first layer Cu atoms in the [10$\bar{1}$0] azimuth.

1. Introduction

In recent years there have been numerous experiments on bimetallic and alloy surfaces due to the fundamental importance of these systems in both heterogeneous catalysis and materials science. From a theoretical point of view, mechanisms of surface segregation and chemisorption are not understood for many alloys. In heterogenous catalysis there are many important mechanisms which need a more profound understanding.

The intermetallic compound Ni_3Al exhibits increasing yield strength with increasing temperature and, in addition, has excellent oxidation and corrosion resistance at elevated temperatures [1,2]. Oxidation properties of this surface at high temperature are therefore of particular interest.

Cu/Ru(0001) system is also interest. There is general agreement now that Cu grows on Ru(0001) forming a pseudomorphic film in the first layer [3,4]. However, the experimental results and their interpretations concerning further growth are still quite controversial. For instance, in an earlier LEED study of the surface system, Christmann et al. [5] found a disordered Cu film in the submonolayer regime and a Cu(111)-like structure for 3D agglomerates at higher coverages. An interesting aspect of this system was later reported by Houston et al. [3] and by Park et al. [4], who agreed on the formation of 2D, pseudomorphic islands only in the first layer and an epitaxial Cu(111) structure for the second layer. Recent STM measurements [6] revealed a layer-by-layer growth at least up to two layers. On the other hand, using low energy alkali ion scattering Overbury et al. [7] suggested that Cu forms well ordered islands which are pseudomorphic for two layers.

In the present work, attempts have been made to study these alloys and bimetal surfaces with the technique of LEIS, which has the advantage of obtaining information concerning the atomic arrangement of the surface layers and adsorption properties. LEIS has been well established as a probe for both surface element composition and surface structure [8-10]. The usual method for LEIS data analysis involves the concept of shadowing or blocking of substrate atoms or adsorbates to explain changes in reflected ion yield. The shadowing positions as a function of incident and azimuthal angles establish the orientation between the atoms on the clean surface or between adsorbate and target atoms. By measuring the scattered or recoiled ion yield at specific scattering and recoiling angles as a function of ion beam incident angle and azimuthal angle to the surface, one can observe variations in structural signals that can be interpreted in terms of the interatomic spacings based on the form of the scattering potential. Low energy alkali ion scattering is an extremely effective technique for surface structural analysis. It is capable of directly measuring the substrate-substrate, adsorbate-substrate and adsorbate-adsorbate interatomic spacings in surfaces, and is complementary to LEED which provides the symmetry of surface long range structure.

An important point in these studies has to be emphasised here. The oxygen adsorption measurements on $Ni_3Al(100)$ and $Ni_3Al(110)$ were carried out with He^+ ions for two reasons. He^+ ions have an extremely high rate of neutralisation during scattering events, ensuring that the detected signal mainly comes from the outermost layer; for Li^+ scattering, O single scattering peaks are superimposed on a large background signal at lower kinetic energies owing to multiple, inelastic scattering of the Li^+ ions from the subsurface. Such a background can totally obscure LEIS O peaks. However, care has to be taken to include the possible angular-dependent neutralisation of the He^+ ions in the interpretation of the experimental results.

2. Experimental

The experimental apparatus has been described previously [11,12]. The experiments were performed in an UHV system with a base pressure of 1×10^{-10} mB. The UHV chamber was equipped with a three-grid LEED optics. The sample could be rotated around two perpendicular axes through the sample surface to control incident and azimuthal angles of the sample, α and ϕ respectively. A spherical electrostatic analyser could be rotated about the sample, allowing variation of the total laboratory scattering angle Θ from 0 to $\pm 130°$. The accuracy of the angles α, ϕ, and θ are $\pm 0.5°$, $\pm 1°$ and $\pm 1°$ respectively.

The analysis of adsorption measurements on $Ni_3Al(100)$ and $Ni_3Al(110)$ was performed with 1 or 2 keV He^+ ions. Typical ion current densities at the sample were between 2×10^{-8} and 8×10^{-8} A/cm². The structural analysis on Cu/Ru(0001) was done by Li^+ ions. The Li^+ ions were generated by a solid alkali charge holder, mass analysed, and well collimated into a beam $\sim$ 2 mm in diameter at a typical current of about $\sim$ 100 pA. The ion scattering was analysed by a spherical electrostatic analyser ($\Delta E/E = 0.02$) with multichannel detection. A typical full energy spectrum was measured in 10 secs, which corresponds to 2×10^{13} ions/cm² of ion dose. No significant effect of ion induced damage and desorption was observed during the measurement. The detailed AES analysis was performed on the scan Auger spectrometer in the Department.

All samples used in this study were aligned using the Laue X-ray diffraction technique. $Ni_3Al(100)$ and $Ni_3Al(110)$ cleaning were achieved with cycles of Ar^+ ion bombardment followed by high temperature ($\sim 900°C$) annealing. The Ru(0001) sample was cleaned by using a standard procedure: Ar^+ ion sputtering between many heating/cooling cycles between 300 and 1200° C in an O_2 pressure of 2×10^{-7} mB. Final cleaning was achieved by heating to 1250° C.

Adsorption measurements were carried out by back filling the chamber with research grade O_2 through a leak valve to the desired pressure. All exposures are expressed in Langmuir units (1 L = 10^{-6} Torr.s = 1.33×10^{-4} Pa.s). After each measurement the Ni_3Al crystal was cleaned and annealed.

High purity Cu was evaporated onto the Ru surface from a resistively heated W basket. The source was carefully outgassed prior to Cu evaporation. The deposition rate was controlled by the voltage difference across the W filament.

3. The initial interaction of O with the $Ni_3Al(100)$ and $Ni_3Al(110)$ surfaces

Previous studies on ordered Ni_3Al alloys [12-14] by LEIS with He^+ and Li^+ ions have revealed that the surface composition of the (100) and (110) faces has 50% Ni-50% Al with some small percentage of Ni enhancement (perhaps 1-2%), and the second layer composition is basically Ni. The surface composition of both surfaces once established, remains constant over a wide temperature range.

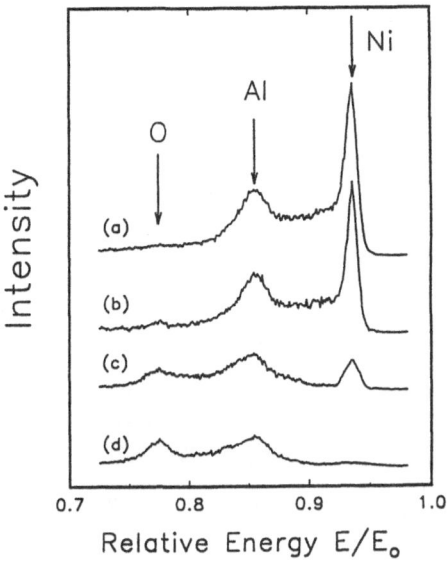

Fig. 1. Energy spectra obtained for 2 keV He$^+$ ions scattered from a Ni$_3$Al(100) surface along the [011] azimuth at $\theta = 2\alpha = 60°$ after different exposures to O$_2$ at 700°C. (a) clean surface, (b) 0.2 L, (c) 3 L and (d) 20 L.

The clean surface of the Ni$_3$Al was exposed to oxygen at a constant pressure for a known time while the crystal was maintained at a temperature of 700°C. Typical He$^+$ energy spectra taken from a Ni$_3$Al(100) surface are shown in fig. 1. An important feature of fig. 1 is that the scattering intensity from Ni atoms decreases much more rapidly than that from Al atoms until Ni is no longer detectable at the longest exposures, in contrast to the presence of a considerable Al peak intensity at all oxygen exposures. A similar result holds for the Ni$_3$Al(110) surface.

For the purpose of obtaining quantitative experimental data, a systematic study of O adsorption as a function of exposure was performed. Example plots for the Ni and Al signals as a function of the O signal at various O exposures are shown in fig. 2. Each data point was normalised to that corresponding to the freshly cleaned component signal intensity. The first observation to be made is that the Ni signal (solid circles) exhibits a linear decline with the adsorbed O intensity, and at an O exposure of about 10 L on the (100) surface (fig. 2a) or about 8 L on the (110) surface (fig. 2b) the Ni signal is no longer detectable by He$^+$ scattering. The assumption made in the following analysis is that the reduction of the Ni peak Intensity is due solely to shadowing of the substrate by adsorbed O overlayers and that the charged fraction of He$^+$ ions scattered from the substrate Ni atoms remains constant. The normalised Ni intensity value, i.e. the ratio of the Ni signal at any oxygen exposure to that of the clean surface is then a direct measure of the number of substrate Ni atoms not covered by O adsorbates. This suggests that the disappearance of the Ni signal coincides with the coverage of the Ni$_3$Al surface by a monolayer of AlO$_x$. Our observations strongly support the formation of an aluminium oxide island on the surface [15], although the mechanism for apparent interchange of the Al atoms to the surface to replace the Ni is not clear in this stage. Experiments along a random (high-index) direction under identical adsorption conditions (not shown) provided the same result as above.

118

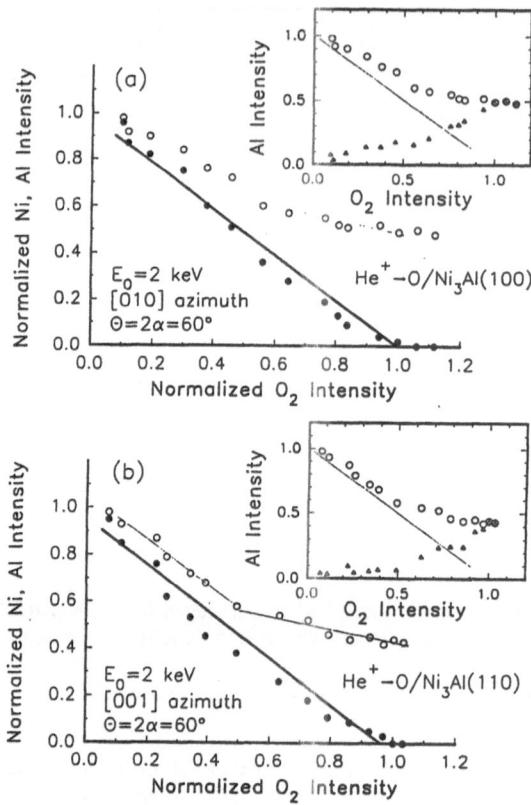

Fig. 2. The decrease in the ion scattering intensity from Ni ($\bullet$) and Al (O) on the Ni$_3$Al(100) and (b) Ni$_3$Al(110) surfaces, compared with the O coverage on the surface. Each data point was normalised to the corresponding freshly cleaned component intensity (dose/clean-surface). The solid and dotted lines are fitted to the experimental Ni and Al results. The insets show the calculated Al intensity of the uncovered substrate (dashed lines), and the difference between the measured Al signal and the calculated Al signal of the substrate ($\vartriangle$) respectively.

When considering the relationship between the Al intensity and the O intensity, it is clear from the data in fig. 2 that the Al signal observed (open circles) appears to involve a two stage process where the slope in the Al intensity changes at a normalised O intensity of about 0.5. As mentioned above, a normalised O intensity of 1.0 is defined as the point of total disappearance of the Ni signal, and interpreted as a monolayer of the AlO$_x$ formed. The results indicate that first part of the change of the Al signal with coverage corresponds to a low coverage of oxygen. In this regime, its variation with coverage is close to parallel to the decrease in the Ni intensity under the same circumstances. With increasing coverage, however, a second component of the Al signal can be seen. This is apparently due to an increase in the number of Al surface atoms. If this is so, it must correspond to an interchange of Ni and Al on the alloy surface. It is also possible that the growth of the AlO$_x$ overlayer is not uniform, and may result from a restructuring which occurs at about half monolayer coverage when the overlayers start to coalesce.

In order to locate the adsorption site for the O uptake, experiments studying the variation of the scattered ion yield from the surface atoms as a function of the angle of incidence along low index azimuths were performed (not shown). The variation of the Ni signal intensity from the partially O-covered surface shows almost the same incident angle dependence as that from a clean surface, indicating that the O at the surface is not influencing the scattering from the Ni atoms. This is further evidence that the O adsorption does not affect the nearest neighbour environment of the Ni atom. It is postulated that the O adsorption promotes an aluminium oxide growth in the form of islands. The Ni exists on the surface only where the surface is clean. In addition, the peak shapes and positions of the Ni signal have been observed not to be affected, again indicating that the uncovered surface from which the Ni signal is derived is clean Ni$_3$Al. The experiment performed above suggests a mechanism whereby oxygen adsorbs at a low pressure at a temperature of 700°C to form a thin oxide film which, depending on the exposure, may be less than one monolayer. Those areas not covered with oxygen or oxide are clean. This is supported by detailed studies of the azimuthal dependence of the scattering of He$^+$ ions from both O-covered Ni$_3$Al(100) and (110) surfaces (coverage of less than one monolayer). These results are shown in figs. 3 and 4 respectively.

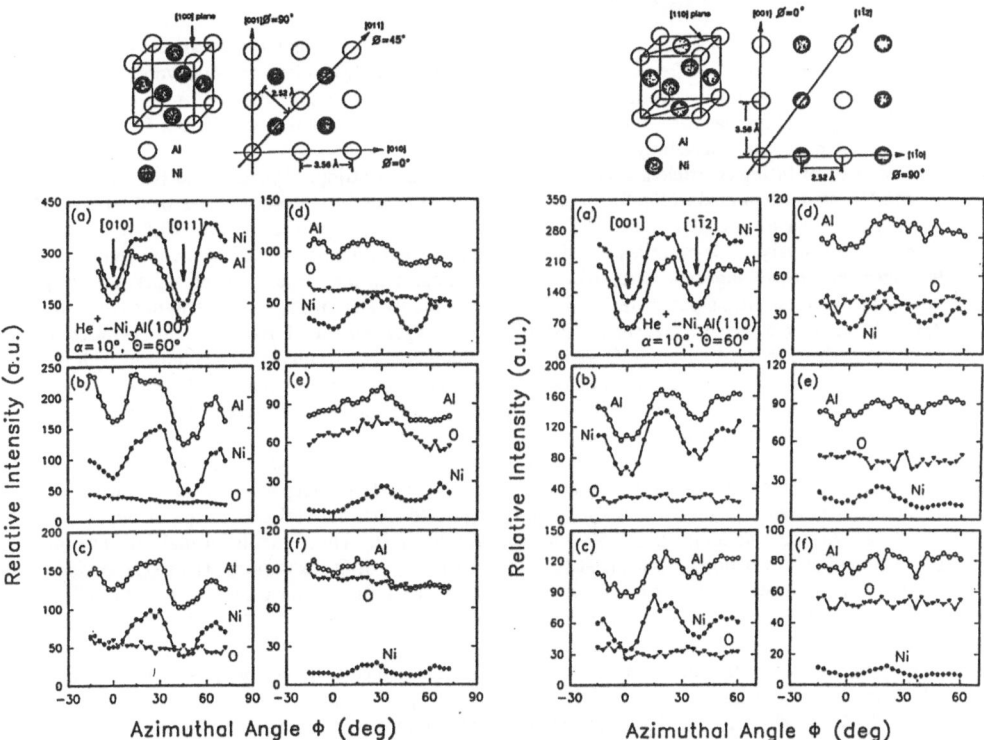

Fig. 3. The azimuthal dependence of the ion scattering from O, Ni and Al from a (100) surface after different exposures to oxygen at 700°C. (a) clean surface, (b) 1 L, (c) 2 L, (d) 4 L, (e) 6 L and (f) 8 L. The azimuthal arrangement of the clean Ni$_3$Al(100) surface is also schematically indicated on the top of the figure.

Fig. 4. The azimuthal dependence of the ion scattering from O, Ni and Al from a (110) surface after different exposures to oxygen at 700°C. (a) clean surface, (b) 1 L, (c) 2 L, (d) 3 L, (e) 5 L and (f) 7 L. The azimuthal arrangement of the clean Ni$_3$Al(110) surface is also schematically indicated on the top of the figure.

In the presence of adsorbed O, the intensity of the ion scattering from the O is independent of the azimuthal angle. This suggests that the adsorbed O atoms may be randomly located in or above the overlayer surface. No shadowing or blocking effects in the overlayers are found. We conclude that the O, and hence the AlO_x film, is not an ordered layer. This is consistent with LEED observations which have shown that a less sharp (1x1) pattern was observed with no additional spots and a high background after O adsorption.

Let us assume that (i) the neutralisation probabilities for He^+ scattering off Ni, Al and O before and after O adsorption are the same, and (ii) any difference in shadowing effects at the edges of the overlayers is negligible, i.e. effectively assuming the island radius is very large in comparison to the overlayer thickness. The model under consideration is that of an AlO_x film (of unknown composition) forming islands on the surface. The uncovered area retains the structure and composition of the clean Ni_3Al surface. Therefore, the scattering from the different components after O adsorption as a function of azimuthal angle ϕ can be described in this model by

$$I_{Ni} (\phi) = (1-\Theta) I_{Ni, uncovered} (\phi) \tag{1}$$
$$I_{Al} (\phi) = (1-\Theta) I_{Al, uncovered} (\phi) + \Theta I_{Al, covered} (\phi) \tag{2}$$
$$I_O (\phi) = \Theta I_{O, covered} (\phi) \tag{3}$$

where $0 \leq \Theta \leq 1$, Θ is the fraction of coverage in monolayers. Using a regression routine for comparison of the azimuthal dependence of the Ni and Al signals from clean and O-covered surfaces, we can deduce a fractional coverage value of Θ during the formation of the first monolayer [12]. This fitting procedure was carried out for the whole azimuthal distribution at different O exposures for each component independently.

The value of fitting coefficient was determined at each scan fitting. Typical fitted results for the Al and Ni scans are plotted in fig. 5a. Figs. 5b and 5c show the fitting coefficient x (and a fraction coverage Θ from the oxygen signal, i.e. 1-x) as a function of exposure for the (100) and (110) surfaces respectively. It is observed that the variations in the fitting parameters of the Ni and Al are the same with increasing exposures, whereas the variations in the Ni and Al intensities are very different. Also, the fitting parameter (which represents the fraction of the surface still uncovered by AlO_x) is complementary to the surface coverage of oxygen measured relative to the saturation coverage. The fractional coverage Θ derived from the Ni and Al results can be compared with the experimental value for the coverage derived from the variation of the oxygen signal with increasing coverage. These are also shown in figs. 5b and 5c (open circles) respectively. Good agreement between these results is obvious.

A further analysis indicates that the standard deviation of the fitting parameter (O coverage) was less than $\pm 10\%$ for most of the experiments. The coverage Θ is independently derived from the azimuthal dependence of the Ni and Al signal strength and from the variation of the O signal with coverage. All three determinations give the same result for Θ. These results strongly support the suggestion of the formation of oxide islands in the initial stages of O adsorption. The oxide, perhaps Al_2O_3, initially grows in the form of islands with minimal influence on clean areas of the Ni_3Al crystal. Finally, following high exposures no bare areas are left on the crystal which then shows the characteristic of the AlO_x films. In addition, the fitted results also show that there seems to be no contribution from the structure of Al in the overlayer. This suggests the formation of an amorphous AlO_x film on the surface.

On the basis of the results described above, we can propose that initially O adsorption destroys the surface order at least within the first atomic layer of Ni_3Al. From a thermodynamic point of view an aluminium oxide island forms in the initial stages of O adsorption, while the Ni remains in its unaffected state. But there is one open question, i.e. what happened to the Ni atoms of the first Ni_3Al layer? As discussed above, the results show no evidence for the formation of Ni-O interactions, i.e. NiO or $NiAl_2O_4$, nor any features which could be attributed to elemental Ni existing somewhere in the region of an AlO_x overlayer. Our interpretation is that segregation of Al occurs during O adsorption due to the chemical driving force, and then the segregated Al atoms cover the surface Ni atoms. The amount of Al which segregates onto the surface compensates for the amount of Al which is covered by oxygen. Therefore, the net effect appears to be a rapid reduction in the amount of surface Ni atoms as a result of the adsorption of oxygen. At elevated temperatures it is known that Ni is easily diffused into the Ni_3Al bulk, so the segregation of Al which has reacted to form AlO_x is counterbalanced by Ni diffusing in the opposite direction towards the Ni_3Al bulk. For example, at 700°C the Ni diffusion coefficient is 5.5×10^{-17} cm^2/s, which is sufficient for an appreciable extent of Ni diffusion into the Ni_3Al bulk.

The similarities in the initial stages of O adsorption on both $Ni_3Al(100)$ and $Ni_3Al(110)$ surfaces are interesting. It appears as if the oxygen strongly interacts with only one constituent of the alloy. This is similar to observations reported for the interaction of O with NiAl(110) and NiAl(111) [16]. Our results provide evidence that exposure to O not only oxidises the surface Al to form an oxide island

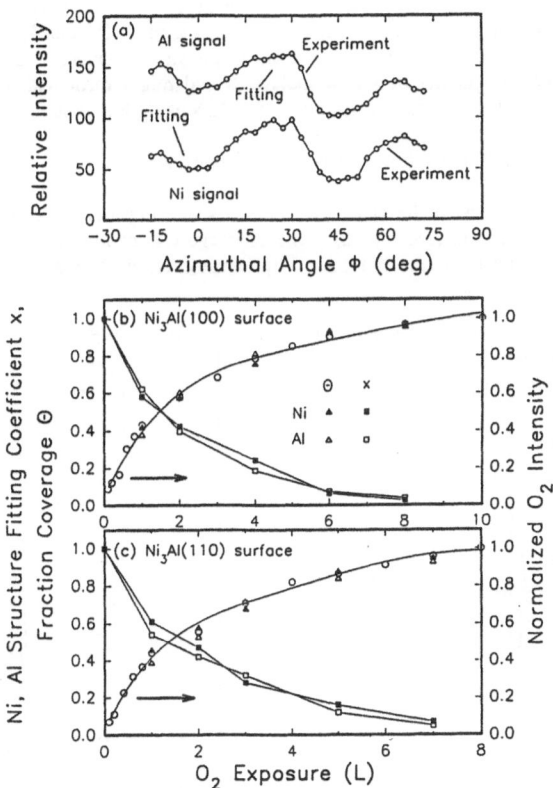

Fig. 5. (a) Example Al and Ni fitting plots (experimental curves from fig. 3c), (b) the fitting coefficient x and a fractional coverage Θ vs O exposure on the $Ni_3Al(100)$ surface. It was done by fitting the whole azimuthal distributions of the Ni and Al (figs. 3b-f) to those of the clean surface (fig. 3a) using a regression routine. The normalised O intensity (open circles) with increasing coverage ($I_0 = \Theta \, I_{O,saturation}$) is also shown for comparison, and (c) similar to b on the $Ni_3Al(110)$ surface (for detail see text).

but also provides an additional driving force causing segregation of Al to the overlayer surface. The Al segregation seems to be more dramatic with increasing O exposures, as shown in fig. 2. These results are in agreement with the lower sublimation energy of Al, affording its segregation to the surface. From thermodynamic considerations the formation of an AlO_x (Al_2O_3) oxide is energetically strongly favoured over the formation of a NiO overlayer because the heat of formation for Al_2O_3 (-400.5 kcal/mol) [17] is much larger than the corresponding value for NiO (-57.3 kcal/mol) [17]. On the other hand, the energy necessary to form elemental Ni and Al from $Ni_{75}Al_{25}$ alloy is only 9.0 kcal/mol based on interpolation of standard data [18]. Therefore, a reaction of the form may occur

$$2\ Ni_3Al\ (s)\ +\ 3/2\ O_2\ (g)\ \rightarrow\ 6\ Ni\ (s)\ +\ Al_2O_3\ (s) \tag{4}$$

where two Ni_3Al layers react with O to form Al_2O_3 and six equivalents of Ni. This would be strongly exothermic. Based on the above enthalpy arguments, this might lead to the formation of another aluminium oxide layer in the second layer, although at present we have no information about the concentration for the deeper layers. In addition, it has been argued [19] that $NiAl_2O_4$ is also a stable oxide, but there is no evidence to support its growth in the surface layer. It is possible that mixed oxides such as $NiAl_2O_4$ might not be planar molecules and might form with Ni atoms below the surface. Unfortunately, this cannot be determined by LEIS using He^+ ions.

4. Cu deposition on Ru(0001) by low energy Li^+ ion scattering

Ru(0001) surface was verified to be clean by He^+ ion scattering and AES measurements, which showed that O, S and C absent (see figs. 6). The coverage was calibrated by comparing our AES and LEED results with those previously reported by Houston et. al [3] and by Park et. al [4].

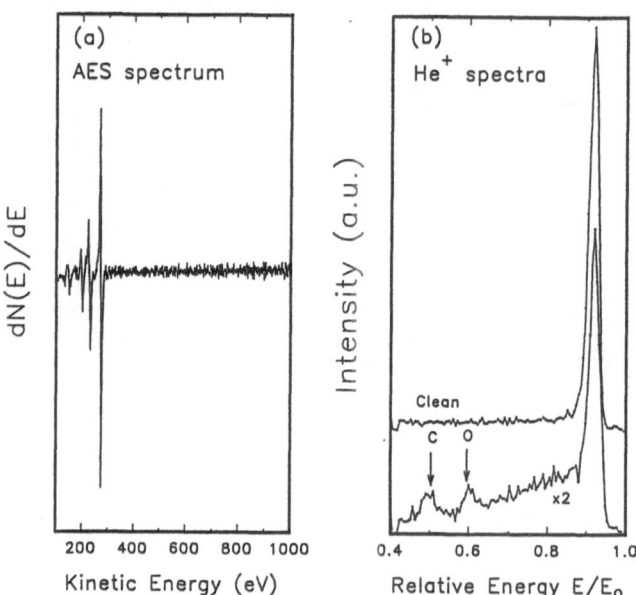

Fig. 6. (a) AES spectra of a clean and well annealed Ru(0001) surface, the negative-to-positive peak intensity ratio in dN(E)/dE Auger spectral feature at 273 eV is 1.37; and (b) energy spectra for 1 keV He^+ ions scattered off dirty and clean Ru(0001) surfaces at a scattering angle of $\Theta=90°$.

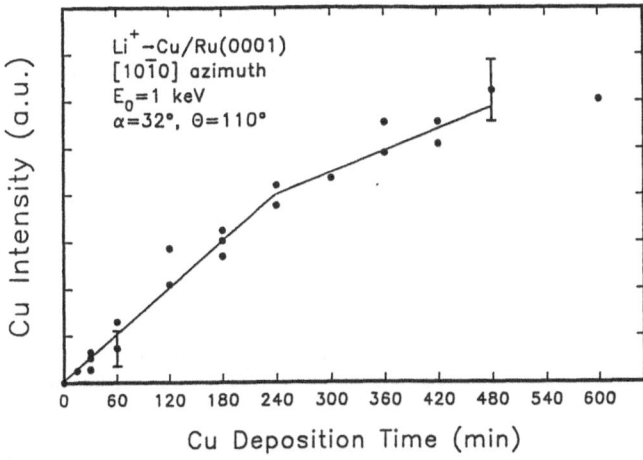

Fig. 7. Intensity of the Li$^+$-Cu single scattering peaks from the Cu covered Ru(0001) surfaces shown as a function of the Cu coverage in the [10$\bar{1}$0] azimuth at $\alpha=32°$ and $\Theta=110°$. The solid lines are intended to guide the eye and representative error bars are indicated.

Our AES measurements by recording the Cu(920 eV)/Ru(272 eV) AES peak-to-peak ratio as a function of Cu coverage were supported by the calculation on the basis of the Gallon mode [20] assuming layer-by-layer (Frank-van der Merwe) growth (not shown). From the AES experiments we specify the coverage scale, which was established based on a linear interpolation between the 0.0 and 1.0 ML and between 1.0 and 2.0 ML points. The experimental error of the Cu coverage should be about ±15% but less than ±0.25 ML. As will be discussed below, the Li$^+$ ion scattering results also give a lower limit of the amount of Cu deposited on the surface.

Our LEED observation showed general agreement with Park et al. [4]. A clean and well annealed Ru(0001) showed the 1x1 hexagonal diffraction pattern with a very low background. For coverage of Cu up to about 1 ML, the sharp integral order 1x1 LEED pattern on a low background intensity remained unchanged, indicating that the first layer Cu formed a pseudomorphic monolayer on Ru(0001). Further growth up to ~2 ML, extra spots appeared and grew in intensity with increasing coverage, indicating layer-by-layer growth at this stage. This pattern was due to multiple scattering between Cu and Ru lattice. At higher coverages the LEED pattern consisted of the superposition of Ru(0001) and Cu(111) patterns.

Cu deposition data for the clean and well-annealed Ru(0001) surface near room temperature are shown in fig. 7, where the Cu signal intensities are given as a function of the coverage. In this experiment the ion beam was incident along the [10$\bar{1}$0] azimuthal direction at an incident angle of $\alpha=32°$ and a scattering angle of $\Theta=110°$. Under these conditions the Li$^+$ ions can be focused onto the first layer Cu through the deposited second layer Cu with increasing coverage (shown below), causing the scattered ion intensity increase if the first two Cu monolayers grow in layer-by-layer model. This increase is observed from the measurements. The Ru signal intensity decreases with increasing Cu coverage (not shown). At the same time, the Li$^+$ single scattering intensity from Cu increases with the Cu coverage. The clear implication from these data is that Cu deposition causes depletion of the scattered Ru intensity along this special geometry due to shadowing effects. The scattered Cu signal increases almost linearly with Cu coverage up to about 1 ML where a distinct breakpoint in the slope is observed. On further increasing the Cu coverage, the Cu signal increases due to the second layer Cu focusing onto the first layer Cu. The almost linear increasing after 1 ML Cu coverage up to about 2 ML indicates layer-by-layer growth at least up to 2 ML. The LEIS data presented in fig. 7 also confirm the coverage calibration established by AES and LEED. The results indicate that the AES data, the model prediction, the LEED observation and the LEIS results are consistent.

124

The intensities of 1 keV Li$^+$ ions scattered off clean and Cu-covered Ru(0001) surfaces as a function of incident angle α are shown in fig. 8. The typical incident angle scan for Ru pattern were taken in the [10$\overline{1}$0] azimuth at a scattering angle of $\Theta = 130°$. A comparison of the corresponding spectra from clean and Cu-covered Ru(0001) shows the following: Firstly, Cu deposition results in attenuation of the first shadowing peak where scattering occurs from only the first Ru layer atoms. Secondly, at coverage of 1.2 ML the new shadowing peak appeared at around $\alpha = 29°$ is observed due to the first Cu overlayer focusing onto the first Ru layer. This confirms pseudomorphic growth of the first Cu overlayer. In other words, that the Cu atoms are in a normal registry position (i.e. there is a continuation of the hcp stacking sequence). Thirdly, the fact that the two Ru shadowing edges at higher incident angles resulting from second and third layer scattering remain almost unchanged for coverages up to 2.2 ML also supports above conclusion. Fourthly, the presence of the second Cu overlayer results in disappearance of the Ru shadowing peak at around $\alpha = 29°$. This is due to shadowing effects. This observation is consistent with the fact that the second Cu overlayer atoms are located at fcc sites. We believe that a Cu(111)-like overlayer has started to develop at this stage. Further analysis regarding the Cu atom sites will be reported in detail somewhere [21].

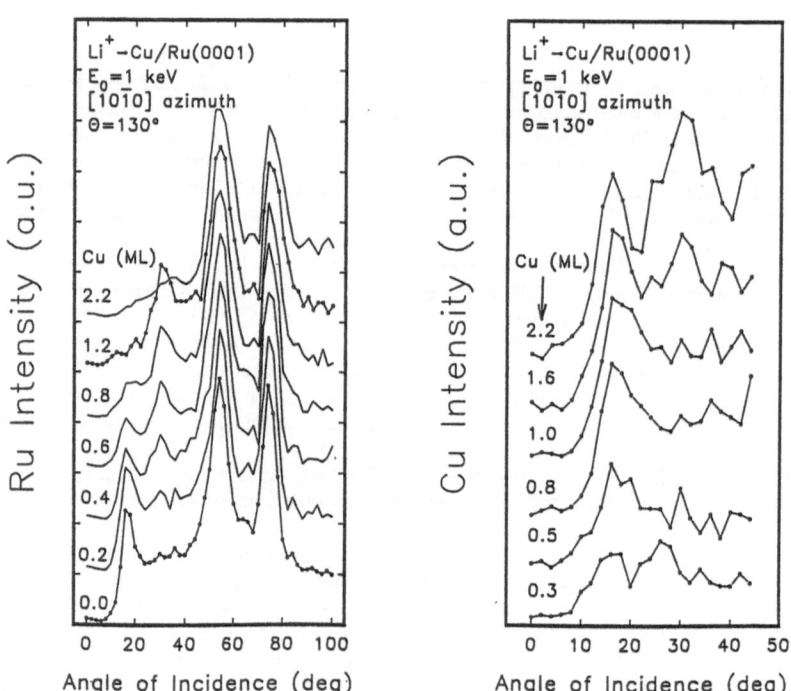

Fig. 8. Intensities of 1 keV Li$^+$ ions scattered from the Ru atoms as a function of incident angle α are shown in the [10$\overline{1}$0] azimuth for various coverages of deposited Cu at $\Theta = 130°$.

Fig. 9. Intensities of 1 keV Li$^+$ ions scattered from the Cu atoms as a function of incident angle α are shown in the [10$\overline{1}$0] azimuth for various coverages of deposited Cu at $\Theta = 130°$.

The incident angle dependence along this azimuthal direction was further tested by collecting Cu signals as a function of Cu coverage. The results of such measurements are shown in fig. 9. The general feature for Cu deposition along the $[10\bar{1}0]$ azimuth results in increasing the first Cu shadowing edge with increasing Cu coverage. This is due to the first layer Cu overlayer shadowing of first layer Cu overlayer. The new shadowing edge at around $\alpha = 29.5°$ appears at Cu coverage of 1.6 ML and becomes pronounced at coverage of 2.2 ML. This is strong evidence that the second Cu overlayer atoms are located at fcc sites. This is in excellent agreement with the observations shown in fig. 8, which shows that the presence of the second Cu overlayer has very little effect up the shape of the Ru single scattering shadowing edges at higher incident angles along the $[10\bar{1}0]$ azimuth.

The structure of the Cu thin films up to several monolayers has been probed by measuring the incident angle dependence along the $[1\bar{1}00]$ azimuth at a scattering angle of $\Theta = 130°$. For comparison, the incident angle dependence for the clean Cu(111) surface along the $[\bar{1}\bar{1}2]$ azimuth under same scattering geometrical conditions has also been measured. This is shown in fig. 10. Both curves are very similar. We believe that an epitaxial Cu(111) structure was fully formed at this stage although a slight broad first shadowing edge was observed.

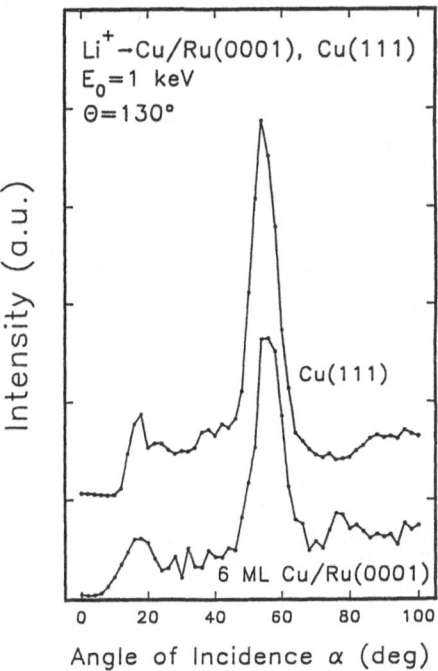

Fig. 10. The incident angle dependence of 1 keV Li$^+$ ions from clean Cu(111) in the $[\bar{1}\bar{1}2]$ azimuth and 6 ML Cu-covered Ru(0001) surface in the $[1\bar{1}00]$ azimuth at a fixed scattering angle of $\Theta = 130°$.

5. Conclusions

Using LEIS it has been possible to obtain some interesting results on the oxygen adsorption and oxide growth on Ni$_3$Al(100) and (110) surfaces. Two conclusions are strongly supported by the results. Firstly, oxygen adsorption activates a segregation of Al to the surface, with the oxygen apparently not bonding to the Ni in the surface plane at least. Secondly, an oxide film of aluminium grows on the surface, with apparently no order relative to the Ni$_3$A matrix and covering part of the surface, the uncovered part retains all of the surface compositional and structural characteristics of clean Ni$_3$Al.

For the surface system of Cu on Ru(0001), Li$^+$ ion scattering results indicate that the first layer Cu atoms are in a normal registry position, and the Cu layer is 2.1 Å above the first layer of the Ru substrate. The second layer Cu atoms have been determined to be at the fcc sites of the Ru substrate by testing the second layer Cu focusing onto the first layer Cu along the [10$\bar{1}$0] azimuth in the angle of incidence scan. The structure of the Cu thin films up to 6 ML has also been probed by measuring the incident angle dependence. The result is consistent with an epitaxial Cu(111) structure by comparing with the results of Cu(111).

Acknowledgements

The authors gratefully acknowledge the financial support of this work by the Australian Research Grants Scheme. One of us (K.W.) also acknowledges support from the agreement for Research at the University of Newcastle.

References

[1] T. Takasugi, N. Masahashi and O. Izumi, Acta Metall. 35 (1987) 381.
[2] D. Farkas and V. Rangarajan, Acta Metall. 35 (1987) 353.
[3] J.E. Houston, C.H.F. Peden, D.S. Blair and D.W. Goodman, Surf. Sci. 167 (1986) 427.
[4] C. Park, E. Bauer and H. Poppa, Surf. Sci. 187 (1987) 86.
[5] K. Christmann, G. Ertl and H. Shimizu, Thin Solid Films 57 (1979) 247;
    J. Catal. 61 (1980) 397.
[6] G.O. Pötschke and R.J. Behm, Phys. Rev. B44 (1991) 1442.
[7] S.H. Overbury, D.R. Mullins, M.T. Paffett and B. Koel, in The Structure of Surface III, edited by S.Y. Tong, M.A. van Hove, K. Takayanagi and X. Xide (Springer, Berlin) (1991) 323.
[8] H. Niehus and R. Spitzl, Surf. Interface Anal. 17 (1991) 287.
[9] S.H. Overbury, Nucl. Instr. Meth. B27 (1987) 65.
[10] Th. Fauster, Vacuum 38 (1988) 129.
[11] Y.G. Shen, D.J. O'Connor and R.J. MacDonald, Nucl. Instr. Meth. B66 (1992) 441.
[12] Y.G. Shen, D.J. O'Connor and R.J. MacDonald, Surf. Interface Anal. 17 (1991) 903; 19 (1992) 729.
[13] D.J. O'Connor, B.V. King, R.J. MacDonald, Y.G. Shen and X. Chen, Aust. J. Phys. 43 (1990) 601.
[14] Y.G. Shen, D.J. O'Connor and R.J. MacDonald, Nucl. Instr. Meth. B67 (1992) 350.
[15] U. Bardi, A. Atrei and G. Rovida, Surf. Sci. 239 (1990) L511.
[16] R.M. Jaeger, H. Kuhlenbeck, H.-J Freund, M. Wuttig, W. Hoffmann, R. Franchy and H. Ibach, Surf. Sci. 259 (1991) 235.
[17] CRC Handbook of Chemistry and Physics, 72th ed. D.R. Lide, (CRC Press, Boca Raton, FL, 1991/1992).
[18] R. Hultgren, P.D. Desai, T.T. Hawkins, M. Gleiser and K.K. Kelly, Selected Values of the Thermodynamic Properties of Binary Alloys (American Society for Metals, Metals Park, OH, 1973).
[19] E.W. Young, J.C. Riviere and L.S. Welch, Appl. Surf. Sci. 28 (1987) 71.
[20] T.E. Gallon, Surf. Sci. 17 (1969) 486.
[21] Y.G. Shen, D.J. O'Connor, J. Yao, K. Wandelt, H. van Zee, R.H. Roberts and R.J. MacDonald, to be published.

Ion Beam Mixing in Metals

B.V. King[1], M.A. Sobhan[1], M.Petravic[2]

[1]Physics Department, University of Newcastle, Callaghan 2308, Australia

[2]Department of Electronic Materials Engineering, Australian National University, PO Box 4, Canberra, Australia

Ion mixing of impurity markers in silver and of isotopic silver bilayers is discussed. The efficiency of mixing is shown to depend on the ion species and energy as well as the substrate temperature and the impurity species. The magnitude of ion mixing broadly agrees with the Koponen and Hautala thermal spike theory of ion mixing.

INTRODUCTION

The process of ion mixing is occurs when an energetic ion impacts on a solid surface and rearranges, or mixes, atoms beneath the surface. The intermixing occurs in different stages. In the first fraction of a picosecond, the energetic primary ion causes displacements of target atoms. These primary recoils collide with other target atoms eventually generating a collision cascade. Damage in the form of a distribution of point defects builds up during this ballistic phase of the cascade which may last a few tenths of a picosecond. Replacement collision sequences during this time generate widely separated vacancy-interstitial pairs. For ion impact energies above a few hundred eV, a distinct spike region may be formed along the ion track containing highly excited atoms resembling a liquid [1]. As the collision cascade "cools", point defects are partially annealed due to spontaneous recombination or clustering forming a core of vacancies or vacancy clusters surrounded by interstitials on the periphery of the original region of energy depsoition. During the cooling or relaxation phase of the cascade, which may last picoseconds in noble metals, extensive mixing may also take place.

In this paper we will discuss the determination of the efficiency with which ion beams cause intermixing in silver. It is useful to study the mechanisms of ion mixing using both high (>100keV) and low (<10keV) ion energies because implantation in each energy regime has advantages and disadvantages. These are outlined below.

(i) In high energy mixing the deposition of energy into displacement processes, which initiates mixing, occurs below the surface and so is spatially separated from the surface where sputtering occurs. Rutherford backscattering (RBS) measurements can then be made of elemental distributions before and after irradiation and the difference ascribed to ion mixing alone. For low energy mixing, the region of high nuclear stopping power, $(dE/dx)_{nucl}$, is close to the surface so the effects of sputtering and mixing are interconnected. Mixing efficiencies can only be found from secondary ion mass spectrometry (SIMS) depth profiles, for example, if sputtering effects e.g. preferential sputtering and the escape depth of the detected species, are well known.

(ii) High energy mixing is a low fluence process whereas to observe changes in elemental concentrations in low energy mixing requires sputter profiling, a high fluence process. With high ion doses come the development of surface topography, especially for inert gas bombardment of noble metals [2], and incorporation of the primary ion into the surface, either causing oxide or alloy formation or segregation leading to, for example, bubble formation. All these processes must be considered in extracting mixing information from sputter profiles. Anisotropic transport processes are relatively more important in low fluence experiments whereas in high fluence experiments, especially if the impurities and matrix have similar atomic masses, anisotropic material transport is cancelled by bulk flow due to density changes [3].

(iii) Above implantation energies of about 20keV, displacement energy deposition does not take place in one connected cascade region. Rather, subcascades are generated by energetic primary recoils along the ion track. Mixing takes place within those subcascades. Quantification of high energy mixing then relies on a knowledge of the subcascade distribution at the depth of interest.

If mixing is assumed to be isotropic then the change in variance, $\Delta\sigma^2$, of a dilute gaussian marker which has been irradiated by a dose ϕ ions cm^{-2} is given by the solution of the simple diffusion equation as $\Delta\sigma^2 = 2Dt = 2k\phi F_D$ where D and t are a diffusion constant and time respectively, k is a constant of proportionality and F_D is the rate of energy loss to displacement processes, $(dE/dx)_{nucl}$. This relation indicates that, to a first approximation, the amount of mixing across an interface is proportional to both the ion dose and the amount of energy each ion deposits at the interface. The constant of proportionality, k, is the mixing efficiency which, as will be shown later, is a function of not only the target species but also the ion species and energy. For low energy mixing [4], the diffusion equation must be modified to take account of a surface receding with a velocity U. In that case

$$\frac{\partial C(x,t)}{\partial t} = \frac{\partial}{\partial x}\left(D(x)\frac{\partial C(x,t)}{\partial x}\right) + U\frac{\partial C(x,t)}{\partial x} \qquad (1)$$

where C(x,t) is the concentration of the marker at a depth x below the surface which has been eroded for a time t. At the surface of the solid the concentration, C(0,t) must obey a boundary condition in terms of r, the ratio of the impurity sputter yield to the matrix sputter yield. Note that the above equation is based on the assumption that any impurities are dilute and that the ion beam is not incorporated into the surface so that U and D(x) are not functions of time. It is assumed that any anisotropic impurity transport, given by the first moment, is cancelled by the reverse material flow which is required to maintain a constant density. The mixing efficiency is then found by determining the value which gives the best agreement between the experimental depth profile and C(x,t) as found from the above equations using calculated values of $F_D(x)$ and r. This process can be simplified for depth profiles of simple structures, e.g. a bilayer or marker, which can be quantified in terms of a single parameter. For example, after long sputter times, when the original interface has been sputtered through, $C(0,t)$ decreases exponentially with sputter time, i.e. $C(0,t) \propto \exp(-Ut/\lambda)$ where λ is the decay length. λ is then found from the slope of the trailing edge of peaks corresponding to sputter depth profiles of buried marker layers. An alternative measure of the amount of mixing for bilayers is the standard deviation, σ, of the error function which describes the mixed interface. In this review paper we will discuss the measurement of λ and σ, the calculation of mixing efficiency and the relevance of this parameter to models of ion mixing. In particular new results for λ for different impurity markers in Ag will be compared to previous measurements of σ for isotopic Ag bilayers.

EXPERIMENTAL

Two types of silver multilayer sample were made. The first type, polycrystalline multilayers made by ion beam sputter deposition , were used for measurement of mixing of dilute impurities (Ti, Ni, Cu, Mo, Zr, Hf and Ta) in Ag. Alternate layers of pure Ag and a dilute impurity in Ag were made by rotating a target comprising pure Ag on one side and a small piece of the impurity element mounted on a pure Ag sheet on the other in front of a 10mA 1keV Ar^+ sputtering beam. The sputtered material was collected onto a silicon wafer at a temperature of 20°C. The second type, for measurement of the intermixing of Ag^{107} and

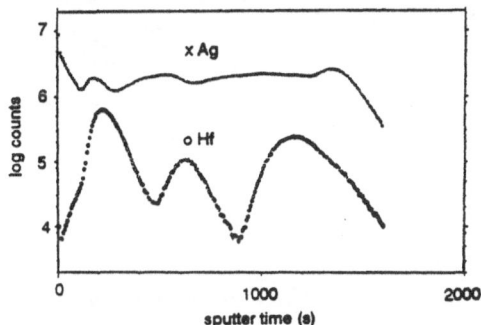

Fig 1. SIMS profile using 3keV Ar of a multilayer of 30nm Ag / 20nm Hf,Ag / 25nm Ag / 5nm Hf,Ag / 40nm Ag / 15nm Hf,Ag / 30nm Ag on Si. The impurity layers were 5% Hf, 95% Ag.

Ag^{109}, used thermal evaporation of the individual isotopes, enriched to 99% abundance, onto either a heated mica substrate covered with a thick film of pure Ag (which is 54% Ag^{107}, 46% Ag^{109}) or onto a silicon substrate which was at room temperature. More experimental details for the second sample may be found elsewhere [5-7]. The films were checked using SEM,TEM, surface profilometry and RBS. Low energy mixing experiments were performed on the isotopic samples [5,6] and on the Ag-impurity samples using SIMS sputter depth profiling, whereas high energy mixing using 200keV Ge^+ or 90keV Si^- was performed only on the isotopic layered sample [6,7].

Figure 1 shows a SIMS depth profile for an impurity-Ag multilayer, which comprises four pure Ag layers alternating with three 95% Ag, 5% Hf layers, a total of about 160nm in thickness. The purpose of the 3 layers is to allow the development of surface topography to be checked. The surface roughness should increase with sputter time whereas mixing and preferential sputtering effects should come to equilibrium after sputtering to a depth equivalent to about twice the ion range, in this case 6nm. Since the top two marker layers are both initially deeper than twice the ion range, the amount of mixing for both and hence λ for both should be the same if topographical effects are negligible. This is the case, because the slopes of the trailing edges of the peaks in fig 1, and thus the values of λ, are the same. However, the Hf^+ peak in fig 1 corresponding to the deepest layer is thicker, possibly due to the development of surface topography, although surface profilometry shows that all SIMS craters are flat to within a standard deviation of 1nm. The Ag^+ signal in figure 1 shows SIMS matrix effects, in that the Ag^+ signal does not change in a complementary way to the Hf^+ signal. No such effect is observed in the depth profile of the Ag^{107}/Ag^{109}/pure Ag multilayer sample (fig 2) which has been profiled with 4keV

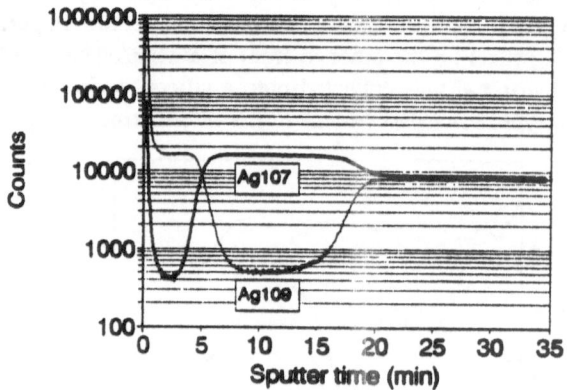

Fig 2 SIMS depth profile of an Ag^{107}/Ag^{109}/Ag multilayer.

O_2^+ [5]. This illustrates the first advantage of depth profiling isotopes of the same element, namely that matrix effects are absent. Apart from an initial transient due to surface contamination and preferential sputtering, three distinct layers are seen, a Ag^{109} rich layer on top of a Ag^{107} rich layer on a thicker pure Ag layer.

The values of decay length, found from fitting an exponenetial decay to the trailing edge of the topmost peak in fig 1, is 6.46nm. Other values of decay length are shown in table 1 and indicate that λ varies from 6.46nm for 3keV Ar^+ irradiation of Hf-Ag multilayers to 12.2nm for 5keV Ar^+ irradiation of Mo-Ag multilayers. λ is seen to increase with ion energy, E, and in fact does scale approximately with $E^{0.5}$ as has been found for impurities in Si [8] and predicted from thermal spike theory [9]. The standard deviation, σ, was found for isotope mixing by fitting the concentration variation across the interface in fig 2 to a sum of two error functions. Calculations were performed for 5, 10 and 15keV O_2^+ give values of σ of 4.5nm, 5.5nm and 7nm respectively. Again σ is seen to increase with energy.

DISCUSSION

The efficiency for low energy mixing may be found from λ or σ by solution of equation 2 with associated boundary conditions. The value of $F_D(x)$ has been determined as the energy deposited only into displacement processes in a full TRIM [10] cascade calculation. Using $F_D(x)$ calculated in this way, λ and σ are plotted as a function of mixing efficiency in figure 3. Figure 3 illustrates that λ is actually a function of r and the mixing efficiency so that to derive the mixing efficiency from λ requires the degree of preferential sputtering to be estimated. In fact, for large values of mixing efficiency λ approaches the value R/r where R is the range over which the ion deposits energy into displacement processes. Unfortunately there are few good measurements of r for impurities in Ag so r

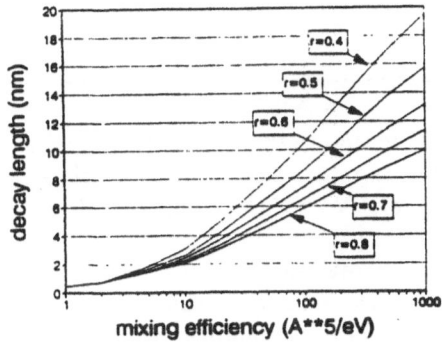

Fig 3   Relationship between the decay length and mixing efficiency as a function of r for 4keV Ar irradiation of Ag.

was calculated from TRIM simulations for 95% Ag- 5% impurity alloys and compared to the analytic expression $r = (M_2/M_1)^{2m} (U_2^s/U_1^s)^{1-2m}$ where $U_{1,2}^s$ and $M_{1,2}$ are the surface binding energies and atomic masses of the impurity and matrix species respectively. For Ag-Cu and Ag-Pd alloys, experimental values of r are 0.71 and 0.67 respectively [11], whereas TRIM simulations give values of 0.97 and 0.83, about 30% higher. For all the impurities considered in this study and m=0.2, equation 7 predicted r values within 20% of the TRIM values. The values of r used for all impurities, which are shown in table 1, have been found by reducing the TRIM estimates by 30% in line with the results above. The mixing efficiencies found from the above calculations are also presented in table 1 and show a range from 50-250 $Å^5eV^{-1}$, a consequence of the almost exponential dependence of mixing efficiency on λ. Given the uncertainty in r, significant uncertainties may be expected in the calculated values of mixing efficiency. However our values appear reasonable when compared to high energy measurements where sputtering is not a factor. For example, mixing efficiencies calculated for 330keV Kr^+ mixing of markers in Ag at 77K also show a wide range

Table 1 Mixing results and thermodynamic parameters for various elements in Ag.

Element	Ion Energy (keV)	λ (nm)	r	$(Dt/\phi F_D)$ ($\text{Å}^5 \text{ eV}^{-1}$)	ΔH_b (eV)	ΔH_m (eV)
Ti	4	7.7	0.58	100	-0.11	-0.22
Cu	8	11.84	0.66	200	-0.05	0.035
Mo	5	12.2	0.52	200	-0.45	0.58
Zr	8	11.25	0.45	70	-0.24	-0.17
Hf	3	6.46	0.4	47	-0.26	-0.15
Ta	3	7.3	0.47	85	-0.48	0.2
Ta	4	8.35	0.47	90	-0.48	0.23
Ta	5	9.27	0.47	90	-0.48	0.23
Ta	8	12.1	0.47	95	-0.48	0.23

from 19 $\text{Å}^5 \text{eV}^{-1}$ for W to 160$\text{Å}^5 \text{eV}^{-1}$ for Pb [12]. Previous work performed using the same impurities as in this work yielded efficiencies of 25$\text{Å}^5 \text{eV}^{-1}$ for Ta using 330keV Kr^+ and 138$\text{Å}^5 \text{eV}^{-1}$ for Cu using 750keV Kr^+.

For isotopic mixing, r is approximately equal to unity, obviating any reliance on crude estimates of r in the calculation of mixing efficiency, a second advantage of profiling isotopes. Oxygen incorporation has been handled [5] by solving for a $Ag_{0.5}O_{0.5}$ target as well as a pure Ag target. The experimental values of σ then correspond to mixing efficiencies ranging from 40-60$\text{Å}^5 \text{eV}^{-1}$ if the target is assumed to be AgO. High energy mixing of Ag isotopic layers yields larger mixing efficiencies, 170$\text{Å}^5 \text{eV}^{-1}$ for 200keV Ge^+ of polycrystalline Ag^{109}/Ag^{107} at room temperature [7] and 210$\text{Å}^5 \text{eV}^{-1}$ for 90keV Si mixing of epitaxial isotope layers again at room temperature [6]. In addition, the mixing efficiency is found, for both high and low energy mixing, to be a weak function of temperature below 300K and to rise rapidly above 400K [6] as radiation enhanced diffusion processes dominate cascade mixing (figure 4).

In summary, mixing efficiencies in Ag vary from about 40-200$\text{Å}^5 \text{eV}^{-1}$, increase with primary ion energy and temperature and are dependent on the impurity species. These experimental values may now be compared to estimates from different theories. TRIM can properly only model the ballistic portion of the cascade because the binary collision approximation used in TRIM breaks down at the low interaction energies chararcteristic of the cooling phase of the cascade. An estimate of ballistic mixing can be obtained by using TRIM to calculate the transport function and then determining D(x). For 5 keV Xe irradiation of Ag the calculated value of D(x), when substituted into equation 1, gives a mixing efficiency of approximately 4Å^5 eV^{-1}, significantly lower than the experimental values.   It would then appear that a purely ballistic model is insufficient to explain observed mixing magnitudes.

Various approaches have been taken to model the effects of the relaxation phase of the cascade. They include molecular dynamics simulation [1] and analytic thermal spike models [13].   The first approach has produced good agreement with experimental mixing efficiencies for 5keV Cu cascades in Cu, although a

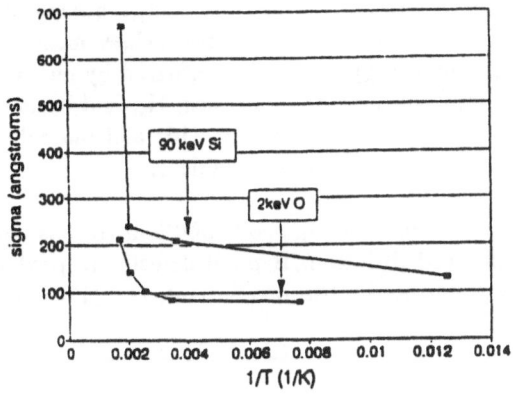

Fig 4   Relative interface width (i.e. 2σ normalised to the interface depth) as a function of target temperature for mixed Ag^{109}/Ag^{107} bilayers [6].

purely repulsive interatomic potential was used [1]. Recent work has extended the ion energy which can be treated up to 25keV [14]. Unfortunately, no results exist for mixing in Ag. A particularly useful analytic theory is that of Koponen and Hautala [14], which models cascade mixing as an interdiffusion of hard spheres in a liquid, since it predicts mixing magnitudes in Ag that (i) increase approximately as $E^{1.7}$ where E, the subcascade energy, ranges from 0.1-10keV and (ii) increase by a factor of 2.5 as the target temperature was raised from 100K to 300K.   These predictions are in reasonable agreement with the experimental results because the high implantation energies (200keV Ge^+ and 90keV Si^-) would be expected to generate a greater proportion of higher energy, and hence more efficient, subcascades than keV bombardment. The value of σ is also found to increase by a factor of 1.8 for 90keV Si^- as the substrate temperature increases from 100-300K (fig 4), in rough agreement with theory. However taking errors into account, there is virtually no change with temperature for low energy mixing, in contradiction to the Koponen and Hautala theory.

The other prediction of thermal spike models which is not confirmed by results in this paper is the variation of mixing efficiency with species. Previously, high energy mixing results in Cu [15] have been linked to impurity diffusion coefficients or, failing any reliable diffusion values, to vacancy-impurity binding enthalpies. If mixing is dominately influenced by interactions within the cooling cascade of energy 1eV or less then it is likely that there should be a correlation between mixing efficiency and thermal diffusion parameters. The correlation between the low energy mixing efficiencies and these parameters has, however, been weak [3] in comparison to high energy experiments where good correlations are seen. Since high energy mixing occurs in low energy cascades it is not clear that there should be any substantive difference between the correlation in the two cases. Perhaps the lack of correlation with low energy mixing efficiencies results from the large uncertainties in the experimental values mentioned above. The lack of correlation for low energy mixing is confirmed by our results in Ag. This is shown by vacancy-impurity binding enthalpies, H_b, and heats of mixing, ΔH_m, calculated from Miedema's heats of solution [16] and the cohesive energies of the impurity and Ag, which are presented in table 1.

The increase in the isotope intermixing above 500K in fig 4 is consistent with the onset of radiation enhanced diffusion. If point defects are predominantly lost by recombination then the enhanced diffusion constant, D_{rad} [17] is given by

$$D_{rad} = 1.22\sqrt{\epsilon Ka_o^2 D_v/32\pi} \qquad (2)$$

where ϵ is the fraction of freely migrating defects, K is the displacement rate, a_o is the lattice parameter and D_v is the diffusion constant for vacancies is given by and ν_o is the vacancy attempt rate and ΔH_v^m is the vacancy migration enthalpy.

$$D_v = a_o^2 \nu_o \exp(-\Delta H_v^m / kT) \qquad (3)$$

For 90keV Si irradiation of Ag, ϵ is about 0.015, K is 0.04dpa s^{-1} and ν_o is 9×10^{13}s^{-1}. Good agreement is found with the experimental values of D of 28Å^{2}s^{-1} and 220Å^{2}s^{-1} at 493K and 543K if ΔH_v^m is taken to be 0.75eV.

CONCLUSION
The efficiencies found for low and high energy ion beam mixing of Ag isotopic bilayers agrees reasonably with the thermal spike model of Koponnen and Hautala. Thermal spike concepts cannot however account for the observed mixing efficiencies of various impurities in Ag, although there is doubt about the measured efficiencies due to a lack of data for preferential sputtering. At temperatures above 500K the mixing rate increases due to radiation enhanced diffusion.

REFERENCES

[1] T. Diaz de la Rubia, R.S Averback , H. Hsieh and R. Benedek, J.Mater.Res. 4 (1989) 579

[2] B.M. Paine, J.Appl.Phys. 53 (1982) 6828

[3] B.V. King, S.G. Puranik, M.A. Sobhan, R.J. MacDonald, Nucl.Instr.Meth. B39 (1989) 153

[4] B.V.King and I.S.T. Tsong, J.Vac.Sci.Technol. A2 (1984) 1443

[5] B.V. King, R.P Webb, V.K.M. Sharma and J.A. Kilner, SIMS VIII, eds A.Benninghoven etal (John Wiley, Chichester,1992) 363

[6] B.V. King, M. Petravic, R.G. Elliman and R.J. Chater, SIMS IX, in press

[7] B.V.King,C.Jeynes,R.P.Webb,J.A.Kilner, Nucl.Instr.Meth. B80/81 (1993) 163.

[8] M. Petravic, R.G. Elliman and J.S. Williams, SIMS VIII, eds A.Benninghoven, K.T.F. Janssen, J. Tumpner and H.W. Werner (John Wiley, Chichester,1992) 367

[9] P. C. Zalm and C.J. Vriezema, Nucl.Instr.Meth.B67 (1992) 495.

[10] J.P.Biersack and L.G. Haggmark, Nucl.Instr.Meth. 174 (1980) 257

[11] Sputtering by Particle Bombardment II, ed R.Behrisch (Springer-Verlag Berlin 1983)

[12] E. Ma, S.-J. Kim, M.-A. Nicolet and R.S. Averback, J.Appl. Phys. 63 (1988) 2449

[13] I. Koponen and M. Hautala, Nucl.Instr.Meth. B69 (1992) 182

[14] T. Diaz de la Rubia and W.J. Phythian, J.Nucl.Mat. 191-4 (1992) 108

[15] S.-J.Kim,M.-A.Nicolet,R.S.Averback, Nucl.Instr.Meth. B19/20 (1987) 662.

[16] A.K. Nissen, F.R. de Boer, R. Boom, P.F. de Chatel and W.C.M. Mattens, Calphad 7 (1983) 51.

[17] A. Müller, V. Naundorf and M.P. Macht, J.Appl. Phys. 64 91988) 3445.

Surface Segregation and Preferential Sputtering
of Binary Alloys

E.Taglauer[1], J. du Plessis[2] and G.N. van Wyk[2]

[1]Max-Planck-Institut für Plasmaphysik, EURATOM Association, D-85748 Garching bei München, Germany

[2]Department of Physics, University of the Orange Free State, 9300 Bloemfontein, South Africa

Abstract. The surface composition of compounds and alloys very generally differs from their bulk concentration. Minimization of the total energy (Gibbs free energy) leads to surface segregation in alloys with an equilibrium composition at a given temperature, characterized by the segregation energy. Ion bombardment of a solid multicomponent material can cause an altered surface concentration whose steady state value depends on the different sputtering yields of the constituents. Bombardment can further support segregation due to radiation enhanced diffusion, expressed by a diffusion coefficient. We report on experimental results, model calculations and numerical simulations concerning preferential sputtering, segregation and their combined effects for various binary systems, i. e. Ta_2O_5, PdPt, CuTi, and FeAl. Detailed analysis requires surface analytical methods with different information depths, as provided by ion scattering spectroscopy (ISS) and Auger electron spectroscopy (AES). From such measurements segregation energies and diffusion coefficients can be extracted.

1. Introduction

Ion bombardment of a solid surface in general creates a situation far from thermodynamic equilibrium. The steady state for a certain experimental condition and the relaxation into equilibrium can be used to to determine specific surface properties [1,2]. It has e.g. been applied to study equilibrium surface morphology [3] and the dynamics of equilibration [4] of elemental metal surfaces. In the case of multicomponent material (e.g. metal alloys, oxides, carbides) the surface composition during or after bombardment is determined by collisional processes and in many cases by surface segregation effects. Investigations of that kind are reported in the literature for a number of alloy systems [e.g. 5-8]. In the following we first deal with collisional processes for the example of Ta_2O_5 and subsequently present measurements and the analysis of combined effects for various alloys (PdPt, CuTi, FeAl) using a theoretical model from which it can be shown that the segregation energy ΔG and the diffusion coefficient D can be uniquely determined.

2. Collisional processes

For systems with large differences in the atomic masses of the constituents, such as Ta_2O_5, the preferential sputtering effects can be described by collisional processes exclusively [9]. This can be seen by the example given in Fig.1: for bombardment with He ions the energy transfer

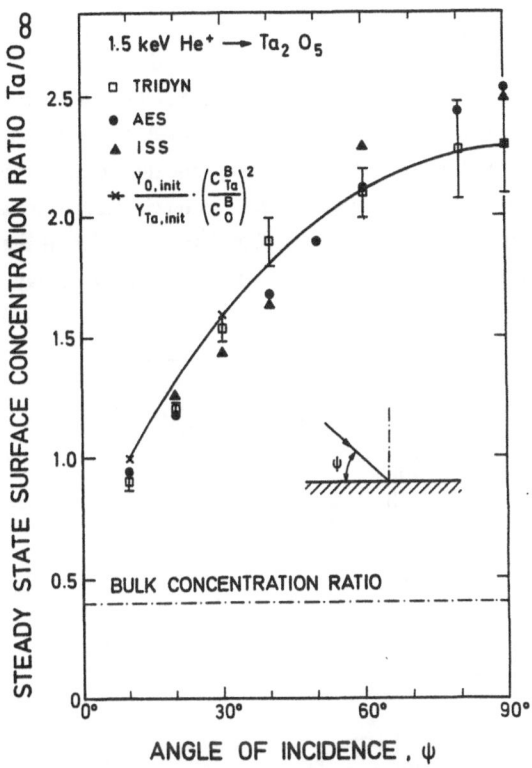

Fig. 1: Ta/O concentration ratio in steady state as a function of the angle of incidence: AES and ISS measurements and TRIDYN simulations, the solid line connects the points of the indicated formula [9].

(as given by two-body collision kinematics) to the oxygen atoms is much more effective than to the Ta atoms (by a factor of 8). Therefore oxygen is preferentially sputtered and the steady state surface concentration ratio Ta/O deviates considerably from the bulk value of 0.4, it is about 1.0 for grazing incidence and increases to about 2.3 for normal incidence. This result can be explained on the basis of mass conservation in steady state [10] which can be expressed in an equation like

$$(C_A / C_B)_{surf} = (C_A / C_B)_{bulk} \cdot Y_B / Y_A \tag{1}$$

138

for the concentrations C of two constituents A and B. In Fig.1 the result from a similar formula is shown that contains the initial values of the sputtering yields Y as a function of the angle of incidence instead of the elemental yields in eq.1. The agreement with these results as well as with the results from numerical calculations using the computer code TRIDYN [11], a dynamical version of TRIM which is based on the binary collision approximation [12], demonstrates that in the case considered here the preferential sputtering effect, i.e. the creation of an altered surface layer, can be explained by collisional processes. In the computer simulation [9] various collision processes with respect to the trajectories of the projectiles and primary and secondary recoils have been identified and their relative contributions to the sputtering yield in the initial and steady state was identified. The dominating process is one in which a primary recoil is directly sputtered by a projectile that penetrates into the target . It has further been shown that the characteristic fluence to reach steady state is determined by the number of oxygen atoms which must be replaced by Ta atoms in the altered layer. The corresponding depth of that layer is given by the range of the energy deposition near the surface as shown in Fig.2 in which the depth distributions of the deposited energy and the

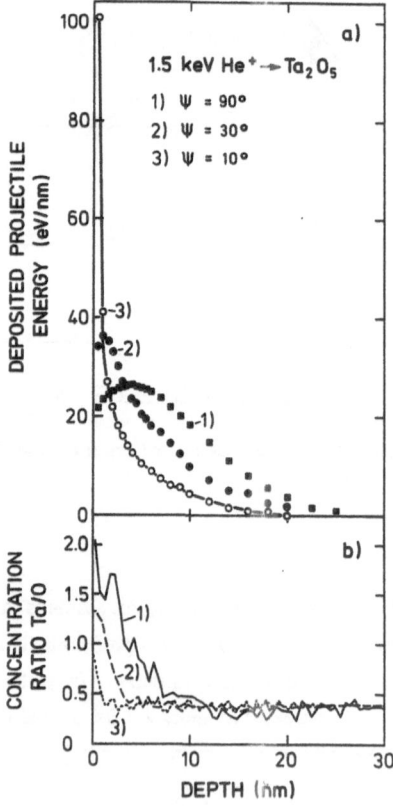

Fig. 2: Depth distributions calculated with TRIM [9] of (a) the average energy deposited per projectile via nuclear collisions and (b) the concentration ratio in the altered layer. Parameter is the angle of incidence relative to the surface.

concentration ratios are compared for three different angles of incidence. The depth of the implanted projectiles is much larger, by about a factor of 6, and therefore not decisive for the depth of the altered layer.

3. Segregation

For metal alloys the purely collisional treatment discussed in the previous section is generally not sufficient, since chemical effects resulting in the segregation of one component on the surface in order to minimize the surface free energy [13,14] have to be considered in addition. This is demonstated in Figs. 3 and 4 which show surface versus bulk concentrations for

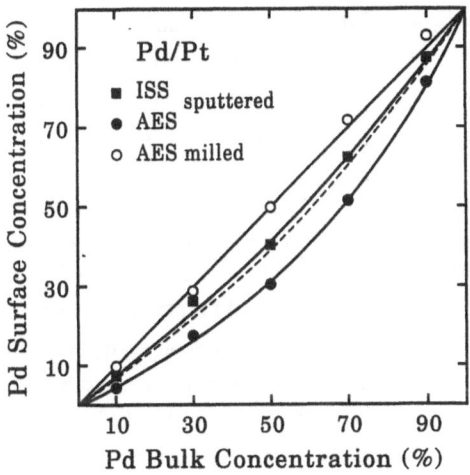

Fig. 3: Pd surface concentration for Ar$^+$-sputtered PdPt alloys (full symbols), showing Pd depletion. Good agreement with bulk values is obtained for in-situ-milled surfaces (open circles). The broken line indicates the surface concentration calculated with eq. (1) [6].

polycrystalline PdPt alloys [6] and amorphous CuTi alloys [15] after Ar$^+$ ion sputtering to saturation (1.5 keV and 2 keV, respectively). In both cases it can be observed that the surface composition is significantly different from the bulk and the values gained from eq.1 are only approximately close to the ISS data. The composition of the top surface layer (ISS) is again different from that surface region sampled by AES (of the order of 5 atomic layers). This difference varies for the two examples, indicating that the combined effects of preferential sputtering and segregation contribute to different extents in the two cases. For PdPt preferential sputtering of Pd can be expected, since its sputtering yield is about a factor of two higher (for the given parameters). The AES data in fact display a substantial Pd depletion. The ISS data, however, show less depletion for the top surface layer and therefore it must be

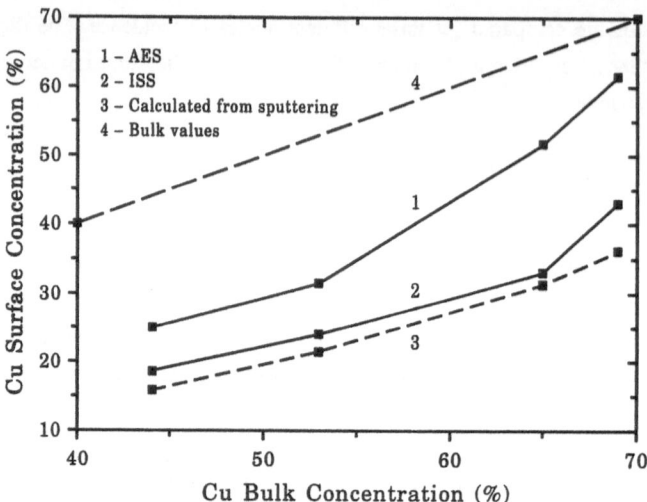

Fig. 4: Cu surface concentrations of sputtered CuTi alloys [15]. AES (1) and ISS (2) measurements and calculations (3) using eq.(1). Curve 4 is the bulk reference line.

concluded that Pd segregation from the altered layer to the surface takes place. This conclusion could be substantiated by the theoretical model described in the following section, giving segregation energies ΔG between 2 and 8 kJ/mole (going from 90% to 10% Pd). Pd segregation is also expected on the basis of present segregation theories [13, 16], a concentration dependence could occur if the strain energy is the driving force, since the PdPt system forms a solid solution with a concentration dependent lattice parameter [17].

In the case of CuTi comparison with the bulk values shows a pronounced Cu depletion due to preferential Cu sputtering, the ratio of the sputtering yields being of the order of 4-6. Here the ISS data exhibit a larger depletion than AES, indicating that Cu segregation is less pronounced than that for Pd in the previous case. Further discussion of the CuTi system is given in connection with the theoretical model presented in the following section.

4. Theoretical model

The model was developed [18] to describe kinetics and equilibrium of Gibbsian segregation in multicomponent alloys and was extended [6,15] to include sputter effects. Along a direction x normal to the surface the sample is divided into a row of cells with distance a, and the atoms are allowed to exchange between these cells by means of three flux terms, the diffusion flux from layer i+1 to layer i, the sputter flux from the surface layer, and the recession flux from layer i+1 to i to keep the receding surface at x = 0. The rate of change of the concentrations in the various cells is then given by a system of nonlinear differential equations:

$$\partial X^{(1)}/\partial t = (MX^{(2)}/a^2) \cdot \Delta \mu^{(2,1)} - I^+ Y_1 X^{(1)}/N_1 + I^+ f_o X^{(2)}/N$$

$$\partial X^{(2)}/\partial t = (MX^{(3)}/a^2) \cdot \Delta \mu^{(3,2)} - (MX^{(2)}/a^2) \cdot \Delta \mu^{(2,1)} + I^+ f_o X^{(3)}/N - I^+ f_o X^{(2)}/N$$

in which $X^{(i)}$ is the concentration of the diffusing species in layer i, M the mobility related to the diffusion constant by D=MRT (taken constant over all layers) and the difference in the chemical potential between layers i+1 and i for species 1 and 2 is

$$\Delta\mu^{(i+1,i)} = \mu_1^{(i+1)} - \mu_1^{(i)} - \mu_2^{(i+1)} + \mu_2^{(i)}.$$

I^+ is the ion flux, N and Ni are the alloy and elemental densities, respectively, and f_o is given by

$$f_o = Y_1 NX_1^{(1)}/N_1 + Y_2 NX_2^{(1)}/N_2$$

with the sputtering yields Y for species 1 and 2. In this model the concentration dependent chemical potetial μ is the driving force and ΔG is the difference in the standard potential for exchanging species 1 and 2 between the surface layer and the bulk. All the parameters except the diffusion cofficient D and the segregation energy ΔG can be taken from the experiment or tabulated values. The system of differential equations can be solved numerically for a given D, ΔG combination and the equilibrium values of the concentrations in the varios cells give a concetration profile of the altered layer. Fitting these concentrations to the ISS results (for the surface layer) and to the AES results for the weighted average over several layers produces a unique pair of D and ΔG values. The results of such calculations for the investigated set of CuTi samples are shown in Fig.5. For the two lower Cu concentrations clear Cu segregation at the surface is obseved, for 65% it is marginal and for 69% there is none. Correspondingly the segregation energies vary (about linearly) from 2.5 kJ/mole to 0. The altered, i.e. Cu depleted layer extends to approximately 15 atomic layers or about 4nm into the bulk. This is in agreement with TRIM calculations for the 44% Cu case as shown in Fig. 6. The distribution of atomic displacements extends over about the same region and thus supports the results from the model calculations regarding the depth range of the radiation induced effects. For the PdPt alloys similar depth profiles as in the two lower distributions of Fig.5 are obtained, the preferential sputtering effect being less and the segregation more pronounced in that case.

The values obtained for the diffusion constants are 2-3 x 10^{-20} m^2/s for the CuTi alloys and 1.2 - 8 x 10^{-20} m^2/s (with increasing Pt concentration, i.e. Pd dilution) for the PdPt alloys. These values ar much higher than those for thermal diffusion at these temperatures and clearly due to radiation enhanced diffusion, in agreement with similar measurements reported in the literature (e. g. [5]).

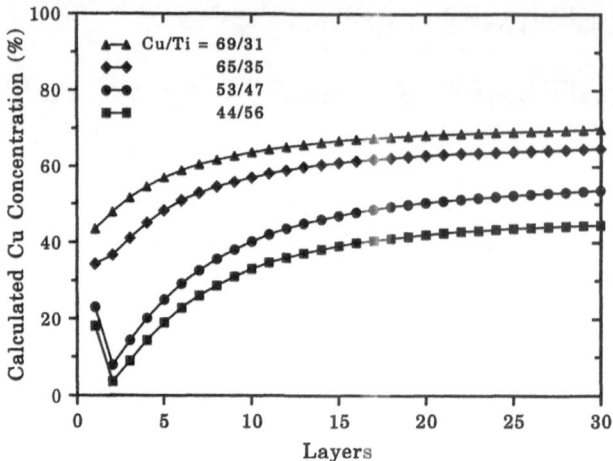

Fig. 5: Cu concentration profiles calculated with the model described in the text for various Ar⁺-bombarded CuTi alloy samples [15].

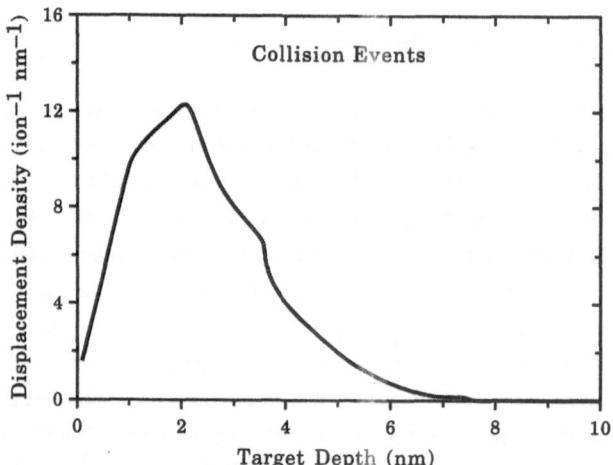

Fig. 6: Atomic displacement density as a function of depth in a 44%Cu 56%Ti alloy bombarded with 2 keV Ar⁺, TRIM calculations [15].

5. Discussion

Regarding the procedure of the calculations the question can be raised whether a unique pair of D and ΔG values can be obtained this way. This has been shown [19] by means of contour mapping and specifically demonstrated for the case of Ne⁺ bombardment of an Fe₃Al alloy single crystal [20]. For this purpose the diffusion coefficient was varied from 1×10^{-24} m²/s to 9×10^{-17} m²/s and the segregation energy between 2 and 50 kJ/mole. The results for the surface

Surface layer concentration   First bulk layer concentration

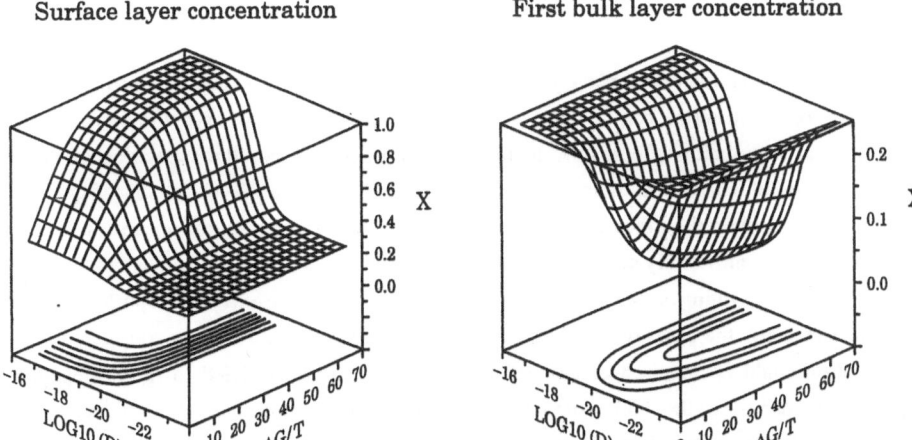

Fig. 7: Surface and first bulk layer Al concentrations X for a Ne$^+$- bombarded Fe$_3$Al alloy in dependence of D and $\Delta G/T$. Bulk level X=0.25. Calculations using the described model [19].

layer and the first bulk layer concentrations are shown in Fig. 7, displaying the Al equilibrium concentrations as a function of D and $\Delta G/T$, the bulk value being X_{Al}= 0.25. For both elements a sputtering yield Y=4 was taken. It can be seen that for D<10^{-20} m^2/s the surface concentration remains at the bulk value and only for higher diffusion coefficients segregation occurs, depending on the segregation energy. The dependence for the second layer concentration is quite different, Al depletion can occur already for low values of D, i.e. the slow replenishment from the bulk cannot balance the fast diffusion flux to the surface driven by the segregation energy. Above D=10^{-20} m^2/s bulk diffusion starts to be fast enough to approach bulk concentration and sputtering is the determining process. Base contours of the three-dimensional plots are also shown in the bottom parts of Fig.7. It can be seen that equivalent lines intersect only in one point when these contour plots are drawn on top of each other. Therefore one set of first and second layer concentrations (as e.g. obtained from different ISS or ISS and AES measurements) yields one unique set of D and $\Delta G/T$ values. Measurements for the Fe$_3$Al alloy mentioned above at three different temperatures gave an average value of ΔG= 34 kJ/mole, i.e. very pronounced Al segregation.

The numerical values given here are deduced from a limited data base and under certain (reasonable) assumptions. They are therefore subject to error margins, but they show that diffusion and segregation parameters can be extracted fom this kind of analysis analysis and the contour diagrams are also useful to delineate the regions in which significant analysis can be made.

6. Concluding remarks

The examples and their analysis given in the preceding sections show that sputtering of multicomponent material can be determined by collisional processes in specific cases, but in general also segregation effects have to be taken into account. For the separation of these effects a careful analysis is necessary, e. g. composition changes upon annealing of a preferentially sputtered surface do not immediately mean that segregation occurs. It is also necessary to use sufficiently surface sensitive techniques in order to establish segregation and identify the segregating species, in some cases contradictory results are reported in the literature (see e.g. the cases discussed in ref.[15]). In the dynamical situation considered here segregation could be manifested although the actual surface is depleted in the segregating species compared to the bulk composition. That clearly demonstrates the necessity for composition analysis with different depth scales as it was applied in the studies reported here by using ISS in various scattering geometries or by combinig ISS with AES. In the special case of the amorphous CuTi alloys segregation studies have to be restriced to the dynamical situation, high-temperature measurements are not possible since the amorphous structure would be lost due to recrystallization.

References

[1]   P. Sigmund: Mat.-Fys. Medd. **43**, 7 (1993)

[2]   E. Taglauer: Mat.-Fys. Medd. **43**, 643 (1993)

[3]   T. Michely and G. Comsa: Nucl. Instr. Meth. Phys. Res. B**82**, 207 (1993)

[4]   K. Kern, I. K. Robinson and E. Vlieg: Surf. Sci. **261**, 118 (1992)

[5]   N. Q. Lam and H. Wiedersich: Nucl. Instr. Methods B**18**, 471 (1987)

[6]   J. du Plessis, G. N. van Wyk and E. Taglauer: Surf. Sci. **220**, 381 (1989)

[7]   D. Marton, J. Fine and G. P. Chambers: Phys. Rev. Letters **61**, 2697 (1988)

[8]   H. J. Kang, J. H. Kim, Y. S. Kim, D. W. Moon and R. Shimizu: Surf. Sci. **226**, 93 (1990)

[9]   B. Baretzky, W. Möller and E. Taglauer: Vacuum **43**, 1207 (1992)

[10] W. L. Patterson and G. A. Shirn: J. Vac. Sci. Technol. **4**, 343 (1967)

[11] W. Möller and W. Eckstein: Nucl. Instr. Meth. B**2**, 814 (1984)

[12] J. P. Biersack and L. G. Haggmark: Nucl. Instr. Meth. **174**, 257 (1980)

[13] T. M. Buck, in "Chemistry and Physics of Solid Surfaces IV" (R. Vanselow and R. Howe, editors, Springer, Berlin 1982) p.435

[14] J. du Plessis, Surface Segregation, in: "Solid State Phenomena", Vol. 11 (Sci.-Tech. Pub. Vaduz, 1990)

[15] G. N. van Wyk, J. du Plessis and E. Taglauer: Surf. Sci. **254**, 73(1991)

[16]   P. M. Ossi: Suf. Sci. **201,** L519 (1988)

[17]   F. J. Kujers, B. M. Tieman and V. Ponec: Surf. Sci. **75,** 675 (1978)

[18]   J. du Plessis and G. N. van Wyk: J. Phys. Chem. Solids **49,**1441 (1988) and **50,** 237
       (1989)

[19]   J. du Plessis and E. Taglauer: Nucl. Instr. Meth. Phys. Res. B**78,** 212 (1993)

[20]   D. Voges, E. Taglauer, H. Dosch and J. Peisl: Surf. Sci. **269/270,** 1142 (1992)

IV.

Properties and Influence of Adsorbates

The Determination of Adsorbate-Substrate Bonding via UPS

Graeme L. Nyberg and Wei Shen

Chemistry Department, La Trobe University, Bundoora, Vic., 3083, Australia
nyberg@latrobe.edu.au

Abstract. Understanding the nature of surface chemical bonding remains one of the central questions of chemisorption, and one which only UPS can answer. Some of the major difficulties in interpreting the adsorbate UP spectrum are: the lack of a comparative gas-phase UP spectrum for a dissociative adsorbate species; clearly identifying the adsorbate-induced changes to the metal substrate (d-band) spectrum; the absence of a well established theoretical framework within which to interpret the results. In tackling these difficulties we have attempted to divide the problem into successive stages: formulate a theoretical model for the dissociative adsorbate species, and thereby identify its 'purely molecular' bands; locate and then identify the lower-BE adsorbate-induced bands; calculate the appropriate symmetry-adapted substrate surface orbitals; formulate a qualitative surface-molecule adsorbate-substrate bonding model. Some techniques which have been of particular value in these procedures are: HF-MO calculations; the use of fractional-difference spectra; the choice of substrate (surface); polarization-dependent ARUPS. Each of these points is illustrated by application to methanethiol (CH_3SH) adsorbed on Cu and Al.

1. Introduction

Virtually the only experimental tool for directly investigating bonding is Ultraviolet Photoemission Spectroscopy (UPS). For maximum effectiveness, though, it needs to be combined with theoretical calculations. In this way it has been very successful both for gas-phase molecules and solids (in association with Molecular Orbital (MO) and band structure calculations, respectively). Interpretation of the adsorbate-induced UP spectral features can be broken down into two stages – identifying those bands which belong purely (or predominantly) to the adsorbate (ie the orbitals are localized within the molecular fragment), and those which are involved in the adsorbate-substrate bonding. While it is the latter which are of essential interest to chemisorption, the determination of the former is a necessary first step.

Even for nondissociative adsorbates there can be difficulties here, either when the majority of the bands are involved in chemisorption (so making difficult the identification of which bands experience a bonding shift), or when the bands are not all individually resolved (so their location is masked). In such cases polarization-dependent (PD) angle-resolved (AR) UPS is required for a full assignment. When we turn to the chemically more interesting cases of dissociative adsorption, the band assignment is very much more difficult, since the adsorbed species would correspond to unstable radicals in the gas phase, and so their spectra are not available. We have tackled this problem by attempting to calculate the MO energy levels of a species which is appropriate to that which is adsorbed. What may not be apparent is that this species is not the radical.

To obtain a complete bonding picture it is necessary to also identify the changes incurred by the substrate. The task of identifying such changes is, however, very difficult on the usual transition metals because of the broad, structureless nature of their d-bands. The value of copper as a substrate in this context, with its relatively narrow d-band components, was demonstrated by several early studies, including those which examined adsorbed sulfur [1], and utilized PD-ARUPS [2,3]. Our own work has recognized this lead and made extensive use of copper as a substrate, with the aim of identifying the adsorbate-induced changes within the d-band region. The other substrate which we have found useful is aluminium, whose UPS is just a flat, featureless, low-intensity, sp-band.

Besides an appropriate substrate, similar consideration needs to be given also to the adsorbate. Here it is desirable for there to be a range of related compounds, all bonding fairly strongly to the substrate, but with some bands which remain relatively unperturbed. Both alcohols and thiols (the sulfur analogs) satisfy these criteria, but the latter have a MO structure (and UP Spectrum) which is somewhat easier to interpret, and so have been employed extensively in this work. As the surface-bonding is common to all homologues, it will be illustrated by the progenitor.

2. Dissociative adsorbates: the 'distorted molecule' model

Fig. 1 shows the gas-phase UP spectrum of methanethiol (CH_3SH; MeSH), together with the difference spectra when it is adsorbed on various copper single-crystal surfaces. In no case would it appear possible to make a simple 1:1 correlation between these bands and the four of the parent molecule. This is consistent with the further observation that dimethyl-disulfide, CH_3SSCH_3, gives rise to the same adsorbate spectrum, which in turn clearly indicates that (in both cases) the adsorbed species is in fact the mercaptide (thiolate), CH_3S, as has been well established [4-8]. Hence it is not the UP-spectrum/electronic-structure of the parent molecule which is relevant, but rather something related to the latter fragment. This problem for band assignment is common to all cases of dissociative adsorption, and one which we have addressed via a theoretical model of an appropriate species.

Fig. 2a gives the MO energy levels (MOEs) of various methanethiol-related species [8]. From this the reason for the unsuitability of the radical electronic structure is immediately apparent. Because it is an open-shell (doublet) species, there are separate sets of levels for each of the alpha and beta spins, giving rise to a doubling of the number of bands. A more appropriate species is the closed-shell (singlet) anion, which does have the right number of levels. Because it is negatively charged, all the levels are raised in energy; this, though, is unimportant. However the fact that the extra charge resides predominantly on the sulfur (rather than being distributed evenly over all the atoms) means that those orbitals with a large sulfur contribution are upshifted more than the rest, so the relative spacing of the levels is altered. What this species does illustrate, though, is that the uppermost level is doubly degenerate. The $S3p_y$ level, which is 'pulled down' in the parent molecule to form the σ-SH orbital, is released with the loss of the thiol hydrogen, and so moves back up to rejoin its $3p_x$ nonbonding ("lone-pair") counterpart.

It is this feature which is incorporated into the third model, the distorted $CH_3S\text{-}H$. When we consider the adsorption process, it is apparent that loss of the thiol H must be preceded by S-H bond elongation and possibly also CSH bond-angle deformation; also, that this is likely to proceed via a minimum-energy transition state on the corresponding potential energy surface. Although in reality this will involve the metal, what we have done is to calculate the potential energy surface with coordinates of S-H bond elongation and CSH angle increase, both of which ultimately lead to the $S3p_{x,y}$-containing orbitals becoming degenerate. The energy levels illustrated correspond to those of the geometry of the minimum distortion (in terms of sums of squares) at which this degeneracy occurs [8].

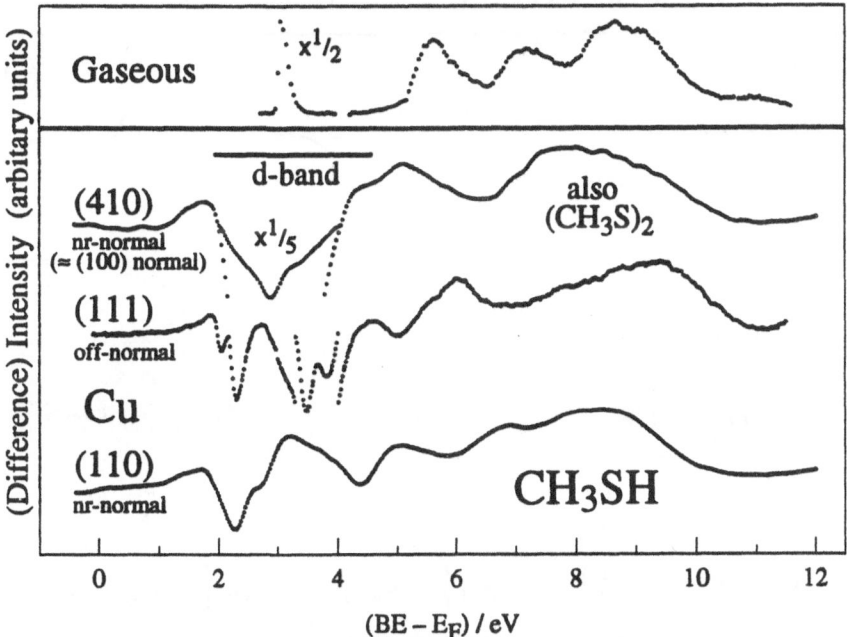

Fig 1. UP Difference Spectra (He I) resulting from the adsorption of CH_3SH (and also $(CH_3S)_2$) on Cu (410), (111) and (110), together with the adsorbate gas-phase spectrum (arbitarily positioned).

This (first) stage of the analysis is independent of the particular substrate. In the second stage, this transitional species will proceed to interact with the latter, so modifying the energies of the surface-bonding subset of orbitals, but leaving the energies of the non-surface-bonding subset largely unchanged. Thus when the upper (S-containing) orbitals interact with the solid, the resulting energies will depend on the strength of this interaction, and so on the particular substrate. This is most clearly evident in the UP spectrum of the same adsorbate on Al (polycrystalline), shown in fig. 2b [9,10]. From this it is clear not only that there are indeed three adsorbate bands, but also that there are bonding shifts in the upper, S-containing MOs, both in accordance with anticipations. The highest BE band can thus be assigned as the doubly-degenerate e(CH3) {or σ'_{CH_3}, σ''_{CH_3}}, the next as the a1(CS) {or σ_{CS}}, and the uppermost as e(S) {or (n'_S, n''_S)}.

While this distorted-molecule model has proved very satisfactory for most of the thiols studied, there are two cases where it has not: the 1,2-ethane- and 1,2-benzene- dithiols adsorbed on copper. Both compounds (the former more than the latter) have an initial adsorption regime where they disruptively dissociate to give adsorbed sulfur (on Cu, but not on Al), but thereafter adsorb molecularly-dissociatively to yield the dimercaptide [9]. However the structure of the resulting higher-BE bands of both adsorbates is different on the two substrates, substantially so for

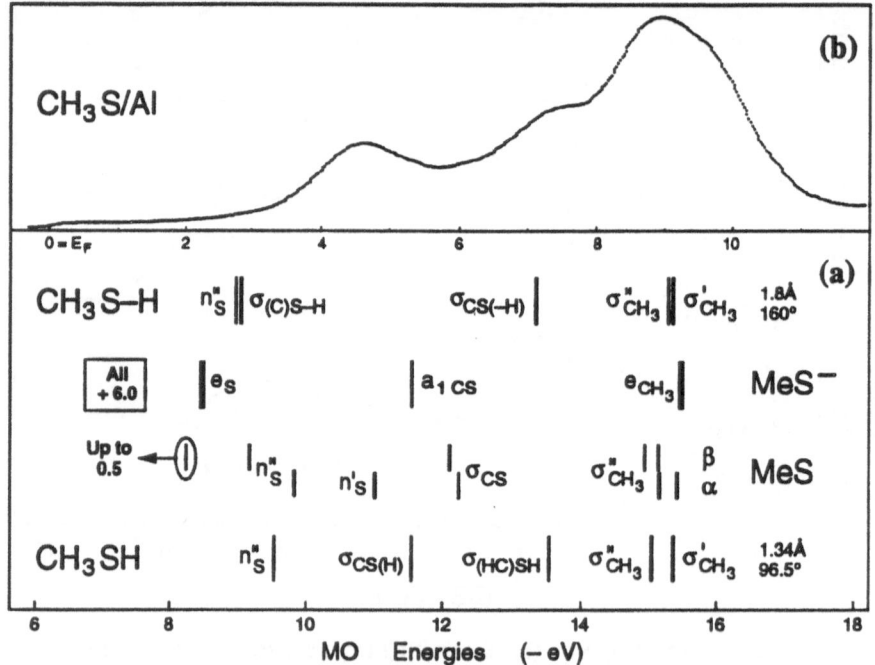

Fig. 2 Molecular Orbital Energies of species related to the Methanethiol adsorbate (a), together with its UP spectrum on Aluminium (b).

EtS_2. Whereas this structure on Al is in line with the distorted CS-H model, thus far we have been unable to reproduce the electronic structure on copper. We assume this is due to a combination of the stronger bonding on the latter, together with geometric and electronic changes in the remainder of the molecule, these being greater for the short, strained, aliphatic chain than its more rigid aromatic counterpart. There is thus no longer a substantial fragment of the molecule which is not significantly affected by the bonding to the surface, and so the simple, locally-distorted-molecule model no longer applies.

3. Further identifying the adsorbate-induced bands

3.1 Choice of substrate

We have already seen that the mercaptide spectrum on aluminium is much easier to interpret than on copper. Such a radical change in substrate may, however, involve quite different surface bonding (or even, in some cases, a different adsorbed species). One method which may be available with some metals for identifying more clearly any adsorbate-induced bands is, though, to employ a different crystal face. This is illustrated in fig. 1 where, in the highest BE region, the broad, molecularly-based band observed on both Cu (410) and (111) is clearly seen on (110) to consist of two components, $e(CH_3)$ and $a_1(CS)$.

In the more difficult d-band region, while the (410)/(100) surface at near-normal emission has a strong, central-band maximum, this is not so for (111), while the (110) surface especially has major, quite narrow, components at the start and end of the d-band region, with only low intensity in between. This 'gap' makes it ideal for identifying any feature which may be induced here by the adsorbate. This is also seen in fig. 1, where the mid-d-band intensity increase at ≥ 3 eV induced by the MeS is one of the clearest observed, and which parallels that of the well-recognized higher-BE adsorbate bands. Similar intensity increases are observed also for the other thiols studied. Being so evident on (110), this further suggests the identification of a similar feature (at ≤ 3 eV) on (111) as also an adsorbate-induced band.

3.2 Fractional difference spectra

One method for aiding adsorbate-induced-band identification is to take the difference spectrum (ie adsorbate-covered – clean-substrate). As can be seen from fig.1 for Cu (410) and (111), however, such full-difference spectra are frequently dominated by the inversion of the substrate d-band, and don't significantly or adequately clarify any structure more centrally within this region. This deficiency can be overcome by using a lesser subtraction, as has been done in the spectra of fig. 3. These fractional-difference spectra correspond to those for which the resulting spectrum-minimum (here within the d-band) just goes to zero. This both reveals as much as possible of the d-band component structure, and is operationally well defined.

The differential reduction in intensity of the d-band component at ≤3 eV indicates that it is preferentially involved in bonding to the adsorbate. The mid-band intensity increase (at $3^{1}/2$ eV in the 'p' normal-emission spectra) is now very distinct. More generally, by taking these fractional difference spectra, features can be identified within what, in the original spectrum, seems to be a fairly structureless d-band region. In addition, by examining a series of related compounds, features which are of uncertain authenticity in an individual spectrum, but which appear consistently in all cases, can be identified with confidence. By this means such structure has been identified even within the d-region of the (410) surface [8].

3.3 Polarization-dependent ARUPS

Complementary to the above technique, but even more powerful for decomposing and identifying bands, is their polarization-dependence, particularly at normal emission where the strongest selection rules apply. Fig. 3 shows the PD normal-emission ARUP difference-spectra of MeSH adsorbed on Cu(111). Now whereas s-polarization has only an x-component (perpendicular to the plane of incidence, but parallel to the surface), p-polarization has both y and z-components (parallel and normal). For this reason we have formed the 'p–s' spectrum, which gives a representation of the 'z-only' photoemission. These spectra confirm that the broad band in the 7-10 eV range indeed has two components – one centred around 9 eV which has (x,y)-symmetry, and one around 8 eV which has z-symmetry. This identifies these bands as corresponding to the $e(CH_3)$ and $a_1(CS)$ orbitals respectively, with the molecule standing upright on the surface.

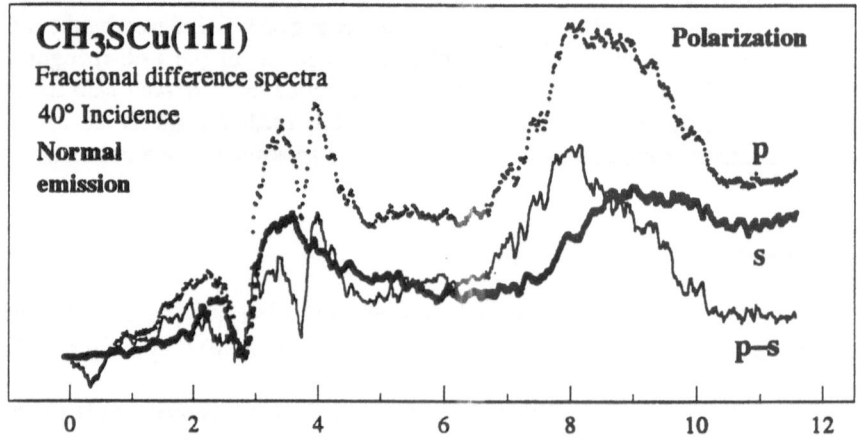

Fig. 3 Fractional difference spectra of MeSH adsorbate on Cu(111), taken with p- and s-polarized He I radiation, and their difference.

Surprisingly, however, the 6 eV band observed off-normal is now absent. Thus it cannot correspond to the $e(S3p_{x,y})$ orbital, as might have been expected, and nor can it have $a_1(z)$ symmetry, because in either case it would be observed at normal emission with p-polarization. Since these are the only two MO-symmetries possible for CH_3S, this means that this band can't in fact be molecularly-based. Moreover at normal emission from the (111) surface (ie along ΓL) the bulk bands are also restricted to these same symmetries. This leads to the conclusion [11] that this 6 eV band must therefore be a substrate-based band which is induced by the presence of the adsorbate, and which has a symmetry which is forbidden at normal emission. We will return to a further consideration of this matter later.

Separation of the z- and (x,y)-dependent bands can be done, just as profitably, also for the lower BE region. While the fractional-difference spectrum allows recognition of peaks at $3^{1}/_{2}$ and 4 eV, the polarization dependence further identifies these as having (predominantly) (x,y)- and z-symmetries, respectively. At lower BE, although only one peak is present in the 'p' spectrum, this is now clearly decomposed into a z-dependent band at ≤ 2 eV, and an (x,y)-symmetry band at $\leq 2^{1}/_{2}$ eV. This information will be subsequently incorporated into an adsorbate-substrate bonding scheme.

4. Eliminating adsorbate umklapp

One of the mechanisms which has been clearly shown to account for some of the d-region intensity increases for Cl adsorbed on Cu is adsorbate umklapp [2,3]. Such a process can occur only when the adsorbate forms an ordered adlayer on top of the ordered substrate, thereby creating a new set of reciprocal lattice vectors. The umklapp peaks all correspond to substrate inter-band transitions, but are seen at different emission directions (though at the same energy), since the original parallel momentum component of the photoelectron is changed by the addition of these adsorbate reciprocal lattice vectors [12]. Experimentally, the conditions for adsorbate surface umklapp should be very critical, as any particular adsorbate superstructure (such as c(2x2), for example) will form only in a specific, narrow, exposure range.

As previously inferred from the spectra of H_2S, MeSH and other thiols on Cu (410), and more recently confirmed on Cu (110) [9], the associated intensity increase in the middle of the d-band does not exhibit any singularity in its exposure dependence, just a monotonic increase until saturation. A similar (though somewhat smaller) intensity increase is also observed from the adsorption of the same molecules on (completely unordered) polycrystalline copper. Both results are totally inconsistent with a surface umklapp mechanism, and so completely eliminate this process in the present adsorbate-substrate systems. More generally, adsorbate umklapp tends to be restricted to atomic adsorbates on annealed samples for which there is a good LEED pattern.

5. Substrate surface orbitals and substitutional site

As the next step in formulating an adsorbate-substrate bonding model, we need to take into account the geometry of the overlayer, ie the adsorption site and coverage. The group of substrate atoms which are potentially involved in bonding with the adsorbate are then those that lie within the Wigner-Seitz surface unit cell. Consideration of the point-group and translational symmetries of this cell then yields the symmetries of the appropriate substrate group orbitals. These are listed for the ($\sqrt{3}$x$\sqrt{3}$)R30° valley, atop, and substitutional arrangements (of, say, Cl, S, or SCH$_3$) on Cu(111) in Table I (for the upper monolayer atoms only).

From this it can be seen (and this is true in general) that some of these substrate combinations are of the same symmetries as those of the adsorbate orbitals, and may therefore interact, and some may not. The latter will thus be unaffected by the presence of the adsorbate, whereas the former will mix (provided they are in the same energy region) and move to new energies (either bonding or antibonding). Note that the original substrate components of a particular symmetry (eg $3d_{xz,yz}$; $E_{(1)}$; Λ_3) will in general be neither completely bonding nor nonbonding, though some (eg the in-plane $3d_{x^2-y^2,xy}$) may well be predominantly nonbonding, due to the minimal overlap of these orbitals with those of the adsorbate.

In the original treatment of this system [11], the four combinations of valley and atop sites with ($\sqrt{3}$x$\sqrt{3}$)R30° and p(2x2) geometries were considered, for both normal and off-normal emission, and the only ones which yielded a band whose allowedness corresponded to that observed for the 6 eV band previously discussed were the atop geometries. This was surprising, since in general a valley site of higher coordination would usually seem to be preferred.

An alternative explanation is that the sulfur occupies a *substitutional* site, formed by removing the central (or 'Inner') substrate atom from the atop site (the relocation energy of a Cu atom, say to a step site, being more than offset by the increased sulfur binding energy). While the occupation of such a substitutional site was determined some time ago for Te on Cu(111) [13], this is the first instance (that we are aware) of such a proposal for a *molecular* adsorbate. Our recent results for thiol adsorption on the more open Cu(110) surface do, moreover, also support incorporation of the sulfur into the first copper monolayer [9]. As seen from Table I, the symmetry-species remain the same as for the atop geometry, and so a substitutional site is consistent with the photoemission results. The difference is that the bonding between the substrate atoms is altered (particularly those which involved strong overlap between the orbitals of the Inner and Outer atoms), so that some of the substrate component bands are likely to appear at different energies. This conclusion is consistent with changes in the substrate density-of-states (DOS) which have been observed previously, both experimentally and in computational studies [eg 2,3; 14]. Note that, although also brought about by the adsorbate, this effect is not kinematic like adsorbate umklapp, but is due to the physical rearrangement (reconstruction) of the substrate atoms (in this specific instance, the removal of one).

Table I. Symmetry species of the adsorbate-atom orbitals, and of the symmery-adapted substrate orbitals, of the surface unit cell for a $(\sqrt{3} \times \sqrt{3})R30°$ lattice on an fcc (111) surface. Bold type indicates those which are forbidden at normal emission (and are also of the wrong symmetry to interact with the adsorbate).

Ads Site Pt Group	valley C_{3v}	atop C_{6v}	substnl C_{6v}
Adsorbate orbitals and symmetries			
$3p_{x,y}$	E	E_1	E_1
$3p_z$	A_1	A_1	A_1
Substrate orbitals and symmetries (& corresp. bulk-band symms)			
Atoms	3xInner	1xIn;2xOut	2xOuter
$3d_{xz,yz}$ (Λ_3)	A_1+A_2+2E	E_1; E_1+E_2	E_1+E_2
$3d_{x^2-y^2,xy}$ (Λ_3)	A_1+A_2+2E	E_2; E_1+E_2	E_1+E_2
4s or $3d_{z^2}$ (Λ_1)	A_1+E	A_1; $A_1+\mathbf{B_1}$	$A_1+\mathbf{B_1}$

6. Adsorbate-substrate bonding: the surface-molecule model

The results from the previous sections are now combined together to form an overall model for the bonding between the adsorbate species and the substrate surface, as illustrated in fig. 4. On the right are the MOEs of RS-H, from fig. 2a, and adjacent is the 'preadsorbed' (or 'physisorbed') species, RS-[M]. The relative orbital spacings of this are identical to those of RS-H, but all the orbital energies are shifted rigidly upwards, corresponding to a uniform relaxation shift of all the orbitals. The magnitude of this shift is empirically chosen to put the purely molecular level(s) at the same energy as observed in the adsorbate spectrum. It is this set of energy levels (which represent the 'transition state' on the surface) with which the substrate interacts. The substrate energy bands, drawn on the left-hand side of the diagram, are simply a schematic representation of the clean ΓL spectrum (appropriate to normal photoemission off the Cu(111) surface). It is assumed that these also approximately represent the energies of the reconstructed surface.

The electronic structure of the resulting adsorption-complex is shown in the adjacent panel. While for aluminium the bonding interactions are relatively straightforward [10], with the copper the situation is more complicated, because of the substrate d-band. One of the main features is the interaction between the adsorbate $(n's,n"s)$ and substrate $3d_{xz,yz}$ (Λ_3) orbitals. While the $d_{x^2-y^2,xy}$-based substrate group orbitals can also have the same π-type $(E_{(1)})$ symmetry, it is apparently the former combination which has the greatest overlap with an (adsorbate) atom (as also on the (100)-type surface [8]). This results in the removal of the latter band from its original energy position, and its redistribution into a pair of π/π^* bonding/antibonding levels, which are associated with the (x,y)-polarized 3.3 and 2.3 eV bands, respectively. In this qualitative model the size of this interaction is purely empirical; its magnitude (~1 eV) is substantially less than that in discrete molecules.

The other interaction involves the σ_{CS} orbital, which needs a substrate combination of the same (A_1) symmetry. For a substitutional site the d-component has to be the d_{z2}, which is in accord with the decrease in this substrate band intensity. This interaction appropriately depresses the σ_{CS} band, and again gives an accompanying antibonding band just above the d-band (z-polarized, at 1.7 eV).

There are two omissions thus far. The first is the interaction with the sp_z-band at E_F, necessary to fill the original e_S vacancy. While an (E-symmetry) interaction with π_S would be most direct, this is possible only for a valley site (Table I), and for a substitutional site would need to operate indirectly, through the σ^* band. This results in a further, vacant, σ^* band above E_F. Note that this lies well below the adsorbate LUMO, which in these compounds is too high above E_F to have any significant part in the bonding scheme, and so has not been included. The second is that the 4 and (off-normal) 6 eV bands are not accounted for in terms of these bonding interactions. Instead, both are attributed to substrate bands ($3d_{z2}$ and 4s respectively) shifted by adsorbate-induced substrate reconstruction (formation of the substitutional site).

Although various uncertainties remain with the present surface-molecule qualitative bonding model, it has the following advantages over the simple 'bonding-shift' model: (i) it takes into account the appropriate symmetries of the substrate group orbitals; (ii) it provides a coherent explanation for the upshift of some bands, as well as the downshift of others; (iii) it offers an explanation as to why some d-band components should be involved in the bonding to so much greater extent than

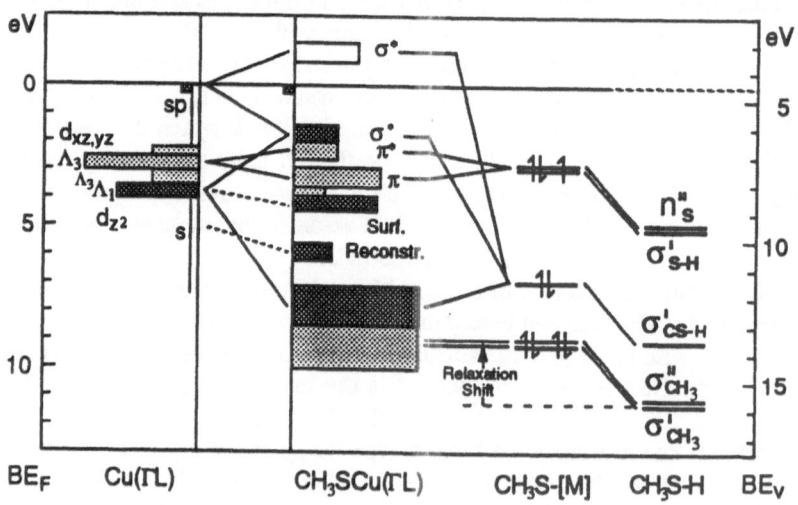

Fig. 4   Surface-molecule bonding model for mercaptide adsorption on Cu(111).

others; (iv) it allows for particular substrate group orbitals to be shifted in energy; (v) it offers a prediction of an increase in the empty DOS above E_F (a prediction which is susceptible to experimental verification by means if Inverse Photoemission); and (vi) essentially similar schemes apply to the other crystal faces.

7.   Conclusion

While many newer techniques have been introduced into surface science over the past several years, and many exciting results have been obtained, it remains true that UPS is still the only experimental 'window' onto adsorbate-substrate bonding. Considering that this technique has been available now for two decades, it could be said that progress has been disappointingly slow. There are probably three reasons for this. One is that readily interpretable results are difficult to obtain. The second is that many investigators (probably as a consequence) have concentrated on the newer techniques, many of which give structural information, which tends to be easier to envisage. The third is the lack of a reliable theoretical model within which to interpret the results. This contrasts with the earlier situations for molecules and solids, for both of which the electronic structure calculations were already well developed, and progress was rapid.

What we have attempted to do in this paper is demonstrate that, by a thoughtful choice of systems studied, and by a careful treatment of the data, it is indeed possible for progress to be made, even for dissociative polyatomic adsorbates. For the specific systems studied, some of the more general results which have been obtained (and partly illustrated herein) are: i) all thiols RSH lose H and form mercaptides RS, which bond through the S atom; ii) there is some initial desulfurization in some systems, but this is self inhibiting [eg 15]; iii) while most molecules stand largely upright, some lie down at low coverages [15]; iv) Al is less reactive than Cu (for these S-containing molecules); v) on copper, the S atom likely incorporates into the first Cu monolayer; vi) the bonding involves both the S3p π and σ electrons, which interact with substrate orbitals of the same symmetries; vii) bonding and antibonding metal-sulfur orbitals result; the former are always filled; the latter are unoccupied on Al, but on Cu lie between the Fermi level and the d-band onset.

Many difficulties remain in advancing the interpretation of bonding effects in UP spectra to the less-tractable transition metal substrates. However with further development and application of the above techniques, together with reliable and generally-available computational methods (probably Density-Functional-Theory-based cluster models) with which to interpret the results, we would anticipate a renewed enthusiasm for adsorbate UPS studies, and a concomitant advance in our understanding of the nature of adsorbate-substrate bonding to that of its progenitors.

8. Acknowledgements

We would like to thank the Australian Research Council for various grants related to this research, and the Department of Industry, Technology and Commerce for making participation by G.L.N. possible in this German-Australian Workshop. W.S. would like to thank La Trobe University for a Postgraduate Research Scholarship.

9. References

1 G.G. Tibbetts, J.M. Burkstrand and J.C. Tracy, Phys. Rev. B, 15 (1977) 3652
2 D. Westphal and A. Goldman, Solid State Commun., 35 (1980) 437;
3 D. Westphal and A. Goldman, Surf. Sci., 131 (1983) 113
4 S.E. Anderson, Ph.D. Thesis, La Trobe University, 1985
5 B.A. Sexton & G.L. Nyberg, Surf. Sci., 165 (1986) 251
6 P.A. Agron, T.A. Carlson, W.B. Dress & G.L. Nyberg, J. Elec. Spec., 42 (1987) 313
7 S. Bao, C.F. McConville & D.P. Woodruff, Surf. Sci., 187 (1987) 133
8 S.E. Anderson and G.L. Nyberg, J. Elec. Spec., 52 (1990) 735
9 Wei Shen, Ph.D. Thesis, La Trobe Univ., 1991
10 Wei Shen and G.L. Nyberg, Surf. Sci., 296 (1993) 49
11 G.L. Nyberg, T. Gengenbach and J. Liesegang, Physica Scripta, 41 (1990) 517
12 J. Anderson and G.J. Lapeyre, Phys. Rev. Letts, 36 (1976) 376
13 F. Comin, P.H. Citrin, P. Eisenberger and J.E. Rowe, Phys. Rev. B, 26 (1982) 7060
14 R. Hoffmann, J. Phys.: Condens. Matter, 5 (1993) A1
15 Wei Shen, G.L. Nyberg and J. Liesegang, Surf. Sci., 298 (1993) 14

Oxygen on Cu(100): From Physisorption to Chemically Induced Reconstruction

K. Baberschke
Institut für Experimentalphysik, Freie Universität Berlin,
D - 14195 Berlin-Dahlem, Germany

Abstract. For approximately half-a-monolayer coverage of oxygen on a single crystal surface of Cu(100) we review four different states of adsorption: (i) a molecular "physisorbed" state below 44 K, (ii) a molecular chemisorbed state between 44 and 100 K with a stretched O-O axis, (iii) a "precursor" atomic adsorption with an unreconstructed Cu surface, and finally (iv) after a dosage of 1200 L O_2 at 500 K the well-known O/Cu(100) state with a sharp ($\sqrt{2}$x2$\sqrt{2}$) R45° LEED pattern, and a reconstructed substrate surface. These four steps present a case study of molecular adsorption via dissociation to the onset of corrosion.

1. Introduction

The Surface version of the Extended X-ray Absorption Fine Structure (SEXAFS) has contributed to the understanding of the bond geometry and the dynamics of adsorbates on metal surfaces [1]. The technique yields complementary information to other surface analytical techniques, in that it is (i) element specific - here we will focus on oxygen with the K-edge at ≈532 eV - and (ii) it is a local probe. No long range order is necessary as needed for diffraction (LEED [2]) and scattering techniques (LEIS). Strictly speaking, when investigating the O-Cu local interaction, the experiment does not distinguish for example between oxygen adsorbed on the surface, Cu oxides, and the O-Cu interaction in the High-T_c-Superconductors (HTSC). For surface physics, the detection of electron yield is favorable, for probing the bulk, it will be of advantage to use O K_α fluorescence detection [3,4]. In both cases the detection channel is proportional to the number of the created 1s-holes in oxygen. Beside the recording of the extended XAFS one can measure the near edge structure (NEXAFS), this is few eV below to ≈30 eV above the edge. In this range transitions from $1s^{-1}$ to unoccupied bound states (Rydberg, LUMO, etc.) and scattering into quasi bound states (σ^* shape resonance) above threshold is recorded [5].

The oxygen - copper interaction attracts some interest: (i) there exist two copper oxides - not so for NiO; (ii) it plays the key role in the HTSC's [6]. (iii) The adsorption of oxygen on a Cu(100) surface is an interesting case study: for a coverage of approximately 1/2 monolayer (ML) at least four different states of adsorption can be studied [7-9]. This happens between 30 and 500 K, depending on the gas dosage. In this overview we will not discuss all the details [9], but rather demonstrate the capability of SEXAFS in general. It will turn out that this technique determines the state of adsorption more in a geometrical way rather than in terms of binding energies.

Tab. 1: Some selected literature for the determination of the adsorption geometry of 1/2 ML of atomic O/Cu(100), listed as a function of time.

1980	LEED	c(2×2) and ($\sqrt{2}$x2$\sqrt{2}$) R45°:	bridge FFH $d_\perp = 0$Å	Onuferku	[10]
1983	PED	c(2×2) and ($\sqrt{2}$x2$\sqrt{2}$) R45°:	FFH $d_\perp = 0.6$Å FFH $d_\perp = 0.3$Å	Tobin	[11]
1984	SEXAFS NEXAFS	c(2×2) and ($\sqrt{2}$x2$\sqrt{2}$) R45°:	FFH $d_\perp = 0.8$Å	Döbler	[12]
1986	LEED	"c(2×2)" ≡ ($\sqrt{2}$x2$\sqrt{2}$) R45°		Mayer	[13]
1987/9	LEED HREELS	"c(2×2)" ≡ ($\sqrt{2}$x2$\sqrt{2}$) R45°		Wuttig	[14]
1989	LEED	($\sqrt{2}$x2$\sqrt{2}$) R45°	and MR	Zeng	[15]
1989	SEXAFS	"c(2×2)":	FFH $d_\perp = 0.8$Å	Wenzel	[16]
1990	PED	"four spot" ⎫ ⎬ ≠ ($\sqrt{2}$x2$\sqrt{2}$) R45°: MR "c(2×2)" ⎭		Asensio	[17]
1990	Theory	"more than one minimum in the potential"		Jacobsen	[18]
1992	SEXAFS	"c(2×2)": and ($\sqrt{2}$x2$\sqrt{2}$) R45°:	FFH $d_\perp = 0.8$Å MR $d_\perp = 0.2$Å	Lederer	[8]

In Sec. 2 we discuss the atomic oxygen [8], this has the longest history (Tab. 1). The molecular adsorption [7], quite new, is presented in Sec. 3. In Sec. 4 the four cases will be discussed under the perspective of the local adsorption geometry.

2. Atomic O on Cu(100)

From 1980 until '85, LEED, PED, SEXAFS, etc. observed 2 structures: the c(2x2) and a ($\sqrt{2}$x2$\sqrt{2}$) R45° (top part of Tab. 1). The first adsorption site varied from bridge to fourfold hollow (FFH). The adsorption geometry of the second was not well clarified. No substrate reconstruction, relaxation, etc. was considered. Theory agreed also with a FFH site and c(2x2) symmetry as a stable configuration [19]. In 1986 a LEED study suggested that the 2 LEED pattern reflect the same adsorption site [13] (middle part of Tab. 1), only in some LEED experiments the fine structure of the ($\sqrt{2}$x2$\sqrt{2}$) R45° pattern was not resolved. This idea was convoyed by a careful HREELS study [14]. The Vancouver group [15] was the first which stated clearly that there happens an adsorbate induced reconstruction of the Cu fcc(100) surface into a missing row along the (110) direction, but they still claimed the existence of one adsorption state only. Several other experiments confirmed that the exposure of oxygen induces a missing row reconstruction of the Cu(100) surface along the [011] direction. The gas dosage ranges up to 1200 L at 500 K adsorption temperature. So, the question still remained: why did the early experiments in 1980, '83 analyze two structures? Recent PED experiments also supported this "new" opinion of two different states [17]. Theory suggested at the same time, that there may be more than one potential minimum [18].

Having all this in mind, the SEXAFS experiments were repeated. To determine the local bonding geometry only from the measure of the n.n. distance and coordination number may be difficult in some cases, as this usually requires already the input of some model structures, and none of the hypothetical structures may be correct. But from a temperature dependent measurement more information can be gained: (i) The correlated vibration of O versus the n.n. Cu atoms damps the amplitude of the EXAFS oscillations. It is the mean square relative displacement (MSRD). In Fig. 1 one sees the stronger damping at high k-values with a Debye-Waller like amplitude factor

$$\exp[-2\ c_2(T) \cdot k^2] \tag{1}$$

In an Einstein model $c_2(T)$ is given by

$$c_2 = \sqrt{(\hbar/\mu\alpha)} \cdot \coth(\Theta_E/2T) \tag{2}$$

In the experiment, only Δc_2 is determined. In Fig. 2 is shown that between 50 and 500 K $c_2(T)$ follows nicely equ. (2). (ii) EXAFS measures the effective pair distribution function (PDF) which is *per se* asymmetric, due to an anharmonic n.n. interaction. This can be expressed as a mean cubic relative displacement (MCRD) and enters in the phase of the oscillations

$$\sin\ [2rk + ... + 4/3\ c_3(T)\ k^3] \tag{3}$$

$$\text{with}\quad c_3 = \frac{\beta\hbar}{a^2\mu}\left[\frac{3}{2}\coth(\frac{\Theta_E}{2T}-1)\right] \tag{4}$$

β and α are defined by $V(r) = \frac{1}{2}\ \alpha\ (r-r_0)^2 - \beta(r-r_0)^3$.

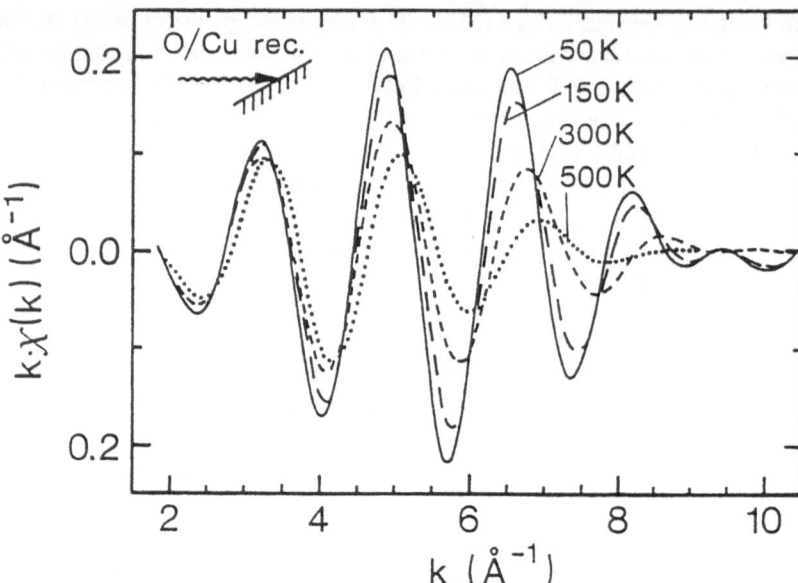

Fig. 1: Fourier back transform $\chi(k)$ for the n.n. peak. The sample was prepared at 500 K and measured (dotted line), and subsequently cooled to 50 K. The data are shown for grazing incidence, probing the interaction of the O -atom with Cu underneath for the MR row reconstruction (see text).

Fig. 1 shows evidently this phase shift towards larger k-values with increasing temperature. This is not a shrink in real distance with increasing T but rather a very direct measure of an asymmetric PDF or an anharmonic effective pair potential (equ. 4).

With this additional information on the dynamics of the O-Cu interaction (for details see [8]), we are quite confident that two distinct different states of the adsorption of 1/2 ML atomic O on Cu(100) do exist (bottom part of Tab. 1). For shortness we describe here the results only [20].

The thermodynamic stable state is the missing row (MR) reconstruction, first discussed in [15]. Oxygen sits on the rim of the MR almost in the first Cu-plane with 3 n.n. in the top layer. But surprisingly the geometry has a great deal of local static disorder, as can be seen in Fig. 3 a. The PDF at 50 K (solid line) is very broad and does not change for 300 K. The sample was prepared at 500 K with 1200 L. Sharp LEED spots were observed.

However there exists a "precursor" state at which the Cu surface remains unreconstructed and the O-atom sits almost in a symmetric pyramid position ≈ 0.8Å above the surface. For this state we observe a much sharper PDF (dashed lines in Fig. 3), which broadens at 300 K. This state is prepared at 300 K substrate tempera-

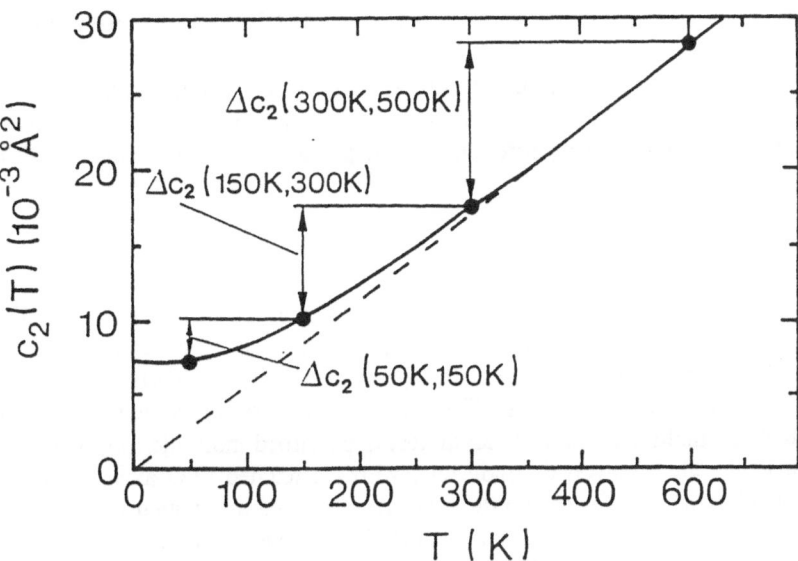

Fig. 2: The EXAFS Debye-Waller factor (MSRD) for O/Cu reconstruction and the same geometry as in Fig. 1. The solid line is equ. 2 with $\Theta_E = 260$ K. The 3 Δc_2 (T_1, T_2) values agree perfectly with Einstein model. No structural change during cooling and heating was observed.

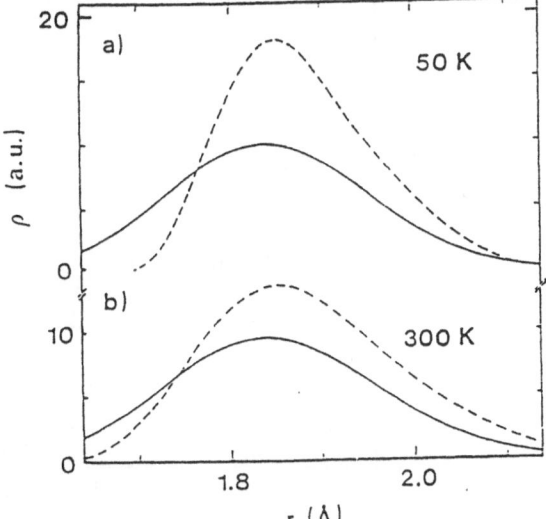

Fig. 3: Pair distribution function (PDF) for the precursor "c(2×2)" (- - -) and the reconstructed ($\sqrt{2}$x2$\sqrt{2}$) R45° (——) state at 50 and 300 K.

ture and 300 L gas exposure. A very poor "c(2×2)" like LEED pattern is observed. The dynamical parameters c_2, c_3 are measured at various temperatures, for grazing and normal incidence photon beam. This measures spring constant α and the anharmonicity β for the motion normal and parallel to the surface, respectively. All parameters show a clear difference for the precursor and the reconstructed case.

3. Molecular O_2 on Cu(100)

In chapter 2 we discussed the exposure of O on the Cu surface at 300 K and 500 K and a subsequent cooling of the sample. If one cools the clean Cu(100) first (i.e. to 20 K) and exposes O_2 then, the gas is adsorbed in molecular form. A thermal desorption (TDS) experiment (Fig. 4) shows three different steps of desorption: At 30 K the multilayer, at 44 K an almost unperturbed molecule (physisorbed), and at 94 K the chemisorbed O_2 molecule, with a stretched O-O axis (the reasoning is given below). For lower exposure/coverage (dashed and dotted line), no multilayer desorption and formation into a chemisorbed state is observed.

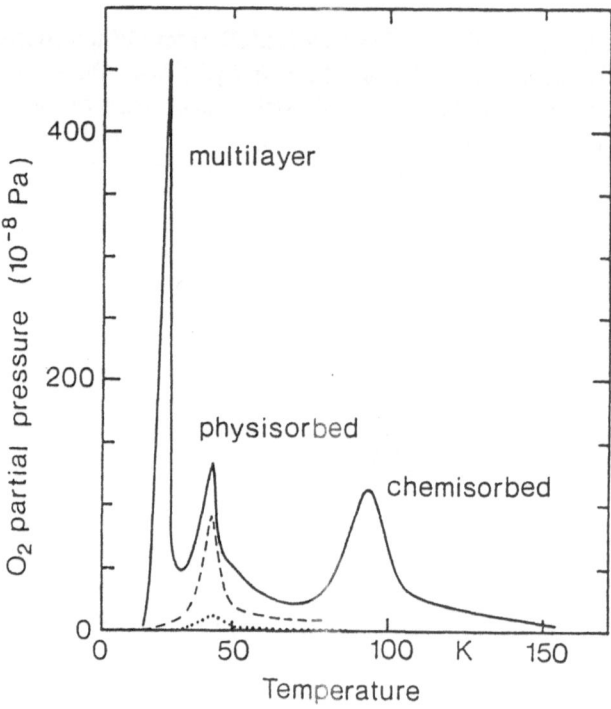

Fig. 4: Thermal desorption spectrum, taken with a heating rate of 2 K/s from 15 K to 160 K. The relative exposure for the full/dashed/dotted lines are 10/2.2/1.0.

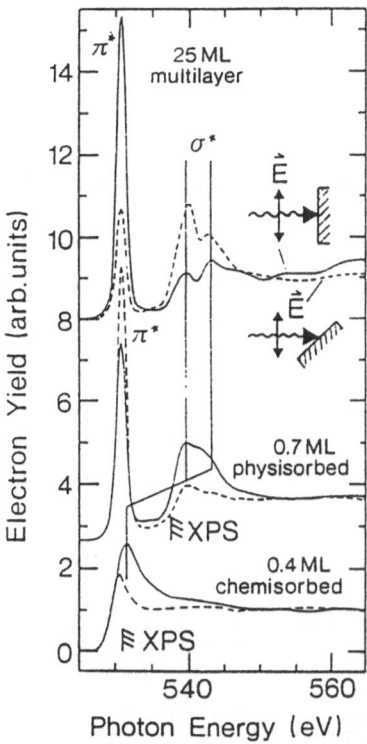

Fig. 5: NEXAFS of O2 /Cu(100) for grazing and normal incidence as indicated in the figure. The coverages are estimated from the edge jump.

Fig. 5 shows the NEXAFS for grazing (dashed) and normal (solid) incidence. The three spectra were taken below the corresponding desorption temperatures at about 20 K, 38 K, and 70 K. The analysis is straightforward and is a nice example for XAFS applications. Firstly we see, that the O_2 molecule indeed exists: The well-known π^* and σ^* resonances at 532, 539, and 544 eV show up in the multilayer and physisorbed state. For the physisorbed state, the σ^* resonances appear at the same energy, that is a strong indication for an "unperturbed" molecule. Secondly the angular dependence determines the orientation of the molecule. For the physisorbed state the molecule lies almost flat on the surface (σ^* intensity vanishes for grazing incidence, where π^* is strongest). The multilayers forming on this first layer have upstanding O-O axis (π^* is largest and σ^* is weakest for grazing incidence), in agreement with the literature.

For the chemisorbed state the analysis is more complicated (for details see [7]). The hybridization of substrate 4s and adsorbate 2p states causes a partial filling of the LUMO, as a consequence the intensity of the π^* resonance almost vanishes. The

σ* resonance is shifted to a lower energy of 532 eV, also the splitting of this resonance disappears. It has been shown previously [21] that for these simple two-
atomic molecules the correlation between stretching of the intramolecular bond
length and the shift of σ* holds. This correlation yields for the present case an O-O
distance of ≈ 1.45 Å.

From Fig. 5 we also see that the XPS binding energy shift by 6.5 eV between
physi- and chemisorption. The SEXAFS analysis goes in the same way as for
atomic O/Cu. The input is an almost flat lying molecule axis and two inequivalent
local positions for the two O-atoms (FFH and bridge site). As a first try the O-O
bond lengths equal to 1.21 Å and 1.45 Å for the physi- and chemisorbed molecule
(see Fig. 6 and [7]) were used. The numbers are given in the inset (O) of Fig. 6: For
both O-atoms, the O-Cu distance is the same (2.01 Å) within experimental error
bars. The coordination numbers are 4 and 2. this gives a small tilt angle of the O-O
axis. As the fourfold symmetry is maintained, four local domains of O-O orientation
will appear.

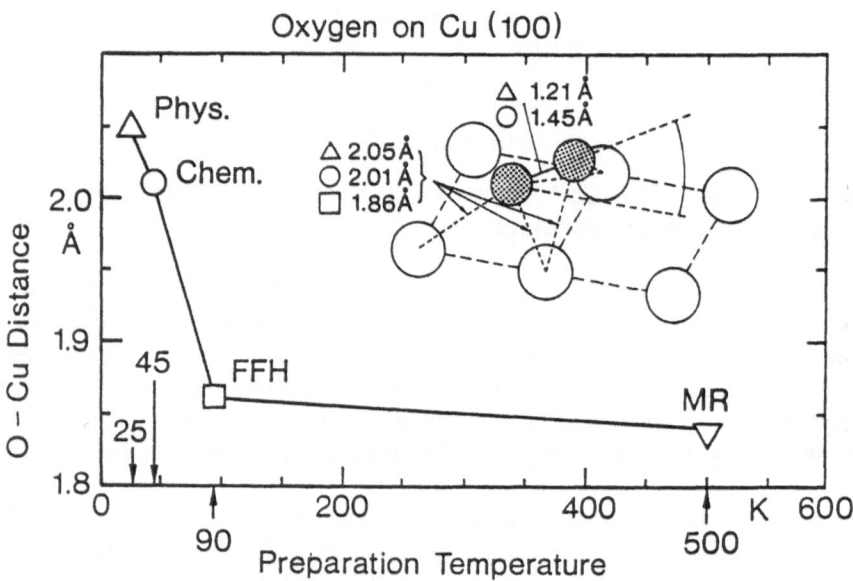

Fig. 6: O-Cu distances determined from SEXAFS for all 4 states of adsorption. At
the x-axis, the preparation temperature is plotted; it should be noted that these temperatures vary, depending on the gas dosage. The inset shows schematically the local orientation of the molecule.

4. Discussion

In Figs. 6 (inset) and 7 the adsorption geometry is drawn schematically.
<u>Physisorbed O_2:</u> An oxygen atom sits in the a fourfold hollow site (FFH) ≈ 1.0 Å
above the Cu surface plane. This is estimated from the measured O-Cu distance of
2.05 Å (Δ, Fig. 6). Within ± 0.04 Å error bar the same distance is measured for the
second oxygen atom. Assuming an unperturbed O-O, this leads to a bridge like position with a tilt angle of about 25 - 30°. Within experimental error bar this agrees
with the angular dependence of the NEXAFS intensities for the π* and σ* resonances.
<u>Chemisorbed O_2:</u> The O-Cu distance is slightly shorter (2.0 Å, see O in Fig. 6),
yielding a shorter vertical shift of 0.9 Å above surface. The second atom sits in the
same bridge site with similar tilt angle. So, the increase of the O-Cu bonding weak-
ens the O-O bonding.
<u>Precursor:</u> Further increase of the O-Cu interaction stretches the intramolecular
bond length more and leads to dissociation. However it is interesting to note that O_2
can stretch up to ≈ 1.45 Å before breaking apart. After dissociation, the O-Cu distance is significantly shorter: 1.86 Å. Every O-atom sits in the same FFH site position (Fig. 7 c) with $d_\perp$ ≈ 0.8 Å. The substrate remains unreconstructed. However
there is no long range periodicity. From the poor LEED spots we assume that the
lateral dimension of the "FFH site islands" is of the order of 20 - 50 Å.

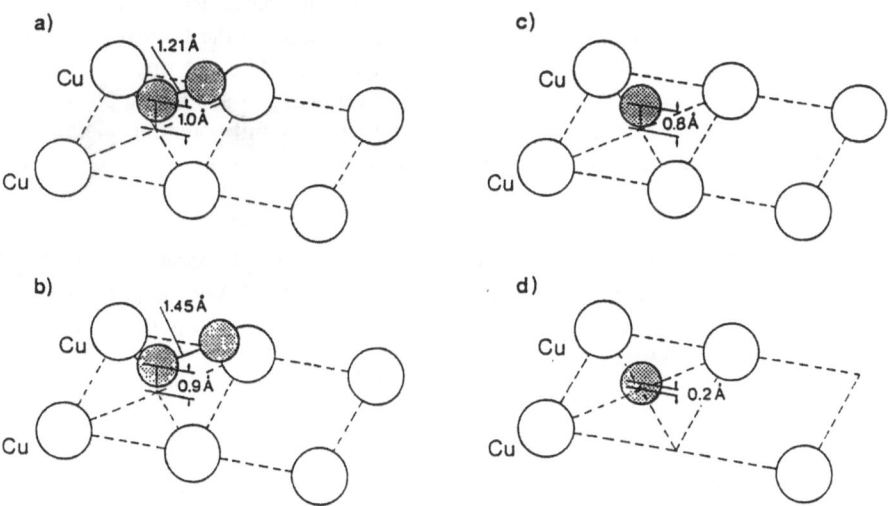

Fig. 7: Schematic drawing of the four adsorption geometries

<u>Substrate reconstruction:</u> With larger gas exposure and higher mobility for the Cu atom of 500 K starts a chemically induced reconstruction with a slightly shorter O-Cu distance of 1.84 Å and (!) a second neighbor of 2.05 Å is seen for grazing incidence - this is the Cu-atom in the second layer [8]. All this leads to a dramatic move of the O-atoms almost in the surface plane($d\perp \approx 0.2$ Å, Fig. 7 d). The O-atom sits almost in the surface plane with three n.n. in the first layer and one in the second. The missing row points along the [110] direction. From the sharp LEED spots one concludes long range (> 100 Å) periodicity. Domains rotated by 90° are possible and have been seen.

The quantitative analysis of the O-Cu bond length correlated to the amount of thermal energy supplied to induce the different adsorption states underlines the local character of the molecular dissociation versus the longer range interactions responsible for a surface reconstruction. Furthermore, information on the static part of the disorder was obtained. It is found that the disordered phase exhibiting only a poor LEED diffraction pattern (broad spots) corresponds locally to a weaker but more ordered n.n. O-Cu bond. Supplying thermal energy to the O covered surface in the "disordered" state, leads to a state of lower energy in the form of a stronger O-Cu bond arranged to form a missing row reconstruction over large areas (sharp LEED spots) at the expense of the order of the individual bonds. It appears that the O-Cu pair can be encountered in many electronic configurations on a surface, in analogy to what is known in the bulk, where two types of oxides exist (CuO and Cu_2O). However the pair-potential analysis indicates that the surface phases correspond to different O-Cu electronic configurations than in the bulk. These results underline the relevance of SEXAFS in connection to the understanding of more fundamental problems as the role of the O-Cu interaction in the occurrence of high temperature superconductivity in ceramic compounds. Here the opportunities offered from surface science, in addressing actual problems of solid state physics, are used in combination to EXAFS which is exploited to its full potential, yielding information beyond the determination of the n.n. distance.

5.   **Acknowledgment**   The work was performed in collaboration with D. Arvanitis. T. Lederer and the other authors of Ref. 1, 7, and 8. We thank B. Schliepe for the preparation of the manuscript. The work was supported in part by the BMFT (Grant No. 05 5KEAAB).

6. References

1. D. Arvanitis, T. Lederer, G. Comelli, M. Tischer, T. Yokoyama, L. Tröger, K. Baberschke, XAFS VII Int. Conf., Jpn. Jnl of Appl. Phys. **32**, 337 (1993)
2. Here we refer to the "poor-mans-LEED"; K. Heinz discusses in this volume how LEED can be used as a local probe.
3. L. Tröger, D. Arvanitis, K. Baberschke, H. Michaelis, U. Grimm, E. Zschech, Phys. Rev. B **46**, 3283 (1992)
4. L. Tröger, T. Yokoyama, D. Arvanitis, T. Lederer, M. Tischer, K. Baberschke, Phys Rev. B **49**, 888 (1994)
5. D. Arvanitis, H. Rabus, T. Lederer, K. Baberschke, Vacuum Ultraviolet Radiation Physics: Proceedings of the 10th VUV conference, p. 347, eds. F.J. Wuillemier, Y. Petroff, I. Nenner, World Scientific, Singapore 1993
6. L. Tröger, D. Arvanitis, H. Rabus, L. Wenzel, K. Baberschke, Phys. Rev. B **41**, 7297 (1990)
7. T. Yokoyama, D. Arvanitis, T. Lederer, M. Tischer, L. Tröger, K. Baberschke, G. Comelli, Phys. Rev. B **48**, 15405 (1993)
8. T. Lederer, D. Arvanitis, G. Comelli, L. Tröger, K. Baberschke, Phys. Rev. B **48**, 15390 (1993)
9. For details see: T. Lederer, Ph. D. Thesis, FU Berlin 1993 (unpublished)
10. J.H. Onuferku, D.P. Woodruff, Surf. Sci. **95**, 555 (1980)
11. J.G. Tobin, L.E. Klebanoff, D.H. Rosenblatt, R.F. Davis, E. Umbach, A.G. Baca, D.A. Shirley, Y. Huang, W.M. Kang, S.Y. Tong, Phys. Rev. B **26**, 7076 (1982)
12. U. Döbler, Ph. D. Thesis, FU Berlin 1982 (unpublished)
13. R. Mayer, C. Zhang, K.G. Lynn, Phys. Rev. B **33**, 8899 (1986)
14. M. Wuttig, R. Franchy, H. Ibach, J. Electr. Spec. 44, 317 (1987), and Surf. Sci. **213**, 103 (1989), ibid. **224**, L979 (1989)
15. Z.H. Zeng, R.A. McFarlane, K.A.R. Mitchell, Surf. Sci. **208**, L7 (1989)
16. L. Wenzel, Ph. D. Thesis, FU Berlin 1989 (unpublished)
17. M. Asensio, M. Ashwin, A. Kilcoyne, D. Woodruff, A.Robinson, T. Lindner, J. Somers, D. Ricken, A. Bradshaw, Surf. Sci. **236**, 1 (1990)
18. K.W. Jacobsen, J.K. Norskov, Phys. Rev. Lett. **65**, 1788 (1990)
19. C.W. Bauschlicher and P.S. Bagus, Phys. Rev. Lett **52**, 200 (1984)
20. Details are given in [1, 8, 9].
21. D. Arvanitis, U. Döbler, L. Wenzel, K. Baberschke, J. Stöhr, Surf. Sci. **178**, 686 (1986)

Inelastic Phenomena of Low-Energy Particle-Surface Interactions

A. Närmann, C. Höfner, T. Schlathölter, and W. Heiland

Universität Osnabrück, FB Physik, D-49069 Osnabrück, Germany

Abstract. In this contribution we summarize recent results on low-energy particle-surface interactions. By scattering under grazing incidence and detecting in specular direction we assure that we probe surface properties only. We used as incident particles hydrogen and helium as well as molecules, such as CO_2. For the simple monoatomic species we present a model that explains the experimental data by accounting for the interrelation between charge exchange and stopping processes. The molecule experiments are compared to chemisorption experiments on clean and alkaline-covered surfaces.

1 Introduction

Charge exchange processes between particles and surfaces are an important part of the plasma-wall interaction in fusion experiments, of chemisorption, of the physics of particle detectors and of radiation damage due to particle bombardment of matter. The understanding of the charge exchange processes was developed within the framework of the physics of the energy loss of swift particles passing through matter. The energy loss, often characterized by the stopping power, i.e. the energy loss per unit length (eV/Å), can be separated into a nuclear and an electronic part. The nuclear part is the elastic loss due to the collisions of the projectile with the nuclei of the solid, gas or liquid. The collisions can be described in a wide energy range classically, limits are set by diffraction at low energies and by nuclear reactions at high energies. For most particles the limits are hence between about 10 eV and a few MeV. The motion of the projectile can be followed as a sequence of binary collions (binary collision approximation, BCA). The total nuclear loss is simply the sum of the 'elastic' losses of the individual collisions. Experimentally problems are encountered at low energies where the traditional scheme using thin foils is troubled by thickness variations, pin-holes, channeling effects, and small angle scattering events. Channeling is the correlated motion of a projectile in single crystals. If a projectile has a small perpendicular momentum with respect to the repulsive potential of a chain or plane of atoms, the projectile will be 'trapped' in channels formed by axes or planes in the crystal. In general the particle beam will be split into a channeled and a 'random' beam. Since there is a critical angle for channeling [1] which increases with descreasing energy channeling effects are a greater nuisance in low energy stopping power experiments. In a number of experiments [2–4] we made use of the channeling effect, i.e. by scattering the particles at grazing incidence off surfaces. In that case the elastic losses are negligible compared to the inelastic loss and the trajectories of the projectiles are well defined. There are cases with rather uniform trajectories [5, 6], with little varation in the outgoing angles and/or the trajectory length.

The inelastic loss or the electronic stopping power is due to elementary excitations of the solid, of invidual atoms in the solid and of the projectile [7, 8]. The excitations of the atoms and the projectile are ionization processes which in turn depend on the actual charge state of the projectile. Hence depending on the velocity of the projectile there will be a charge state distribution of the projectiles. In gaseous targets this charge state distributions can be estimated from capture and loss processes [9]. In solids at low particle velocities the charge state of the projectile is defined by the dynamical screening by the valence electrons of the solid [10]. The charge exchange processes contribute to the energy loss and to the energy loss straggling [11]. The major contributions to the energy loss are however the excitations of plasmons and electron-hole pairs. These lead to a linear dependence of electronic stopping power on the projectile velocity [12]. In sum we face a problem where the charge state and the losses of the projectile scattering from a surface are both intrinsically connected with solid state properties, e.g. the response of the solid to a moving charge. The properties in question are in general the band structure and the dielectric function of the solid.

This contribution discusses in some detail our present understanding of the inelastic loss and energy loss straggling of H and He at low energies (below v_B, the Bohr velocity). This discussion includes the charge exchange processes. The results are compared to experiments. In a second section we will discuss the results of experiments using molecular particles where charge exchange processes are observed. The charge exchange includes the formation of negative molecular ions which are also observed in chemisorption experiments as precursors of dissociative chemisorption or associative desorption.

2 Scattering of Ions and Atoms

As mentioned before, charge state and energy loss of particles interacting with matter are closely related. A particle being scattered off a surface at grazing incidence interacts only with the electrons extending beyond the uppermost atomic layer. The interaction between this particle and the electrons can be expected to be much stronger if this particle is an ion compared to the case when it is neutral.

An ion in front of an 'electron see' gives rise to an image charge which in turn attracts the ion further to the surface (see e.g. [13]). For a neutral particle this image charge does not exist. The probability of exciting an electron-hole pair or a plasmon (and thereby losing the corresponding amount of kinetic energy) is higher for ions than it is for neutral particles; and similarly one expects it to be higher for highly-charged ions than for singly-charged ions.

During the interaction the particle usually does not stay in a single charge state but is subject to electron exchange processes altering its charge state. In such a case the strength of interaction is varying with the charge state.

Aiming at a theoretical description we need to know a) what the probability is for a particle to switch from a given state to another and b) what is the rate of energy loss in a given state. Equipped with this set of numbers, we are able to arrive at a final expression that delivers an energy loss distribution for a particle starting out in, say, in some charged state and leaving the interaction region in the ground state.

2.1 Helium incident on a metal surface

This case was dealt with in detail in Ref. [4] for He^+ incident on a Ni(110) surface and will here be summarized shortly. The description for helium is actually simpler than for hydrogen because in our velocity range there are only two charge states (He^+ and He^0) involved for helium versus three (H^+, H^0, and H^-) for hydrogen.

The Auger-neutralization to neutral He^0 is very efficient at low velocities and the transition rate for reionization is very low, as was shown in [4]. From this we conclude that only one charge-changing event takes place during the interaction, so that the rate of change of the He^0-yield can be expressed by

$$\frac{dn(He^0)}{dt} \simeq \frac{1}{\tau_A} n(He^+).$$

where $\frac{1}{\tau_A}$ is the Auger transition rate.

Taking into account that the friction coefficient (strength of interaction) changes with the charge state of the He particle, we write for the energy loss before neutralization

$$Q_+(x) = \int_0^x \gamma^+ v ds \quad \text{and after neutralization} \quad Q_0(x) = \int_x^L \gamma^0 v ds$$

so that the total energy loss Q is given by

$$Q = Q_+(x) + Q_0(x) = (\gamma^+ - \gamma^0)vx + \gamma^0 vL, \quad x \in [0, L] \tag{1}$$

where $\gamma^{+,0}$ are the friction coefficients for He$^+$ and He^0, respectively, v is the velocity of the particle, x the free path for the particle being an ion, and L the trajectory length. Transforming variables finally yields the energy-loss spectrum for the reflected neutral particles:

$$\frac{dn(He^0)}{dQ} \propto \exp\left(-\frac{Q-Q_0}{(\gamma^+ - \gamma^0)v\lambda_A}\right) \Theta(Q-Q_0)\Theta(Q_+ - Q), \qquad (2)$$

where $\Theta(x)$ is the step-function, being 1 (0) for positive (negative) argument, $Q_0 = Q_0(0)$, $Q_+ = Q_+(L)$, and $\lambda_A = v\tau_A$.

In eq.(2) the straggling of the energy loss [14] has been neglected. We include the straggling by convoluting the right hand side of eq.(2) with a Gaussian straggling distribution function of width Ω. This gives

$$\frac{dn(He^0)}{dQ} \propto \exp\left(-\frac{Q-Q_0}{(\gamma^+ - \gamma^0)v\lambda_A}\right) \left\{1 + \frac{2}{\sqrt{\pi}}\int_0^y \exp\left(-t^2\right) dt\right\} \qquad (3)$$

where

$$y = \frac{1}{\sqrt{2}}\left(\frac{Q-Q_0}{\Omega} - \frac{\Omega}{(\gamma^+ - \gamma^0)\,v\lambda_A}\right).$$

The calculated spectrum depends on the following parameters: τ_A, γ_s^+, γ_s^0, and the straggling parameter Ω^2. The Auger-lifetime τ_A has been calculated in [4]; the ratio γ_s^+/γ_s^0 has been calculated in linear theory for an electron gas as a function of the electron density [15]. Q_0 and Ω were also calculated in [4] using a local-density approach by means of

$$Q_0 = \int_{-\infty}^{\infty} \gamma_s^0(l)v\,dl \text{ and } \Omega^2 = \int_{-\infty}^{\infty} W(l)\,dl$$

where γ_s^0 (Fig.3 in [12]) and W (Fig.3 in [16]) were obtained locally assuming that they take in each point the values associated with the corresponding electronic local density. Note that these are nonlinear density functional calculations.

Putting everything together, we compare in fig.1 the experimental curve for 4.9 keV He$^+$ $\rightarrow$ Ni(110) with the theoretical curves with and without straggling.

We find, that with inclusion of straggling we can match the experimental data well. For further details the reader is referred to [4].

2.2 Hydrogen incident on a metal surface

In [17] we reported similar experiments performed with incident hydrogen (H$^+$ as well as H^0) on a Ni surface using the same geometry as before. The reader is referred to this paper for a more detailed discussion of the H$-$metal-surface interaction. Apart from H$^+$ and H^0 we have to deal with a third charge state.

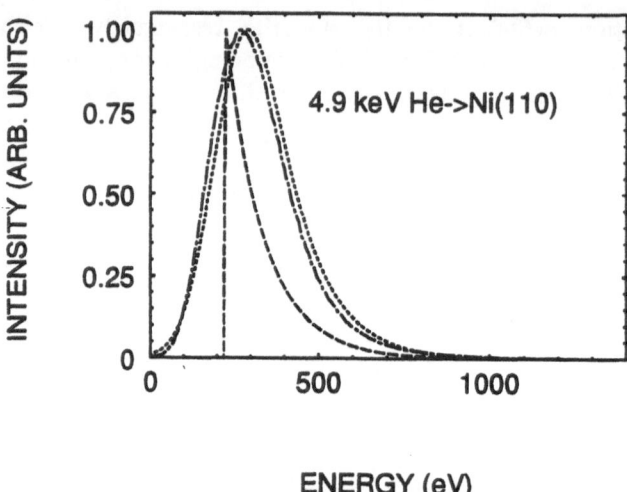

Figure 1: Energy-loss spectra for emerging He^0 with incident He$^+$ scattered off Ni(110) for an incident energy of 4.9 keV (angle of incidence 5°, scattering angle 10°). The dash-dotted line is the experimental result, the dotted line is the theoretical result including straggling, and the dashed line is the theoretical result excluding straggling.

Nonlinear density functional calculations show that the selfconsistent solution for the screening of a static proton in an electron gas at metallic densities ($r_s \gtrsim 2$) gives a ground state configuration in which the bound state around the proton is doubly occupied [18]. We call this the H$^-$ configuration. If we force this level to be singly occupied or empty one gets solutions that represent excited states of the total system that we call H^0 or H$^+$ configuration, respectively. From these potentials the frictions coefficients were determined [7].

Knowing that the H$^-$ state is stable inside a free electron gas, we use a simplified set of rate equations by allowing electron exchange in one direction only: H$^+ \longrightarrow$ H$^0 \longrightarrow$ H$^-$. We neglect the reionization process H$^0 \longrightarrow$ H$^+$ because of the theoretical estimates [19–21] which show that this process has a cross section which is more than an order of magnitude lower.

Then the set of rate equations is given by

$$\frac{d}{dt}n^+ = -\frac{1}{\tau_N}n^+ \tag{4}$$

$$\frac{d}{dt}n^0 = \frac{1}{\tau_N}n^+ - \frac{1}{\tau_C}n^0 \tag{5}$$

$$\frac{d}{dt}n^- = \frac{1}{\tau_C}n^0. \tag{6}$$

The transition rate for a specific process is given by $1/\tau_i$ where i is substituted by N (neutralization $H^+ \longrightarrow H^0$) and C (electron capture $H^0 \longrightarrow H^-$).

As an estimate for the time constant of the H^- formation via resonant tunneling, we take the calculation from [22], based on an Anderson-Newns hamiltonian. In [23] we showed that for neutralization of H^+ the Auger transition rate and resonant tunneling rate at a metal surface are about equal. The time constants are of the order of $10^{-15}s$, which is smaller than the time of interaction T.

For an H^- particle at the end of the trajectory we expect the excess electron to be lost on the outward path. These particles will show up in that part of the spectrum where one would expect neutrals. The H^+ part is separated experimentally by applying a post-acceleration voltage. This implies, that the total energy loss spectrum we are interested in is given by the energy distribution of those particles that arrive at the detector in the *neutral* state. We assume that all detected neutral particles were in the H^- state near the surface.

Following the same line as in the He case, we finally obtain an energy loss spectrum

$$\frac{dn^-}{dQ} = \frac{sgn(\gamma^+ - \gamma^0)}{Q_A}\left(\exp\left(-\alpha(t'_{min}, Q)\right) - \exp\left(-\alpha(t'_{max}, Q)\right)\right)\Theta(Q-Q_1)\Theta(Q_2-Q).$$

(7)

Here $sgn(x) := x/|x|$, $Q_A := v^2\tau_N(\gamma^+ - \gamma^-) - v^2\tau_C(\gamma^0 - \gamma^-)$, $\alpha(t'_{min}, Q) = (Q - Q^-)/(\gamma^+ - \gamma^-)v^2\tau_N$, $\alpha(t'_{max}, Q) = (Q - Q^-)/(\gamma^0 - \gamma^-)v^2\tau_C$, $Q_1 = Q^- := \gamma^- v^2 T$, and $Q_2 = Q^0 := \gamma^0 v^2 T$, with v the particle's velocity and T the time of interaction, which was determined from MARLOWE calculations [4, 24] via the trajectory length.

In Fig.2 we compare the experimental data for the case of 1.5 and 4.8 keV incident H^+ on Ni(110) along a random, i.e. a high indexed direction, with the calculated curves. The broken line represents the energy loss distribution without accounting for straggling (Eq. 7), the solid line the energy distribution including straggling by convoluting with a Gaussian of width Ω, which was used as a fit parameter. Note, that the experimental maximum of the energy loss distribution is reproduced without any adjustable parameters. Only for high energies there is a slight mismatch. This is due to the larger penetration depth of the particles which leads to a higher electron density seen by the particles. That, in turn, affects the friction coefficient. In the inset of Fig.2b we show the curves after calculating the spectra with $\gamma^- = 0.35a.u.$ instead of $0.3a.u.$.

2.3 Statistics

Up to now we considered only two- and three-state systems, where we further simplified the description by introducing appropriate approximations in due course. We were, so to say, lucky in that the physical situation allowed us to introduce these approximations.

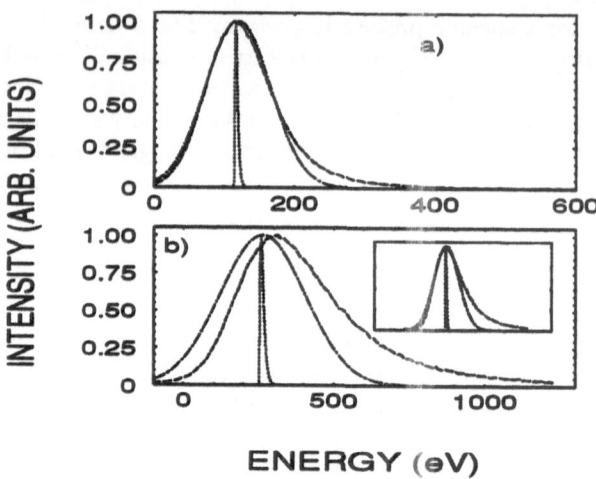

Figure 2: Experimental data (dots) and theoretical calculation without straggling (broken line) and including straggling (solid line). The incident energies are 1.5 keV (a) and 4.83 keV (b). The inset in b) shows the result obtained with $\gamma^- = 0.35 a.u.$.

There are cases where one has to resort to a description of the system under investigation that involves more than three states and where one possibly also has to account for more complicated charge-changing events.

Recently a general formalism has been developed for computation of energy-loss spectra of penetrating particles in the presence of charge exchange [11, 25, 26]. The input to this formalism consists of a $n \times n$-matrix σ, where n is the number of states needed to describe the system properly and $Nv\{d\sigma_{IJ}(T)/dT\}$ is a set of differential transition rates between accessible states I and J. N is the density of scattering centers and v the projectile velocity.

The output generated by the formalism is a transfer matrix $\mathbf{F}(\Delta E, x) = \{F_{IJ}(\Delta E, x)\}$. Here $F_{IJ}(\Delta E, x)d(\Delta E)$ is the probability for a projectile occupying state I at $x = 0$ to occupy state J after a pathlength x and to have lost kinetic energy $(\Delta E, d(\Delta E))$ by an arbitrary sequence of events. In [25, 11] it was shown that $\mathbf{F}(\Delta E, x)$ obeys a generalized Bothe-Landau formula

$$\mathbf{F}(\Delta E, x) = \frac{1}{2\pi} \int_{-\infty}^{\infty} dk e^{ik\Delta E} \exp\left(Nx(\mathbf{Q} - \sigma(k))\right), \qquad (8)$$

where $Q_{IJ} = \int d\sigma_{IJ} - \delta_{IJ} \sum_L \int d\sigma_{IL}$ and $\sigma_{IJ}(k) = \int d\sigma_{IJ}(1 - e^{-ikT})$. Eq. (8) assumes the individual events to be statistically independent and transition rates $d\sigma_{IJ}(T)/dT$ to be independent of time.

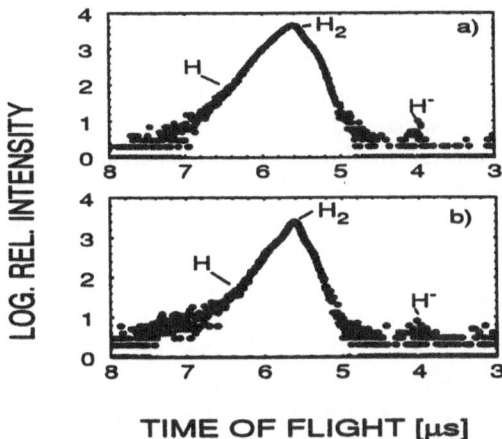

Figure 3: a) Time of Flight (TOF) spectrum of ionic H_2 scattered from Pd(110) at a grazing angle of 5° into scattering angle of 10°. The detector aperture is 1.2°. A positive acceleration voltage is applied to separate the negative ions from the neutrals. The primary energy is 580 eV, equivalent to 5.41 μs. Note that the energy scale runs in the plot from left to right. The ion yield is of the order of 10^{-4}. The surviving neutral H_2 can not be separated from the dissociated neutral H. b) as in a) but for incident neutral H_2.

This formula has been used to extract experimental parameters such as charge state distributions, as well as mean energy loss, straggling, and skewness characterizing the energy spectrum [11, 26]. For simple systems analytical expressions for the energy loss spectra were derived that go beyond equations 3 and 7 [27].

3 Scattering of Molecules

In 1975 it was found that 5 keV D_3^+ ions survive the scattering from a Au surface at grazing incidence [28]. Even when considering the perpendicular momentum only, the 'perpendicular' energy $E_\perp = E_0 \sin\phi = 20eV$ is larger than the binding energy of the molecular ion, $2.56eV$ only. The angle ϕ is the glancing angle of incidence. Later [29] it was found that not only ions survive but that a large fraction of the incident ions, H_2^+ and N_2^+, can capture an electron and survive the collision with the surface as neutralized molecules. These findings were corroberated for many different ionic species and also for different surfaces [30–33]. A new experimental finding were negative molecules, first O_2^- scattered from Ag(111), which was interpreted as an 'harpooning' event [34]. The incident particles were O_2^+ ions such that the molecules capture in fact two electrons.

Similar capture processes occur with NO, C_2, and CO_2 [35–37] More recently we found negative ion formation also in the case of neutral species incident on a metal surface, i.e. CO_2 scattering from Pd + K at low energies (400 eV) and grazing incidence. In this case the neutral molecule captures one electron and leaves the surface as a CO_2^- species [38]. 'Harpooning' is used in atom-molecule scattering events to describe the jumping of an electron from, e.g., an alkali-atom on to, e.g., a halogene molecule [39]. The electron transfer causes a change of the interaction cross section between the atoms and the molecules involved. In surface scattering events the electron transfer will in general cause a $T - V$-transfer, i.e. the transfer of kinetic energy (T) to potential energy (V) [40, 41]. It is interesting to note that in all cases where negative molecules are found in molecule-surface scattering discussed here, there exist negative molecular precursors in the corresponding associative desorption or dissociative chemisorption processes [39–49]. The fast beam experiment may provide direct evidence for those precursors if they are the same species and if they survive the collision. In the following we will discuss some experiments in greater detail and we will discuss the models which are presently used to describe the survival or the dissociation of the fast molecules. Figure 3 shows a typical experimental result of a time of flight experiment (TOF) using a molecular ion or a neutral molecule as the primary particles [53].

The experimental technique is identical with the one used for energy loss experiments with atomic particles discussed in the previous section. The yield is given in a logarithmic scale, the time scale is inversed such showing essentially an energy spectrum. The yield of ions is very small from clean high-work-function metal surfaces. This is the case for both, negative and positive scattered ions. The neutral peak contains molecules and atoms. The evidence for the dissociated atoms follows quite obvious from the finding that there are particles with energies above the primary beam energy E_0. From conservation of energy and momentum and from the conversion from the center of mass to the laboratory system follows that the energy of dissociated particles is given by $E = \frac{1}{2}E_0 + \frac{1}{2}E_D \pm \sqrt{E_0 E_D}\cos\alpha$, where E_D is the dissociation energy and α the angle between the molecular axis (of a diatomic molecule) and the direction of the incident beam. From the half width of the peak of the neutrals a mean dissociation energy of a few eV can be estimated (Fig.3). It is obvious that the peak of the dissociated particles has to be symmetric and hence extends to the low energy side as much as to the high energy side. A de-convolution of the peak of the dissociated particles in order to evaluate e.g. the potential energy surface the particles are dissociating from, is impaired by the additional elastic scattering and inelastic energy losses along their trajectories. Under certain assumption it is possible to simulate the spectra [54]. As an input a repulsive potential was used, qualitively related to the $b^3\Sigma_u$ of H_2. Hence it is assumed that the dissociation is due to a resonant electron capture process. All possible vibrational and rotational excitations have been neglected.

The small peak in the TOF spectra on top of the broad distribution of the dissociated particles (Fig.3) is identified as surviving neutral H_2. In case

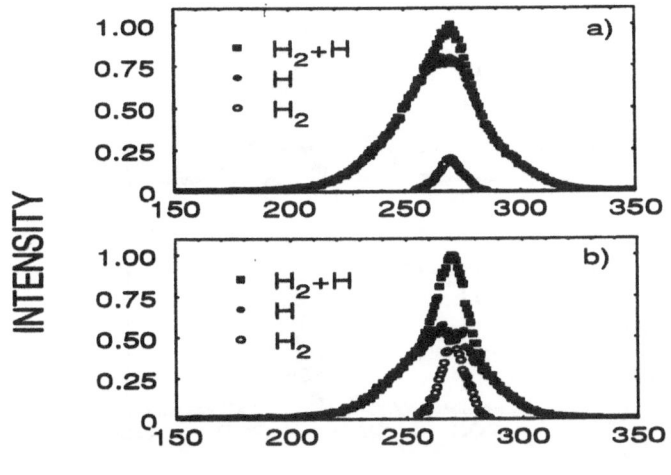

Figure 4: a) The same data as in Fig.3a converted into energy spectra. Here the total spectrum (squares) is separated into molecules (open circles) and atoms (filled circles) as described in the text. Here the intensity scale is linear. b)the same data as in Fig.3b converted into energy spectra in linear scaling. Symbols as in a), squares=total spectrum, open circles=atoms, and filled circles=molecules.

of incident ions it is assumed that it is formed by an Auger electron capture process as discussed above for atomic ions. With neutrals as incident particles these particles are survivors indeed.

In Fig.4 we show the spectra converted into energy spectra with a linear energy scale. For the interpretation of the neutral, dissociated peak in the region of the 'hat' of neutral molecules, we use either an experimental spectrum obtained from atoms at the same speed, or, if available, the ionic peak of dissociated particles, whenever there is enough statistics. In most cases the ionic peak fits the neutral peak quite well. This experimental finding stresses the point that the spectra are dominated by scattering and straggling independent of the dissociation mechanism since we cannot assume that the neutrals and the ions are dissociated by the same mechanism. Or, if the dissociation mechanism is e.g. a resonant charge capture, then there has to be an ionizing collision following the dissociation event. In either case there is little 'memory' of the incident state in the energy spectra obtained under identical experimental conditions but from the low work function surface of a monolayer of K on Pd. The changes in the spectra are obvious: the survival of molecular species increased and there is now an appreciable amount of negative atomic H^-. The results are summarized in Table 1.

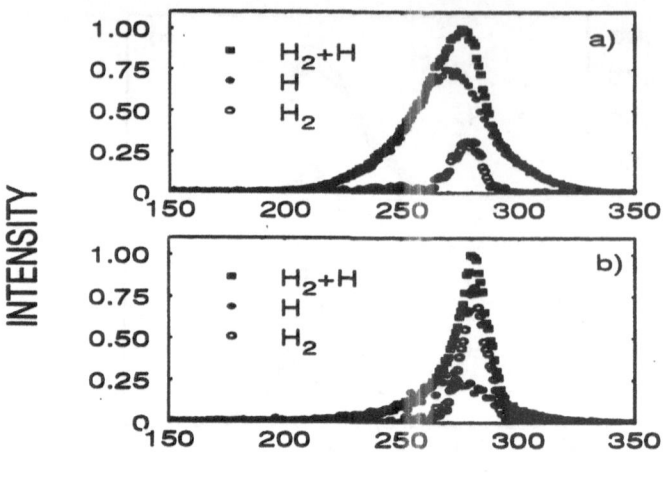

Figure 5: a) Energy spectra as in Fig.4 but for Pd(110) with a coverage of 1 monolayer of K. Symbols, energy, and angles as before. The incident particles are H_2^+ ions. b) Energy spectra as in a) but for incident neutral H_2.

	$Y_m(H_2^+)$	$Y_m(H_2^0)$	$Y_{H^-}(H_2^+)$	$Y_{H^-}(H_2^0)$
Ni(110)	0.10	0.33	$\ll 10^{-3}$	$\ll 10^{-3}$
Pd(110)	0.12	0.34	$\ll 10^{-3}$	$\ll 10^{-3}$
Pd(110) + K	0.24	0.63	0.02	7×10^{-3}

Table 1: Molecular survival fractions Y_m and negative ion fractions Y_{H^-} for 580 eV H_2^+ and H_2^0 scattering at grazing incidence ($\phi = 5°$) from Ni, Pd, and Pd +K.

The yields are the respective peak integrals divided by the sum of all particles. We note that K causes a decrease of the dissociation of both neutral and ionic incident H_2. In chemisorption experiments it was found that K decreases the dissociative chemisorption of H_2.

There are essentially two major pathways for the dissociation in the experiments discussed here: dissociation by electron capture (or loss) and dissociation by ro-vibrational excitation.

3.1 Dissociation by Electron Capture (or Loss)

The electron capture events in its simple version are adapted from the 'Hagstrum-model' of the neutralization of atomic ions (Fig.6).

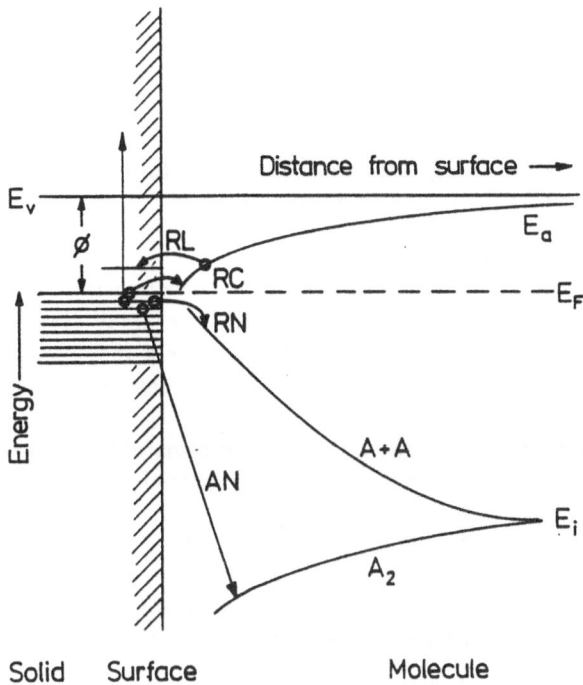

Figure 6: Scheme for the charge exchange processes between a surface and a molecular ion. RL and RC are resonant loss and resonant capture, respectively, RN is resonant neutralization and AN is Auger neutralization. E_a is the electron affinity level, E_F the Fermi energy, E_i the ionization energy at infinity.

This is an adiabatic model which does not include screening effects that are obviously important as discussed above. Calculations based on such a model [55] gave satisfactory agreement with experimental findings [56]. The model predicted a dependence of the dissociation on the orientation of the incident H_2 molecule due to the conservation of the total symmetry [57]. This was qualitatively found experimentally [58]. The second type of charge exchange models is based on the evolution of potential energy surfaces [42]. The theoretical treatment is rather difficult such that the models used for the metal surface are not very realistic yet, i.e. two Cu-atoms [59]. Experimentally, insight into these models is partly obtained from the analysis of both the angular and the TOF distribution of the dissociated particles [60–63]. Still another suggestion in the range of charge capture models is the qualitative transfer of the 'screening model', successful in the atomic scattering world, to the molecular world. This suggestion is stimulated by the experiments using neutral molecules as primary particles (Figs. 3 – 5) and by trajectory calculations. Due to the geometry used, the particles probe essentially only the electron distribution above the surface. The calculations show that the trajectory of these particles at the surface is

	P_0^D	P_{AC}	P_{RC}^D
K	0.37	0.38	0.62
Ni	0.67	0.30	0.70
Pd	0.66	0.35	0.65

Table 2: Dissociation and charge capture probabilities for H_2^0 and H_2^+ interacting with K, Pd, and Ni at grazing incidence ($\phi = 5°$) and 580 eV.

deep and long enough to support the idea that the dissociation is essentially caused by the screening of the H. We know that the 'final' state of H in any metal is a screened proton. So the survival of H_2 is a matter of penetration depth and time spent in the electron distribution. This model has yet to be based on theoretical calculations.

If we analyse our H data (Figs. 3 – 5) on the basis of resonant and Auger-capture events we can write down the following probabilities for the different events. The dissociation probability by resonant capture is given by $P_{RC}^D = (1 - Y_m(H_2^+)) - P_{AC}P_0^D$ where $Y_m(H_2^+)$ is the molecular neutral yield for incident H_2^+, P_{AC} the probability for the Auger capture and P_0^D the dissociation probability of neutral H_2. Naturally, $P_0^D = 1 - Y_m(H_2^0)$ and $P_{AC} + P_{RC}^D = 1$ [38]. From the experimental yield (Table 1) we obtain then the probabilities as summarized in Table 2.

The 'blocking' of the resonant channel by the K adsorption is qualitatively understood by inspection of Fig 6. Since the electron distribution is energetically shifted closer to the vacuum level, the resonance condition with the Franck-Condon region of the $b^3\Sigma_u$ state is destroyed. However, Fig.6 is not very useful for the interpretation of the data of incident H_2. In other words, the memory of the initial charge state is not only destroyed by scattering and straggling. Independent of the initial charge state the screening process, which includes neutralization, is the important mechanism, and only a minor part of the dissociation on the incoming trajectory may be due to the resonant capture into $b^3\Sigma_u$ directly.

3.2 'Mechanical' Dissociation

The other class of models, i.e. the 'mechanical' models, have been introduced by Bitenskij and Parilis (Fig.7 [64]).

The model has been elaborated and compared to experiments extensively [65–69]. In Fig.8 we show results of the analysis of the scattering of H_2 from Ag(111) where only the scattered ions have been analysed. The analysis is based on the assumption of a ro–vibrational excitation which leads to the dependence on $E_0\Theta^2$. The line shown is the calculation which agrees with the experimental data not shown here. The dissociated fraction f_{dis} has been calculated from $f_{dis} = 1 - D_0/(mE_0\Theta^2 y_{max}^2)$ where D_0 is the binding energy of H_2 and $y_{max}^2 \simeq$

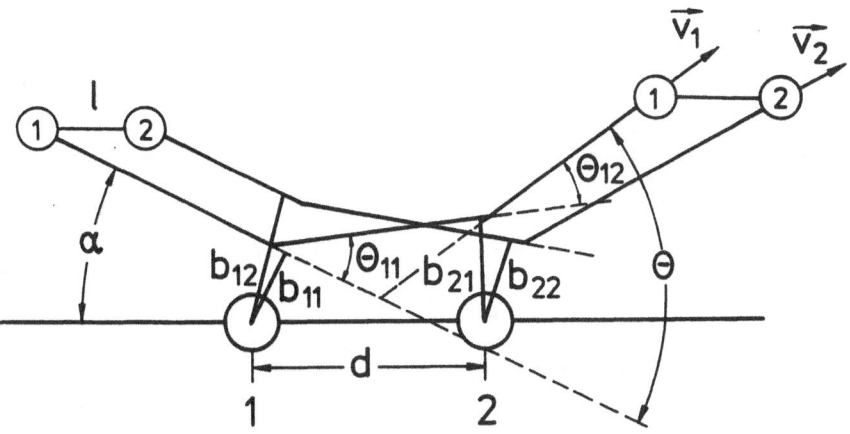

Figure 7: Multiple scattering scheme for the dissociation of molecules at a surface [64]. The constituents of the molecule are separated by l, the lattice parameter in the surface is d. Θ is the total scattering angle. Θ_{ii} and b_{ii} are the individual scattering angles and impact parameters, respectively. Dissociation occurs when the relative energy of the two atoms surpasses the binding energy after the scattering event.

0.79 a parameter, where y is the bond angle between the molecule and a surface atom at the point of impact [68]. We compare this calculation with our data of Pd. The difference is experimentally, that in one case we use ionic H_2^+ but analyse the scattered neutrals and in the other case we use incident neutrals as in the Ag case but again analyse the scattered neutrals. The qualitative evidence from this comparison may be that at higher $E_0\Theta^2$ values the dissociation is indeed mainly due to ro-vibrational excitation, however at lower $E_0\Theta^2$ values we have a contribution from the resonant capture to $b^3\Sigma_u$. With incident neutral H_2^0 we see less dissociation and a rather linear dependence on $E_0\Theta^2$. Θ is the scattering angle which is in our experiment is $2\times\phi$, ϕ being the angle of incidence. For small angles, here $\phi = 5°$, $E_0\Theta^2 \simeq E_\perp$, i.e. the dissociation decreases with penetration depth into the electron distribution above the surface of the solid. This observation supports the screening model. We note that the chemisorption experiments with D_2 on Pt and Pt + K are interpreted by the fact that K causes a spatial extension of the electron distribution perpendicular to the surface and hence an effective increase of the barrier for the dissociation [47]. In our screening model that would correspond to a different screening of H_2 in the case of Pd compared to the case of Pd + K.

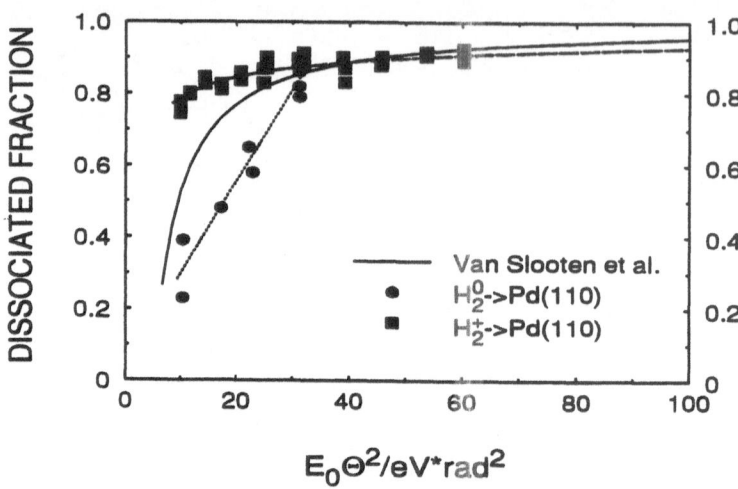

Figure 8: Dissociated fraction for the cases H_2^0 and H_2^+ on Pd(110) as a function of the energy times the total scattering angle squared. The data are compared to a calculation of van Slooten et al. [68] assuming ro-vibrational induced dissociation. These data agree with experiments using neutral H_2 incident on Ag(111) but detecting dissociated ions and surviving ions, not the neutrals as in our experiment.

3.3 Negative Molecular Ions

As an example for the experiments where negative molecular ions are found [31–37] we will discuss the scattering of CO_2 from Pd(110) and Pd(110) + K [38]. Figures 9 and 10 show the TOF spectra of CO_2^+ and CO_2 scattered at low energies, 580 eV, and grazing incidence, $\phi = 5°$ into a scattering angle of $10°$.

The velocity of CO_2 is $v = 5 \times 10^{14} \text{Å} s^{-1}$ and the velocity component perpendicular to the surface $v_\perp = 4 \times 10^{13} \text{Å} s^{-1}$. In Fig.10b the CO_2 peak is clearly identified. In the case of CO_2^+ incident on the K surface we find complete dissociation whereas from the high work function Pd surface and incident CO_2 the TOF spectrum is comparable to those of N_2^+ or CO^+, i.e. the dissociation probability is very low. We conclude therefore that at the low velocities and the preparational state of the surface used here the ro-vibrational dissociation plays a minor role only. Hence, when we change from the neutrals to ions or from the clean surface to the K-case, we assume that charge state effects determine the fate of the molecules. Nevertheless we can a priori not exclude that a charge capture event, e.g. into a vibrationally excited state by a 'harpooning-like' event [40], will make the molecule prone for a ro-vibrational dissociation. So, if we analyse the yields obtained from the experimental results (Table 3) assuming Auger and resonant capture as in the case of H, we have to add here a probability P_-^D. This is the probability for dissociation caused by the formation

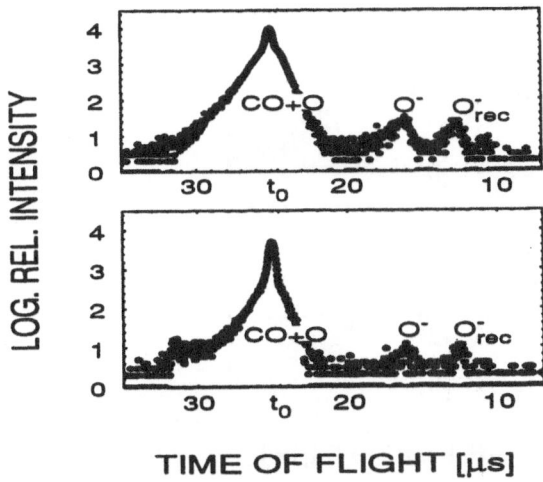

Figure 9: a) TOF spectra of CO_2^+ ions scattered from clean Pd(110) at 580 eV, geometry as in the case of H_2, also with positive post-acceleration. b) As in a) but for neutral CO_2^0.

	$Y_{CO_2}(CO_2^+)$	$Y_{CO_2}(CO_2^0)$	$Y_{O^-}(CO_2^+)$	$Y_{O^-}(CO_2^0)$	$Y_{CO_2^-}(CO_2^0)$
Pd(110)	0.52	0.82	7×10^{-4}	5×10^{-4}	0
Pd(110) + K	0	0.55	0.17	0.08	6×10^{-3}

Table 3: Molecular survival fraction Y_{CO_2}, negative oxygen yield Y_{O^-}, and negative molecular ion yield $Y_{CO_2^-}$ for CO_2^+ and CO_2^0 scattering at grazing incidence ($\phi = 5°$) and 580 eV.

of negative ions. We can not identify the actual electronic state since there is a host of negative molecular CO_2^- states [70]. There is some probability that the negative state is the bend CO_2^- which has been identified in chemisorption experiments [50]. This bend species is naturally vibrationally excited. Also for the case of the resonant capture into a dissociative 'triplet' state of the CO_2 neutral system we can not identify the actual electronic molecular state. Notwithstanding these uncertainities, the analysis reveals the relative importance of the possible mechanisms quite clearly for the four experimental cases of Figs. 9 and 10 (Table 4).

In the case of CO_2^+ on K the filling of dissociative states, either triplet or negative, is so efficient that the Auger neutralization does not come into play at all. This finding is in agreement with the estimate of the time spent on the ingoing part of the trajectory from the point where the interaction (charge capture) may start. If we assume 20 a.u. as this start point and a collision

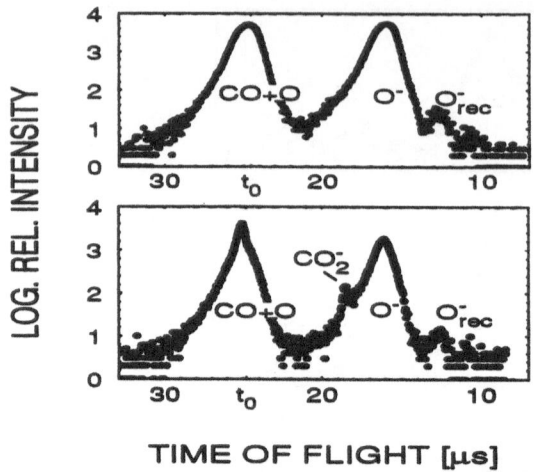

Figure 10: a) TOF spectra of CO_2 ions scattered from Pd(110) + K(1 monolayer) b) TOF spectra of CO_2^0 neutrals scattered from Pd(110) +K(1monolayer).

	P_-^D	P_{AC}	P_{RC}^D
Pd	0.18	0.63	0.37
Pd +K	0.45	0	0.55

Table 4: Dissociation and charge capture probabilities of CO_2^+ and CO_2^0 interacting with Pd and Pd+K at grazing incidence.

distance of about 5 a.u., the molecules spend about $10^{-12}s$ in the approach region. This is indeed enough time to go through a whole sequence of charge capture followed by dissociative processes. Dissociative processes proceed on a time scale of $10^{-12}s$ comparable to vibrational time scales [70]. In the case of neutral CO_2 incident on Pd+K the dissociation can practically only proceed via negative states. These states are not necessarily the same states as in the case of CO_2^+ because the first capture in that case is very likely not into the ground state of the molecule but into an excited state and hence the following second capture will form an excited CO_2^- state. With the much slower and larger CO_2 compared to H_2 we have not to invoke more complicated models than the adiabatic model of Fig.6 for a qualitative understanding of the experimental results. We conclude with the observation that the CO_2 data agree with the findings in chemisorption experiments that K causes an increase of the dissociative chemisorption of CO_2 [50].

Acknowledgements

We thank the DFG, DAAD, and SNF for financial support. Helpful discussions with H. J. Freund (Bochum), V. Kempter (Clausthal), F. Flores (Madrid), P. M. Echenique (San Sebastián), and P. Sigmund (Odense) are gratefully acknowledged.

References

[1] J. LINDHARD. *K. Dan. Vidensk. Selsk. Mat. Fys. Medd.* **34**(14) (1965).

[2] A. NÄRMANN, H. DERKS, W. HEILAND, R. MONREAL, E. GOLDBERG, AND F. FLORES. *Surf. Sci.* **217**, 255 (1989).

[3] A. NÄRMANN, W. HEILAND, H. DERKS, S. SCHUBERT, F. FLORES, E. GOLDBERG, AND C. MONREAL. In P. VARGA AND G. BETZ, editors, "Proc. Symp. Surf. Sci.", p. 183, Kaprun (1988).

[4] A. NÄRMANN, W. HEILAND, R. MONREAL, F. FLORES, AND P. M. ECHENIQUE. *Phys. Rev. B* **44**, 2003–2018 (1991).

[5] H. DERKS, A. NÄRMANN, AND W. HEILAND. *Nucl. Instrum. Methods B* **44**, 125 (1989).

[6] C. HÖFNER, A. NÄRMANN, AND W. HEILAND. *Nucl. Instrum. Methods B72* **72**, 227–233 (1992).

[7] P. M. ECHENIQUE, R. M. NIEMINEN, AND R. H. RITCHIE. *Solid State Commun.* **37**, 779 (1981).

[8] P. M. ECHENIQUE, I. NAGY, AND A. ARNAU. *Int. J. Quantum Chemistry* **23**, 521–543 (1989).

[9] S. K. ALLISON. *Rev. Mod. Phys.* **30**(4), 1137–1168 (1958). (E) 31(1959)839.

[10] P. M. ECHENIQUE, F. FLORES, AND R. H. RITCHIE. *Nucl. Instrum. Methods 33* **33**, 91 (1988).

[11] P. SIGMUND. *Nucl. Instrum. Methods B* **69**, 113–122 (1992). This paper also contains an extensive bibliography about charge exchange and energy loss.

[12] P. M. ECHENIQUE, R. M. NIEMINEN, J. C. ASHLEY, AND R. H. RITCHIE. *Phys. Rev. A* **33**, 897 (1986).

[13] H. WINTER, C. AUTH, R. SCHUCH, AND E. BEEBE. *Phys. Rev. Lett.* **71**, 1939 (1993).

[14] M. A. KUMAKHOV AND F. F. KOMAROV. "Energy loss and ion ranges in solids". Gordon and Breach, New York (1981).

[15] T. L. FERRELL AND R. H. RITCHIE. *Phys. Rev. B* **16**, 115 (1977).

[16] J. C. ASHLEY, A. GRAS-MARTÍ, AND P. M. ECHENIQUE. *Phys. Rev. A* **34**, 2495 (1986).

[17] A. NÄRMANN, K. SCHMIDT, C. HÖFNER, W. HEILAND, AND A. ARNAU. *Nucl. Instrum. Methods B* **78**, 72–76 (1993).

[18] E. ZAREMBA, L. M. SANDER, H. B. SHORE, AND J. H. ROSE. *J. Phys.* **F7**, 1763 (1977).

[19] F. SOLS AND F. FLORES. *Phys. Rev. B* **30**, 4878 (1984).

[20] F. SOLS AND F. FLORES. *Nucl. Instrum. Methods B* **13**, 171 (1986).

[21] F. SOLS. "Procesos dinámicos en la interacción de partículas en movimiento con la materia". PhD thesis, Universidad Autónoma de Madrid, Spain (1985).

[22] J. H. RECHTIEN. Untersuchung des Adsorptionssystems $O_2/Si(001)$ durch Molekülionenstreuung. Master's thesis, Universität Osnabrück, Germany (1988).

[23] K. J. SNOWDON, R. HENTSCHKE, A. NÄRMANN, AND W. HEILAND. *Surf. Sci.* **173**, 581 (1986).

[24] A. NÄRMANN, R. MONREAL, P. M. ECHENIQUE, F. FLORES, W. HEILAND, AND S. SCHUBERT. *Phys. Rev. Lett.* **64**, 1601 (1990).

[25] P. SIGMUND. In A. GRAS-MARTÍ, H. M. URBASSEK, N. R. ARISTA, AND F. FLORES, editors, "Interaction of Charged Particles with Solids and Surfaces", pp. 73–144, New York (1991). NATO ASI Series B271, Plenum Press.

[26] A. NÄRMANN AND P. SIGMUND. *Phys. Rev. A* (1994). submitted.

[27] A. NÄRMANN. (1994). in preparation.

[28] W. ECKSTEIN, H. VERBEEK, AND S. DATZ. *Appl. Phys. Lett.* **27**, 52 (1975).

[29] B. WILLERDING, W. HEILAND, AND K. J. SNOWDON. *Phys. Rev. Lett.* **53**, 2031 (1984).

[30] B. WILLERDING, K. J. SNOWDON, AND W. HEILAND. *Z. Phys. B* **59**, 435 (1985).

[31] S. R. KASI, M. A. KILBURN, C. S. SASS, AND J. W. RABALAIS. *Surf. Sci. Rep.* **10**, 1 (1989).

[32] A. W. KLEYN. *J. Phys. Cond. Matt.* **4**, 8375 (1992).

[33] W. HEILAND. In H. H. BRONGERSMA AND R. A. VAN SANTEN, editors, "Aspects of Heterogenious Catalysis Studied by Particle Beams", p. 113, New York (1991). NATO ASI Series B265, Plenum Press.

[34] P. HAOCHANG, T. C. M. HORN, AND A. W. KLEYN. *Phys. Rev. Lett.* **57**, 3035 (1986).

[35] P. H. F. REIJNEN, A. RANKEMA, U. VAN SLOOTEN, AND A. W. KLEYN. *Surf. Sci.* **253**, 24 (1991).

[36] H. FRANKE, K. SCHMIDT, AND W. HEILAND. *Surf. Sci.* **269/270**, 219 (1992).

[37] S. SCHUBERT, U. IMKE, W. HEILAND, K. J. SNOWDON, P. H. F. REIJNEN, AND A. W. KLEYN. *Surf. Sci.* **205**, L793 (1988).

[38] K. SCHMIDT, H. FRANKE, T. SCHLATHÖLTER, C. HÖFNER, A. NÄRMANN, AND W. HEILAND. *Surf. Sci.* in press.

[39] D. R. HERSCHBACH. *Adv. Chem. Phys.* **10**, 319 (1966).

[40] J. W. GADZUK AND S. HOLLOWAY. *Physica Scripta* **32**, 41 (1985).

[41] J. W. GADZUK AND J. K. NØRSKOV. *J. Can. Phys.* **81**, 2828 (1984).

[42] H. J. FREUND AND R. P. MESSUER. *Surf. Sci.* **172**, 1 (1986).

[43] C. BACKX, C. P. M. DE GROOT, AND P. BILJOEN. *Surf. Sci.* **104**, 3 (1981).

[44] C. T. CAMPBELL. *Surf. Sci.* **174**, L641 (1986).

[45] W. WURTH, J. STÖHR, P. FECHNER, X. PAN, R. BAUCHSPIESS, Y. BABE, E. HUDEL, G. ROEKER, AND D. MENZEL. *Phys. Rev. Lett.* **65**, 2426 (1990).

[46] A. C. LUNTZ, M. D. WILLIAMS, AND D. S. BETHUNE. *J. Chem. Phys.* **89**, 4381 (1988).

[47] J. K. BROWN, A. C. LUNTZ, AND P. A. SCHULZ. *J. Chem. Phys.* **95**, 3767 (1991).

[48] J. WAMBACH, G. ODÖRFER, H. J. FREUND, H. KUHLENBECK, AND M. NEUMANN. *Surf. Sci.* **209**, 159 (1989).

[49] S. WOHLRAB, D. EHRLICH, J. WAMBACH, H. KUHLENBECK, AND H. J. FREUND. *Surf. Sci.* **220**, 243 (1989).

[50] B. BARTOS, H. J. FREUND, H. KUHLENBECK, M. NEUMANN, H. LINDNER, AND K. MÜLLER. *Surf. Sci.* **179**, 59 (1987).

[51] M. KISKINOVA, G. PIRUNG, AND H. P. BONZEL. *Surf. Sci.* **136**, 285 (1984).

[52] M. SCHLAYEGAN, J. M. CAVALLO, R. E. GLOVER, AND R. L. PARK. *Phys. Rev. Lett.* **53**, 1578 (1984).

[53] K. SCHMIDT, T. SCHLATHÖLTER, A. NÄRMANN, AND W. HEILAND. *Chem. Phys. Lett.* **200**, 465 (1992).

[54] W. HEILAND, U. BEITAT, AND E. TAGLAUER. *Phys. Rev. B* **19**, 1677 (1977).

[55] U. IMKE, K. J. SNOWDON, AND W. HEILAND. *Phys. Rev. B* **34**, 41 and 48 (1986).

[56] U. IMKE, S. SCHUBERT, K. J. SNOWDON, AND W. HEILAND. *Surf. Sci.* **189/190**, 960 (1987).

[57] G. H. DUNN. *Phys. Rev. Lett.* **8**, 62 (1962).

[58] W. TAPPE, A. NIEHOF, K. SCHMIDT, AND W.HEILAND. *Europhys. Lett.* **15**, 405 (1991).

[59] J. HARRIS AND S. ANDERSSON. *Phys. Rev. Lett.* **55**, 1583 (1985).

[60] J. H. RECHTIEN, W. MIX, D. DANAILOV, AND K. J. SNOWDON. *Surf. Sci.* **271**, 501 (1992).

[61] J. H. RECHTIEN, R. HARDER, G. HERRMANN, AND K. J. SNOWDON. *Surf. Sci,* **269/270**, 240 (1992).

[62] J. H. RECHTIEN, R. HARDER, G. HERRMANN, AND K. J. SNOWDON. *Surf. Sci.* **272**, 240 (1992).

[63] A. NESBITT, R. HARDER, G. HERRMANN, A. GOLICHOWSKI, AND K. SNOWDON. *Faraday Discuss.* **96** (1993).

[64] I. S. BITENSKIJ AND E. S. PARILIS. *Nucl. Instr. Meth. B* **2**, 384 (1984).

[65] M. M. JAKAS AND D. E. HARRISON. *Surf. Sci.* **149**, 500 (1985).

[66] P. J. VAN DEN HOEK AND A. W. KLEYN. *J. Chem. Phys.* **91**, 4318 (1989).

[67] P. J. VAN DEN HOEK, T. C. M. HORN, AND A. W. KLEYN. *Surf. Sci.* **198**, L335 (1988).

[68] U. VAN SLOOTEN, D. ANDERSSON, A. W. KLEYN, AND E. A. GISLASON. *Surf. Sci.* **274**, 1 (1992).

[69] J. SCHINS, R. B. VRIJEN, W. J. VAN DER ZANDE, AND J. LOS. *Surf. Sci.* **280**, 145 (1993).

[70] H. MASSEY. "Negative Ions". Cambridge University Press (1976).

Adsorption on Epitaxial Oxide Films as Model Systems for Heterogeneous Catalysis

S. Wohlrab, F. Winkelmann, J. Libuda, M. Bäumer, H. Kuhlenbeck and H.-J. Freund

Department of Physical Chemistry, Ruhr-Universität Bochum
Universitätsstraße 150, D-44780 Bochum

Abstract. We have studied the interaction of Pt with $Al_2O_3(111)/NiAl(110)$ and the adsorption of CO on the platinum covered oxide film for different Pt coverages and deposition temperatures. At 300 K the platinum forms a twodimensional layer which is partially ordered. No order within the metal deposit could be observed for deposition at 100 K. On the Pt deposit prepared at room temperature CO dissociation occurs, in contrast to non dissociative CO adsorption on Pt single crystals. With TDS (**T**hermal **D**esorption **S**pectroscopy) an unusual CO desorption signal is observed at $T \approx 150$ K. Upon annealing at elevated temperature the Pt seems to diffuse into the oxide film or through the oxide film into the NiAl substrate.

1. Introduction

Oxide surfaces are interesting research objects for different reasons. From a more basic point of view oxides represent a class of materials which is quite different from metals. Whereas at least for the transition metals the electronic properties are well known today this is not the case to this extent for oxides. Even for one of the most studied oxides, NiO, there have been controversial discussions about the electronic structure only some years ago [1, 2, 3, 4]. This is connected with the fact that oxides are ionic materials and one has to deal not only with delocalized but also with localized electronic levels so that multi electron effects may be quite important [1, 3, 4, 5, 6, 7].

Apart from this basic interest in oxides there is also a more practical interest. Oxides play an important role in catalysis as support materials or as catalytically active components of catalysts and therefore their interaction with gases and metals has been an object of research for a long time [8, 9]. Most of these studies, however, deal with oxide powders. Due to experimental problems one has begun to study well defined oxide surfaces only a few years ago [10, 11, 12]. One problem is that oxides are usually insulators and therefore they tend to charge which hinders the application of electron spectroscopy. For this reason we study oxide films of limited thickness which do not charge.

Al_2O_3 is a typical support material in catalysis. We have prepared thin Al_2O_3 films by oxidation of NiAl(110). In previous studies we have investigated the electronic and geometric structure of this film [13, 14]. The surface is chemically quite inert with respect to CO adsorption as indicated by desorption temperatures between 38 K and 65 K [15, 16]. In this paper we present results of electron spectroscopic studies for platinum deposits on the Al_2O_3 film. In addition, the adsorption of CO on the Pt deposits has been studied. Platinum interacts strongly with the alumina film, destroying the long range order of the oxide even for coverages far below a monolayer [17]. At least two different types of CO adsorption sites exist on the Pt covered surface. As deduced from TDS, the first type of site is to be assigned to metallic Pt. The nature of the other type of site is not fully understood yet. It leads to a CO desorption temperature of $T \approx 150$ K and is only observed for Pt deposition at room temperature. We tentatively assign it to Pt which has changed its charge state or to oxide sites which are electronically modified by the interaction with Pt. For Pt deposition at $T \approx 300$ K also decomposition of CO is observed.

2. Experimental

The experiments have been performed in three different UHV chambers. All chambers contain equipment for LEED (Low Energy Electron Diffraction), AES (Auger Electron Spectroscopy), TDS, residual gas analysis with a quadrupole mass spectrometer, and crystal cleaning with an ion gun. The sample holders allow for cooling down to $T \approx 90$ K and sample heating by electron impact or heat radiation. One chamber is equipped with an HREELS (High Resolution Electron Energy Loss Spectroscopy) spectrometer which is capable of resolutions down to about 5 meV. Another one contains a XPS (X-ray Photoelectron Spectroscopy) spectrometer. The XPS spectrometer is equipped with a monochromatized X-ray source and an unmonochromatized X-ray tube. The third chamber contains an electron monochromator and a rotable electron analyzer for angular resolved EELS (Electron Energy Loss Spectroscopy) and ARUPS (Angle Resolved Ultraviolet Photoelectron Spectroscopy).

The NiAl(110) samples were cleaned by standard sputtering and heating sequences. After cleaning the oxide film was prepared by admitting 1200 L ($1 L = 10^{-6}$ Torr×sec) of oxygen at elevated temperature ($T \approx 550$ K) with subsequent annealing at $T \approx 1200$ K as described elsewhere [13]. The quality of the resulting thin oxide film was checked by LEED. For the final oxide layer a manifold of sharp LEED spots with low background intensity was observed, indicative of a well ordered oxide structure.

The metal films were produced by a metal evaporation source as described elsewhere [18]. Thickness control was established by a quartz microbalance. For this purpose the evaporation rate of the metal source was determined by evaporating a thick film of platinum. This value was used to calculate the evaporation time needed to prepare a platinum film with a given thickness.

3. Results and Discussion

3.1. Pt/Al$_2$O$_3$(111)/NiAl(110)

In a previous study we have established a structural model for the thin aluminium oxide film on NiAl [13]. According to this model the structure of the oxide film is similar to that of γ-Al$_2$O$_3$(111) with the surface most likely terminated by oxygen ions. The film has a thickness of two Al–O bilayers which corresponds to 5 Å. Due to the large unit cell of the oxide film a multitude of spots is observed in the LEED pattern. The oxide is chemically rather inert with respect to adsorption of CO [15] as alluded to in the introduction, so that adsorption of CO on the pure oxide did not occur in the investigations presented here which have been performed at temperatures T ≥ 100 K.

Evaporation of a small amount of platinum results in a strongly increased background intensity in the LEED pattern and a marked reduction of the intensity of the oxide spots [17]. Upon evaporation of 0.5 ML of platinum the intensity of the specularly reflected electron beam in an HREELS experiment decreases by two orders of magnitude and the elastic beam is spread out over a large angular range [19]. All these data are indicative of a loss of long range order within the oxide film, implying a strong interaction of platinum with the

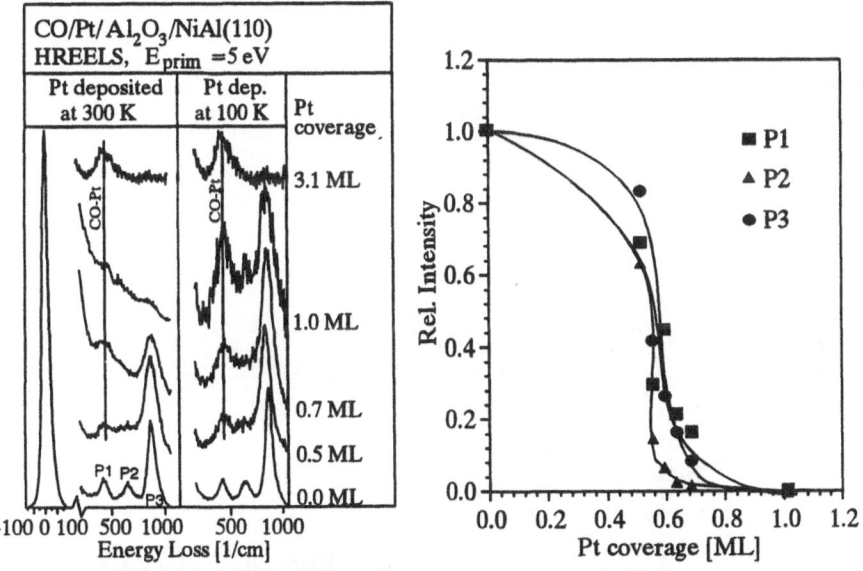

Fig. 1: Left: Series of HREELS spectra for CO/Pt/Al$_2$O$_3$(111)/NiAl(110) as a function of coverage. The two sets of spectra show data taken for two different Pt deposition temperatures. Right: Intensity of the Al$_2$O$_3$ phonons as a function of Pt coverage for a Pt deposition temperature of 300 K.

aluminium oxide film. A hexagonal LEED pattern, indicative of the formation of Pt(111) is observed when a Pt layer with a thickness of several monolayers is prepared at a substrate temperature of $T \approx 500\,K$ [17].

In fig. 1 (left panel) a set of HREELS spectra is shown for CO on a platinum covered Al_2O_3 film as a function of the Pt coverage. The two sets of spectra show data for deposition temperatures of $T \approx 100\,K$ and $T \approx 300\,K$, respectively. Obviously the oxide phonons are damped when Pt is deposited. At a platinum coverage of 0.8 ML the phonons are nearly fully suppressed for deposition at room temperature whereas they are still quite intense when Pt is deposited at $T \approx 100\,K$. This points towards differences in the structures of the Pt layers.

In the right panel of fig. 1 the intensity of the oxide phonons vs. Pt coverage is depicted for platinum deposition at room temperature. As is obvious from these data, the intensity of the phonons strongly decreases in the coverage range between 0.5 and 1 monolayer of platinum. Similar results have been obtained in other studies where metal deposition on dielectric substrates was investigated [20, 21]. The phonon damping may be explained by dielectric theory as to be due to the interaction between the oxide phonons and the plasmons within the metal overlayer [20]. Within this theory the quenching of the phonon intensities at about 1 ML of Pt requires the first Pt layer to form a continuous film or at least a film consisting of small, highly dispersed clusters distributed over the surface so that the free spaces between the clusters are rather small. STM (Scanning Tunneling Microscopy) data of 0.1 ML of Pt on

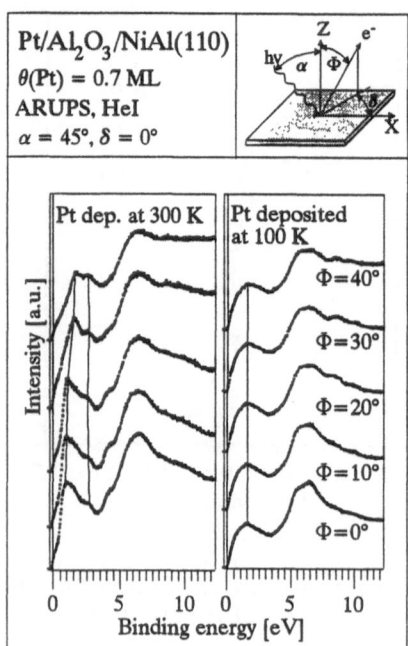

Fig. 2: ARUPS data of 0.7 ML of Pt on $Al_2O_3(111)/NiAl(110)$ taken as a function of the detection angle. Left panel: Pt deposition at $T \approx 300\,K$. Right panel: Pt deposition at $T \approx 100\,K$.

Al$_2$O$_3$(111)/NiAl(110) support this type of growth mode, i.e. they show the formation of small Pt islands with a height of one atomic layer [22]. For Pt deposited at T $\approx$ 100 K the growth mode must be different as indicated by the less pronounced damping of the oxide phonons. This is likely due to a hindered diffusion of Pt atoms at this temperature.

The differences between the Pt growth modes for deposition at T $\approx$ 100 K and T $\approx$ 300 K are also documented by ARUPS spectra. Sets of data taken at these deposition temperatures are shown in fig. 2. A Pt valence emission is found near to the Fermi level where Al$_2$O$_3$ has no electronic states. As may be seen from these data there is dispersion of the electronic levels of Pt as a function of the detection angle when Pt is deposited at T $\approx$ 300 K. This means that the Pt layer must be at least partially ordered for deposition at room temperature. No dispersion of the Pt bands is observed for Pt deposition at T $\approx$ 100 K so that in this case the Pt deposit must be less ordered.

In fig. 3 (left panel) two sets of HREELS spectra are shown. This figure compares spectra taken at deposition temperatures of T $\approx$ 100 K and T $\approx$ 300 K,

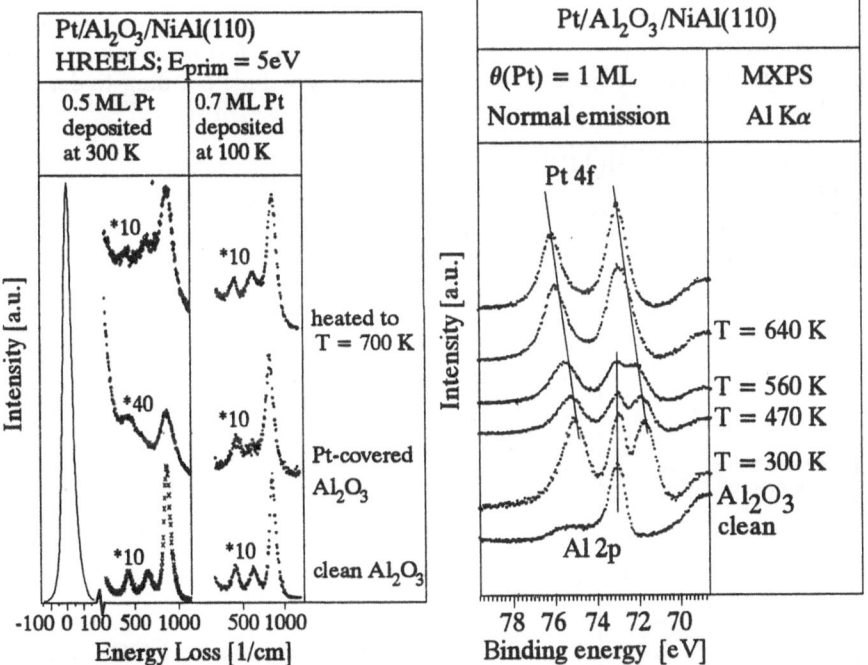

Fig. 3: Left: Comparison of HREELS spectra of Pt/Al$_2$O$_3$(111)/NiAl(110) taken at Pt deposition temperature with spectra taken after annealing at T $\approx$ 700 K. Right: XPS data of 1 ML of Pt on Al$_2$O$_3$(111)/NiAl(110) taken after annealing at different temperatures.

respectively, with spectra recorded after annealing at $T \approx 700$ K. For both deposition temperatures the intensities of the substrate phonons recover after the annealing process which means that something must have happened to the Pt layer.

The XPS data shown in the right panel of fig. 3 display the Pt 4f levels of 1 ML of Pt after annealing at different temperatures. These levels do not vanish when the layer is annealed at elevated temperatures so that the platinum can not be desorbed. The binding energy shifts may indicate that the Pt atoms change their charge states upon annealing. Since ISS (Ion Scattering Spectroscopy) data (not shown here) indicate that the amount of Pt at the surface decreases upon annealing we suppose that the platinum migrates into the oxide or through the oxide into the NiAl substrate.

3.2. CO/Pt/Al$_2$O$_3$(111)/NiAl(110)

In fig. 4 a set of TDS spectra for CO/Pt/Al$_2$O$_3$(111)/NiAl(110) is compared with spectra for two different Pt single crystal surfaces [23, 24]. Most of the CO desorbs in the temperature range between $300\,K > T > 550\,K$ from the Pt covered oxide film. Desorption peaks in this temperature range are to be attributed to CO which bonds to metallic platinum. The fine structure of these peaks is due to different bonding sites which might include bonding to defects,

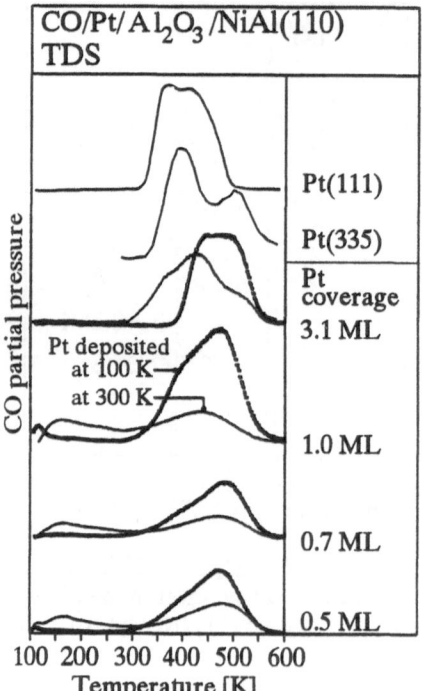

Fig. 4: Comparison of TDS spectra of CO on Pt/Al$_2$O$_3$(111)/NiAl(110) with spectra of CO on Pt single crystal surfaces. Spectra for two different deposition temperatures are shown for each Pt deposition on the oxide surface.

steps and regular Pt(111) sites [19, 23, 24]. For Pt deposited at $T \approx 300\,K$ the structures between $300\,K > T > 550\,K$ are weaker, indicating the presence of less available metallic Pt sites. The small desorption peak at $T \approx 110\,K$ in the spectrum for a Pt coverage of $1\,ML$ is to be attributed to CO_2 desorption.

For a Pt deposition temperature of $T \approx 300\,K$ an additional CO desorption signal at $T \approx 150\,K$ is observed. This result may be compared with CO desorption spectra from transition metal oxides. It is often observed that CO desorbs from transition metal oxides in the temperature range between $100\,K > T > 200\,K$ [25, 26]. Therefore, a possible explanation for the signal at $T \approx 150\,K$ might be that part of CO bonds to oxide sites which are modified by the interaction with Pt or to Pt atoms which have changed their charge state. Further information may be deduced from the data shown in fig. 5. This figure displays series of CO desorption spectra as a function of Pt coverage. In each case the first desorption spectrum from the respective layer is compared with the second one. An obvious difference between the first and the second TDS spectra is that the high temperature CO desorption signals in the second

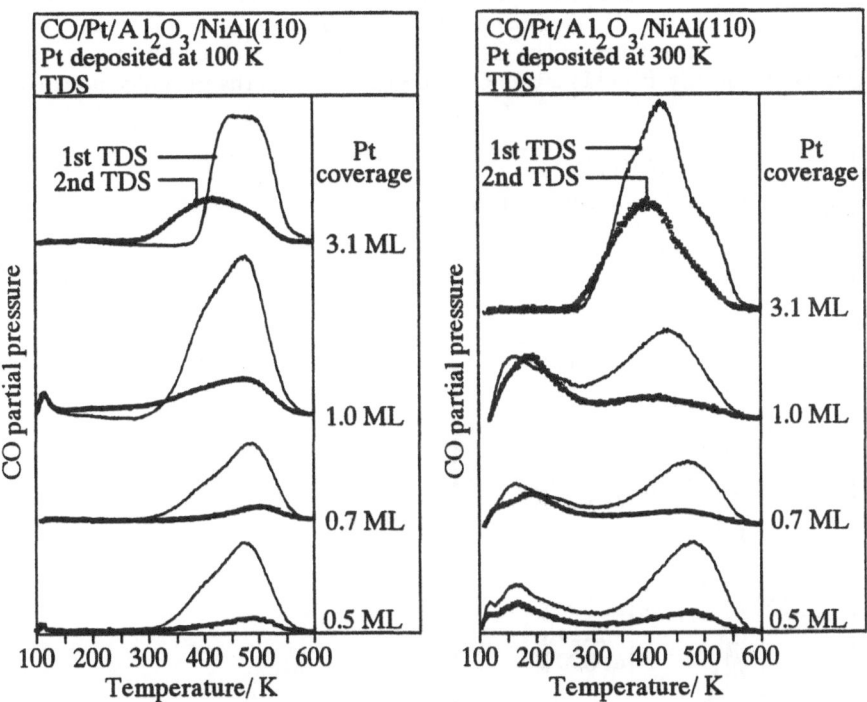

Fig. 5: Series of TDS spectra of CO on $Pt/Al_2O_3(111)/NiAl(110)$ for Pt deposition temperatures of $T \approx 100\,K$ (left panel) and $T \approx 300\,K$ (right panel). In each case the first TDS spectrum from the respective film is compared with the second one.

spectra are weaker than in the first ones. This may be explained as to be due to migration of Pt into the oxide layer or through the oxide layer into the NiAl alloy during the heating process needed to record the first spectrum.

The left panel of fig. 5 reveals that the intensities of the desorption peaks at $T \approx 150$ K are not much different for the first and the second desorption spectra. This indicates that the number of adsorption sites leading to this peak is not considerably influenced by the annealing process. In particular, these sites can not be generated by heating a Pt deposit which has been prepared at $T \approx 100$ K to $T \approx 700$ K as concluded from the data shown in the left panel of fig. 5.

When CO is adsorbed on Pt which was deposited at $T \approx 300$ K, decomposition of part of the carbon monoxide occurs. This is illustrated in fig. 6. The spectrum at the bottom displays the C1s level of CO adsorbed at $T \approx 120$ K on 0.8 ML Pt/Al$_2$O$_3$(111)/NiAl(110) which was prepared at room temperature. The second spectrum displays the C1s level after annealing at $T \approx 800$ K. Since CO is desorbed at this temperature, the remaining signal must be due to the C1s level of carbon generated by CO dissociation. A comparison with the spectrum at the bottom shows that most of the CO is already dissociated at room temperature. The situation is different when Pt is deposited at $T \approx 120$ K. In this case most of the CO remains in its molecular form as indicated by a comparison with the spectrum of the C1s level of CO on a thick Pt film (top) which offers preferentially Pt(111) sites. On Pt(111) CO adsorbs in molecular form.

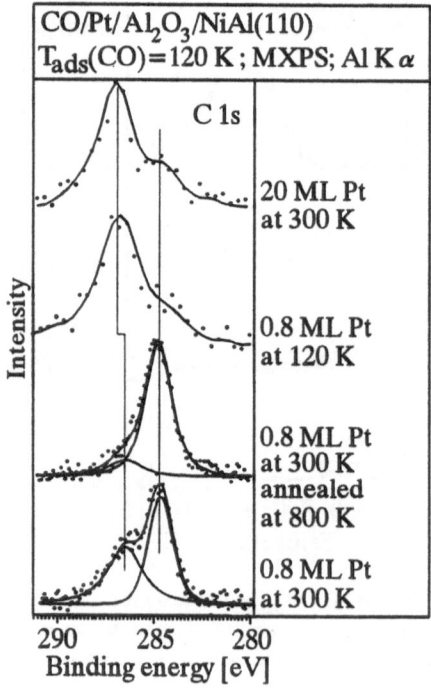

Fig. 6: C1s-XPS spectra of CO on Pt on Al$_2$O$_3$(111)/NiAl(110) taken under different conditions.

4. Summary

In this paper we have shown that platinum on γ-Al$_2$O$_3$(111)/NiAl(110) represents a system with interesting properties. Pt interacts strongly with the oxide film, thereby destroying the long range order within the film. The Pt/Al$_2$O$_3$ system exhibits an unusual adsorption behaviour with respect to CO when Pt is deposited at room temperature. Besides bonding of CO to metallic Pt, also a bonding state, which leads to desorption at T $\approx$ 150 K, is observed. Also, dissociation of CO takes place, in contrast to nondissociative CO adsorption on bulk Pt. These properties and the good applicability of electron spectroscopy at all temperatures make the Pt/Al$_2$O$_3$/NiAl(110) a promising model system for a catalyst. Further studies, including the interaction of NO and CO on this system will be performed in near future.

Acknowledgements

This work has been supported by the Deutsche Forschungsgemeinschaft (DFG), the Ministerium für Wissenschaft und Forschung des Landes Nordrhein-Westfalen and the Fonds der Chemischen Industrie.

References

[1] H. Kuhlenbeck, G. Odörfer, R. Jaeger, G. Illing, M. Menges, Th. Mull, H.-J. Freund, M. Pöhlchen, V. Staemmler, S. Witzel, C. Scharfschwerdt, K. Wennemann, T. Liedtke, and M. Neumann, Phys. Rev. **B43**, 1969 (1991).

[2] Z.-X. Shen, C. K. Shih, O. Jepsen, W. E. Spicer, I. Lindau, and J. W. Allen, Phys. Rev. Lett. **64**, 2442 (1990).

[3] S. Hüfner, P. Steiner, and I. Sander, Sol. Stat. Commun. **72**, 359 (1989).

[4] G. van der Laan, J. Zaanen, G. A. Sawatzky, R. Kernatak, and J.-M. Esteva, Phys. Rev. **B33**, 4253 (1986).

[5] A. Freitag, V. Staemmler, D. Cappus, C. A. Ventrice, K. Al-Shamery, H. Kuhlenbeck, and H.-J. Freund, Chem. Phys. Lett. **210**, 10 (1993).

[6] G. Pacchioni and P. S. Bagus, in *Adsorption on Ordered Surfaces of Ionic Solids and Thin Films*, editors H.-J. Freund and E. Umbach, volume 33 of *Springer Series in Surface Sciences*, page 180, Berlin (1993), Springer Verlag.

[7] G. J. M. Janssen and W. C. Nieuwpoort, Phys. Rev. **B38**, 3449 (1988).

[8] H. H. Kung, *Transition Metal Oxides: Surface Chemistry and Catalysis*, Elsevier, Amsterdam (1989).

[9] A. A. Davydov, *Infrared Spectroscopy of Adsorbed Species on the Surfaces of Transition Metal Oxides*, Wiley&Sons, Chichester, United Kingdom (1990).

[10] G. Heiland and H. Lüth, in *The Chemical Physics of Solid Surfaces and Heterogeneous Catalysis*, editors D. A. King and D. P. Woodruff, volume 3, Elsevier, Amsterdam (1982).

[11] V. E. Henrich, Rep. Prog. Phys. **48**, 1481 (1985).

[12] H.-J. Freund and E. Umbach (Editors), *Adsorption on Ordered Surfaces of Ionic Solids and Thin Films*, volume 33 of *Springer Series in Surface Sciences*, Springer Verlag, Berlin (1993).

[13] R. M. Jaeger, H. Kuhlenbeck, H.-J. Freund, M. Wuttig, W. Hoffmann, R. Franchy, and H. Ibach, Surf. Sci. **259**, 235 (1991).

[14] F. Winkelmann, S. Wohlrab, J. Libuda, M. Bäumer, D. Cappus, M. Menges, K. Al-Shamery, H. Kuhlenbeck, and H.-J. Freund, Surf. Sci. , in press.

[15] R. M. Jaeger, H. Kuhlenbeck, and H.-J. Freund, Chem. Phys. Lett. **203**, 41 (1993).

[16] R. M. Jaeger, J. Libuda, M. Bäumer, K. Homann, H. Kuhlenbeck, and H.-J. Freund, J. Electr. Rel. Phen. **64/65**, 217 (1993).

[17] J. Libuda, M. Bäumer, and H.-J. Freund, J. Vac. Sci. Technol. (1994), in press.

[18] F. Winkelmann, PhD thesis, Ruhr-Universität Bochum, Bochum, FRG, in preparation.

[19] S. Wohlrab, PhD thesis, Ruhr-Universität Bochum, Bochum,FRG (1993).

[20] L. H. Dubois, G. P. Schwartz, R. E. Camley, and D. L. Mills, Phys. Rev. **B29**, 3208 (1984).

[21] C. A. Ventrice, Jr, D. Ehrlich, E. L. Garfunkel, B. Dillmann, D. Heskett, and H.-J. Freund, Phys. Rev. **B46**, 12892 (1992).

[22] Th. Bertrams, F. Winkelmann, H.-J. Freund, and H. Neddermeyer, to be published.

[23] H. Steininger, S. Lehwald, and H. Ibach, Surf. Sci. **123**, 264 (1982).

[24] J. S. Luo, R. G. Tobin, D. K. Lambert, G. B. Fisher, and C. L. DiMaggio, Surf. Sci. **274**, 53 (1992).

[25] C. Xu, B. Dillmann, H. Kuhlenbeck, and H. J. Freund, Phys. Rev. Lett. **67**, 3551 (1991).

[26] D. Cappus, H. Kuhlenbeck, and H.-J. Freund, to be published.

Kinetic Oscillations on Single Crystal Surfaces

R. Imbihl

Institut für Physikalische Chemie, Universität Hannover
Callinstr. 3-3A, D-30167 Hannover

Abstract. Kinetic oscillations in heterogeneously catalyzed reactions have been studied for over a decade using single crystals as catalysts. This paper presents a few selected examples in order to illustrate some of the most interesting aspects of these studies: the formulation of oscillation mechanism based on experimental facts, the mathematical analysis of kinetic models using bifurcation theory and spatiotemporal pattern formation on the catalyst surface.

1. Introduction

In reaction systems which are far from thermodynamic equilibrium new phenomena may occur which are not permitted under equilibrium conditions (1). Instead of approaching its equilibrium state monotonically, a chemical reaction may exhibit kinetic oscillations, spontaneous spatial pattern formation or the system may even display chaotic dynamics. The prototype of an oscillatory chemical reaction has for decades been the famous Belousov-Zhabotinskii (BZ) reaction. However, since the discovery of kinetic oscillations in catalytic CO oxidation by Wicke et al. (2) heterogeneously catalyzed reactions have also been a field of very active research. Most of the progress which has been achieved in this field can be attributed to single crystal studies performed under isothermal low pressure conditions ($p < 10^{-3}$ mbar) (3-6).

Using the analytical tools of surface science it has been possible to establish experimentally verified oscillation mechanism for practically all reaction systems which have been investigated so far. The oscillatory catalytic reaction studied on single crystal surfaces comprise essentially two classes of reactions: catalytic CO oxidation on Pt and Pd surfaces and catalytic NO reduction with various reducing agents (CO, H_2, or NH_3) on Pt(100) and Rh(110) surfaces. Having established an oscillatory mechanism one can formulate mathematical models resulting in a set of coupled differential equations (DE's). With these DE's one can either simulate the reaction system by numerical integration or one can analyze the stability of possible solution to these DE's by conducting a so-called bifurcation analysis.

The properties of an individual oscillator, however, do not suffice to characterize an oscillating surface reaction. In order to achieve macroscopic variations of the reaction rate it is necessary to synchronize a large number of local oscillators either by diffusional coupling or by coupling via the gas-phase. Complete synchronization, i.e., a spatially homogeneously oscillating surface, is only rarely seen, but one typically observes spatiotemporal pattern formation (6,7). These patterns can exhibit quite strongly varying degrees of complexity ranging from simple reaction fronts and rotating spirals to soliton-like behavior and spatiotemporal turbulence. This aspect of an oscillating surface reactions is currently in the focus of intense research and it is in this field where most of the interesting discoveries have been made.

An attempt to give a complete overview of the field of oscillatory surface reactions would be quite unreasonable in view of the large body of material that has been gathered in the last ten years. For this reason only some selected examples are presented here illustrating the various aspects of oscillatory surface reactions mentioned above.

2. Oscillation mechanisms

2.1 CO oxidation on Pt surfaces

Kinetic oscillations in catalytic CO oxidation have been studied under low-pressure conditions ($p < 10^{-3}$ mbar) on the low-index planes Pt(100) and Pt(110) and on high-index planes of the (001)-zone including the Pt(210) surface (5-10). A typical example showing the recording of kinetic oscillations during catalytic CO oxidation on a Pt(110) surface is displayed in Fig. 1. One notes that the work function change $\Delta\varphi$ follows exactly the variation of the reaction rate r_{CO_2} (being proportional to p_{CO}). This relation is essentially a consequence of the fact that oxygen adsorption is rate limiting under the conditions where rate oscillations occur and since the dipole moment of the oxygen adsorbate complex is much higher than that of adsorbed CO, the reaction rate will be proportional to $\Delta\varphi$.

The mechanism of catalytic CO oxidation follows a simple scheme known as Langmuir-Hinshelwood (LH) mechanism which incorporates the following steps:

$$CO \quad + \quad * \quad \rightleftarrows \quad CO_{ad}$$

$$O_2 \quad + \quad 2* \quad \rightarrow \quad 2O_{ad}$$

$$O_{ad} \quad + \quad CO_{ad} \quad \rightarrow \quad CO_2 \quad + \quad 2*$$

(* denotes a vacant adsorption site).

The LH scheme predicts multistability, i.e., a hysteresis in the parameter dependence of the reaction rate which has been found experimentally (4). The LH mechanism alone, however, does not suffice to create oscillatory behavior. In order to produce rate oscillations one needs an additional mechanism which periodically modulates the catalytic activity of the surface. On Pt surfaces this additional feedback step is provided by the mechanism of an adsorbate-induced surface-phase transition (SPT). How this mechanism can cause rate oscillations is demonstrated by the 1x1 $\rightleftarrows$ 1x2 SPT of Pt(110) depicted in Fig. 2 (5,6).

The clean Pt(110) surface exhibits a 1x2 reconstruction of the "missing row" type which upon CO adsorption can be reversibly converted into the non-reconstructed 1x1 surface. Since the sticking

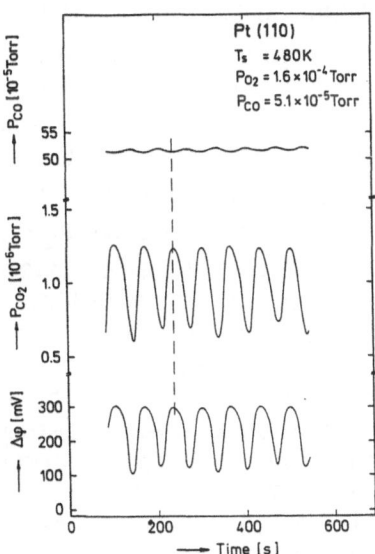

Fig. 1: Typical example of kinetic oscillations in catalytic CO oxidation on a Pt(110) surface.

CO covered   **Pt (110)**   clean

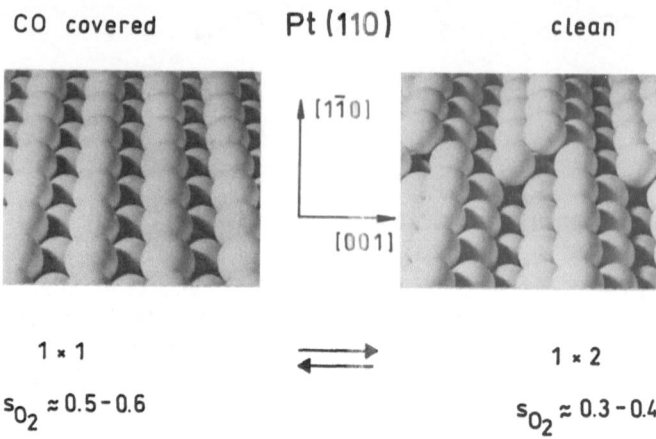

1 × 1   1 × 2

$s_{O_2} \approx 0.5 - 0.6$ $s_{O_2} \approx 0.3 - 0.4$

Fig. 2: Ball model illustrating the CO-induced $1 \times 1 \rightleftarrows 1 \times 2$ SPT of Pt(110). The different s_{O_2}'s of the two phases are responsible for rate oscillations during catalytic CO oxidation. The model also demonstrates how the mass transport of Pt atoms that is involved in the SPT creates an atomic step on the surface.

coefficient of oxygen s_{O_2} is higher on the 1×1 phase (s_{O_2} estimated ≈ 0.6) than on the 1×2 surface ($s_{O_2} \approx 0.3 - 0.4$), a simple oscillatory cycle can be constructed. Starting with the reactive CO covered 1×1 phase, a high O_2 adsorption rate and hence high reaction rate will cause the CO coverage to decrease. Below a critical CO coverage $\theta_{CO,crit} \approx 0.2$, the surface reconstructs. On the less active 1×2 surface, CO adsorption will dominate against O_2 adsorption and cause the CO coverage to increase above $\theta_{CO,crit}$. Thus, the initial configuration of a CO covered 1×1 substrate is reestablished.

An analogous mechanism can be constructed for Pt(100) where CO adsorption causes a lifting of the quasi-hexagonal ("hex") reconstruction of the surface. The proposed mechanism could be verified for both orientations, Pt(100) and Pt(110), by in-situ LEED experiments (8,9). These measurements demonstrated that the oscillations in the reaction rate are in fact accompanied by periodic intensity variations of the extra beams belonging to the reconstructed surface. For high-index planes such as Pt(210) the proposed mechanistic scheme has to be modified slightly. The clean Pt(210) surface does not reconstruct, but under reaction conditions the Pt(210) surface facets into (110) and (310) orientations (10). In this way low-index planes are created upon which the same mechanism can then proceed which has already been verified for Pt(100) and Pt(110).

2.2 CO oxidation on Pd surfaces

Kinetic oscillations in catalytic CO oxidation on Pd catalysts were first observed with polycrystalline material under high-pressure conditions (p > 1 mbar) (11). Under isothermal low-pressure conditions rate oscillations in CO oxidation have been studied on a Pd(110) single crystal surface (12-16). The oscillations on Pd(110) exhibit several characteristic features: they require a minimum pressure of 10^{-3} mbar, the ratio p_{O_2}/p_{CO} has to be large – of the order of 10^2 to 10^3, and they only occur at relatively low temperature between ≈300 K and 450 K.

Since clean Pd surfaces do not reconstruct, it is evident that the concept of rate oscillations driven by a reversible SPT which had been verified for Pt surfaces cannot be applied directly to Pd surfaces. Pd surfaces, on the other hand, have a strong tendency to incorporate oxygen atoms leading to subsurface-oxygen formation. This property of Pd is presumably responsible for the reversal of the usual clockwise (cw) hysteresis one observes in the reaction upon variation of p_{CO} (p_{O_2} and T being kept fixed) into a counterclockwise (ccw) hysteresis at high p_{O_2} (13). The origin of the ccw hysteresis can be explained by a filling of the subsurface oxen reservoir taking place at low p_{CO} when the surface is oxygen covered followed by a depletion of subsurface oxygen when at high p_{CO} oxygen atoms diffuse back to the CO covered surface and react to form CO_2.

Since the cw hysteresis upon increasing p_{O_2} turns continuously into a ccw hysteresis one obtains a characteristic cross-shaped phase diagram as one plots the turning points of the hysteresis in a p_{O_2}/p_{CO}-diagram (14). This diagram is reproduced in fig. 3a. Kinetic oscillations whose existence range is marked by the shaded area in the plot do only occur under conditions of the ccw hysteresis, i.e., they are found above the critical oxygen pressure, $p_{crit} \approx 10^{-3}$ mbar, at which the crossing of the two hysteresis branches takes place.

Apparently the formation of subsurface oxygen, O_{sub}, is essential for the rate oscillations on Pd(110). One can thus formulate an oscillation mechanism by simply adding a fourth equation

$$O_{ad} \quad \underset{\leftarrow}{\overset{\rightarrow}{}} \quad O_{sub}$$

to the equations of the LH mechanism. During oscillations the catalytic activity of the Pd(110) surface is periodically modulated by a filling and depletion of the subsurface oxygen reservoir since O_{sub} leads to a strong decrease of s_{O_2}.

Putting the reaction scheme of the LH mechanism plus the reversible formation of subsurface oxygen into kinetic equations, one obtains a set of three coupled DE's whose numerical integrations yields rate oscillations similar to the ones observed in the experiment (15,16). Instead of showing the results of these simulations only the resulting bifurcation diagram (also called phase diagram or stability diagram) is reproduced in fig. 3b. This simulated diagram is to be compared with the experimental diagram in fig. 3a.

The term bifurcation simply denotes a qualitative change in the character of the solution as a parameter is varied such as, e.g., a transition from a stable steady state to oscillatory behavior. Bifurcation analysis examines the stability of steady states and of periodic solutions of a system of differential equations and thus allows one to compare the predictions of a mathematical model with experimental data. Similar to first and second order phase transitions various types of bifurcations can be classified and a bifurcation diagram like the one displayed in fig. 3b indicates the various kinds of transitions one encounters upon a parameter change. The different types of bifurcations that appear in the diagram of Fig. 3b are relatively easy to understand and can be looked up in textbooks (17).

2.3 NO reduction on Pt(100)

The reduction of NO on noble metal catalysts is of considerable technological importance due to the key role NO_x emissions play in air pollution. Besides the technological importance, these reactions also exhibit an interesting dynamical behavior as one finds multistability, kinetic oscillations and spatial pattern formation (6,18,19). Kinetic oscillations in the NO+CO reaction were first studied on polycrystalline Pt catalysts at $p \geq 10^{-4}$ mbar before they were investigated intensively on a Pt(100) surface. On the same orientation one also observes kinetic oscillations if one replaces CO by H_2 or NH_3 as a reducing agent (6). All three oscillatory systems display a very similar dynamical behavior whose origin presumably lies in the autocatalytic behavior of NO dissociation. In order to dissociate NO, a vacant site ($*$) is required:

$$NO + * \rightarrow N_{ad} + O_{ad}$$

If more vacant sites are produced in the subsequent product forming steps than are consumed in the above step, an autocatalytic increase in the number of vacant sites takes place:

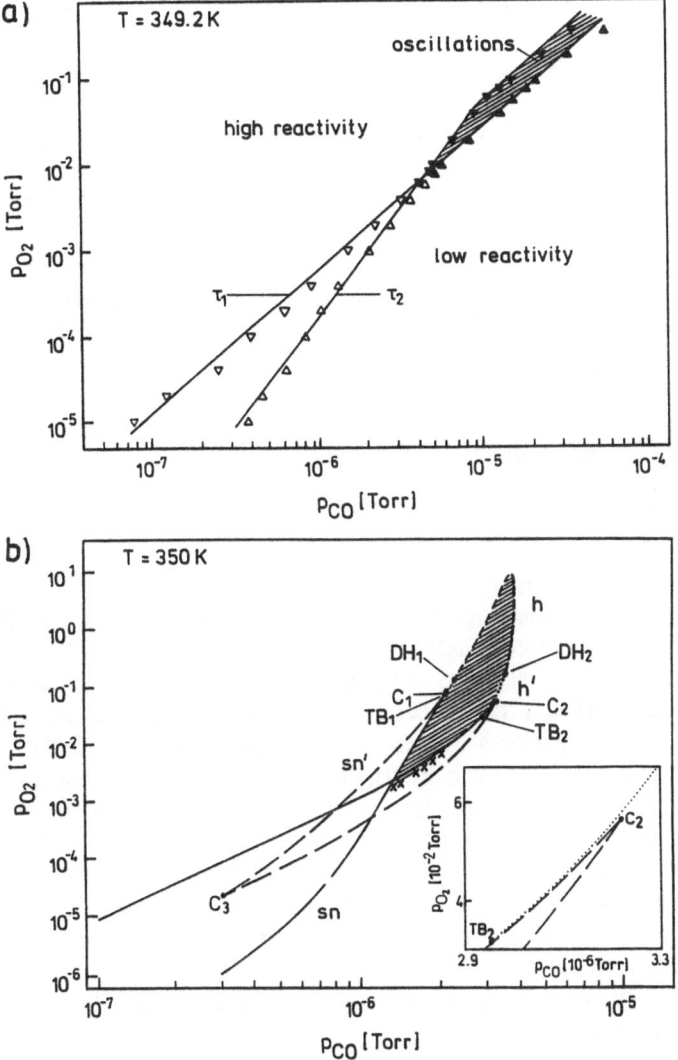

Fig. 3: Cross-shaped bifurcation diagram for catalytic CO oxidation on Pd(110) showing the various regions of monostability, bistability and oscillatory behaviour in p_{CO}, p_{O_2} parameter space at T=350 K.

(a) Experimental bifurcation diagram for catalytic CO oxidation on Pd(110) at T=349.2 K τ_1 and τ_2 represent the transition points of the rate hysteresis. From (14).

(b) Calculated bifurcation diagram in p_{O_2}, p_{CO} parameter space.

The shaded area represents the existence range for oscillaions. The capital letters denote codimension-2 bifurcation points, the small letters codimension-1 bifurcation lines: h = Hopf bifurction, sn = saddle node bifurcation. From (16).

$$O_{ad} \quad + \quad CO_{ad} \rightarrow \quad CO_2 \quad + \quad 2*$$

$$N_{ad} \quad \rightarrow \quad \tfrac{1}{2} N_2 + \quad *$$

This mechanism explains the occurrence of a so-called "surface explosion", i.e., the formation of extremely narrow product peaks (FWHM ≈ 2 - 3 K) if a layer of coadsorbed CO and NO is heated up in temperature programmed reaction experiments. As was demonstrated by mathematical modeling, the same autocatalytic mechanism is also the principal driving force for the rate oscillations in the NO+CO reaction while the 1x1⇄ hex SPT of Pt(100) only plays a minor role (18).

3. Spatiotemporal pattern formation

3.1 General background

An oscillating catalyst surface may be envisioned as being composed of individual oscillators which need to be coupled in order to achieve macroscopic rate variations. Under isothermal low-pressure conditions spatial coupling is provided by two basic mechanisms, surface diffusion via a mobile adsorbate and via partial pressure changes in the gas phase (5-8). The latter simply arises due to mass balance in the reaction since the oscillations in the product rate have to be accompanied by corresponding variations in the educt partial pressures. These depend on experimental parameters like pumping rate and chamber volume, but typically amplitudes of the order of ≈ 1 % of the educt partial pressures are observed as demonstrated by the example displayed in fig. 1.

The two coupling modes compete in their influence as one can see by constructing two limiting cases. If gas-phase coupling were to dominate, a homogeneously oscillating surface would result in the simplest case since the partial pressure variations in the gas phase affect all parts of the surface in the same way and practically without any delay (τ < 1 ms). On the other hand, the coupling of a surface reaction via diffusion of a mobile adsorbate gives rise to propagating reaction fronts. These reaction fronts can create spatial patterns with varying degrees of complexity such as target patterns, spiral waves or even give rise to turbulent behavior.

The principle mechanism which gives rise to a propagating reaction front consists of an autocatalytic reaction according to the scheme

$$A \quad + \quad mX \quad \rightarrow \quad n\,X \quad \text{with} \ \ n>m$$

which is coupled to the diffusion of the autocatalytic component X. These two prerequisites suffice to create a propagating reaction zone, and, as a very simple example, one may consider the spreading of a grass fire. It can be shown quite generally that the velocity c_f of such a wave front can be written in the form:

$$c_f \quad \sim \quad \sqrt{D_X \, k_r}$$

where D_X denotes the diffusion constant of the autocatalytic component and k_r summarizes the effect of the reaction.

Aside from the chemical wave pattern generated by the mechanism described above, a second type of spatial pattern exists in reaction-diffusion systems – the so-called Turing patterns denoted after A. M. Turing who first discussed them in a theoretical paper in 1952 (20). In contrast to chemical wave patterns, these patterns are stationary but they exhibit a periodic variation of the concentration profile along the spatial coordinates. Turing patterns are rarely observed in oscillatory chemical reaction systems, but the faceting of Pt(110) in catalytic CO oxidation represents one of the very few examples (in fact the first one) in which a Turing structure has been identified in experiment.

3.2 Chemical wave patterns

Chemical wave patterns, i.e., propagating reaction fronts of various forms, have been observed in practically all oscillatory reactions investigated so far on single crystal surfaces. With the newly developed photoemission electron microscope (PEEM) an ideal tool is available to image the laterally varying adsorbate concentration in surface reactions with high spatial (=0.1 - 1 μm) and temporal resolution (20 ms) (6,7). This method is based upon the principle that the yield of photoelectrons depends sensitively on the work function if one illuminates the sample with photons whose energy is just above the threshold for excitation of photoelectrons.

Probably the richest variety of spatial patterns so far has been found in catalytic CO oxidation on Pt(110) (7). At lower temperature (T ≤ 500 K) spiral and target patterns are observed, but at elevated temperatures (500 K ≤ T ≤ 550 K) one finds new and interesting phenomena. Solitary pulses ("soliton-like" behavior) and standing wave patterns were detected. A large variety of different wave patterns has also been found in the NO reducing reactions, i.e. in the NO+CO reaction and in the NO+NH$_3$ reaction on Pt(100) (6). Here just one particular example shall be discussed which are the unusual chemical wave patterns that are observed in the NO+H$_2$ reaction on Rh(110) (21).

3.3 Turing patterns

If one uses a ball model of the Pt(110)-1x1 surface as displayed in Fig. 2 and tries to rearrange the atoms such that a 1x2 structure is produced, one realizes that this cannot be accomplished without creating an atomic step. The mass transport of Pt atoms that is associated with the 1x1⇄1x2 SPT therefore necessarily involves a certain roughening of the surface. By choosing suitable reaction conditions this roughening may even turn into a real faceting of the surface, giving rise to a splitting of the integral order LEED beams (9,22). As was demonstrated by an investigation with a high-resolution LEED instrument, the facets on Pt(110) form a regular pattern in which facets of uniform orientation and size are arranged in the shape of a symmetric sawtooth characterized by a lateral periodicity of ≈200 Å.

Since the periodic facet pattern only forms under reaction conditions, it has been classified as a dissipative structure of the Turing type. This interpretation could be confirmed in a Monte Carlo simulation, the steps of which were based solely on the LH mechanism of catalytic CO oxidation and on experimentally known properties of the 1x1⇄1x2 SPT of Pt(110) (23). As demonstrated by the surface profile displayed in Fig. 5, the same sawtooth-like profile develops under reaction conditions in the simulation as was observed experimentally. The essential mechanism which gives rise to the formation of a Turing pattern has been shown to be provided by the coupling between kinetic instabilities and the mass transport of Pt atoms via the 1x1⇄1x2 SPT.

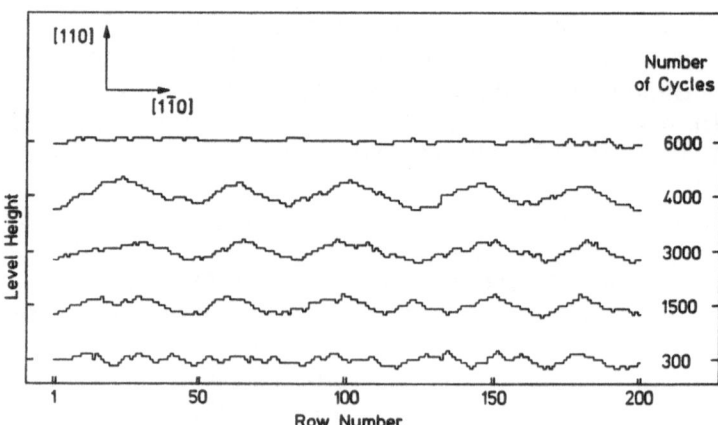

Fig. 5: Monte Carlo simulation showing the development of a regular facet structure during catalytic CO oxidation on a Pt(110) surface followed by a restoration of the flat surface after stopping the gas flow at 4000 cycles. From (23).

One of the principally different properties of surface reactions as compared to spatial pattern formation in fluid phase is that of a diffusional anisotropy being a consequence of the fixed lattice geometry on which the diffusion of the adparticles takes. From the geometry of the unreconstructed Rh(110) surface one expects diffusion to be fast along the (110)-oriented troughs and this is what is seen in experiment as demonstrated by the elliptically shaped target pattern in Fig. 4a. Upon a change of temperature, however, the shape of the target pattern changes drastically and one obtains the square-shaped chemical waves displayed in fig. 4b. The unusual shape of the pattern is a consequence of the large structural variability of the Rh(110) surface. This surface exhibits a variety of adsorbate-induced reconstructions leading to an effective anisotropy of the front velocity along two main crystallographic directions.

(a)   (b)

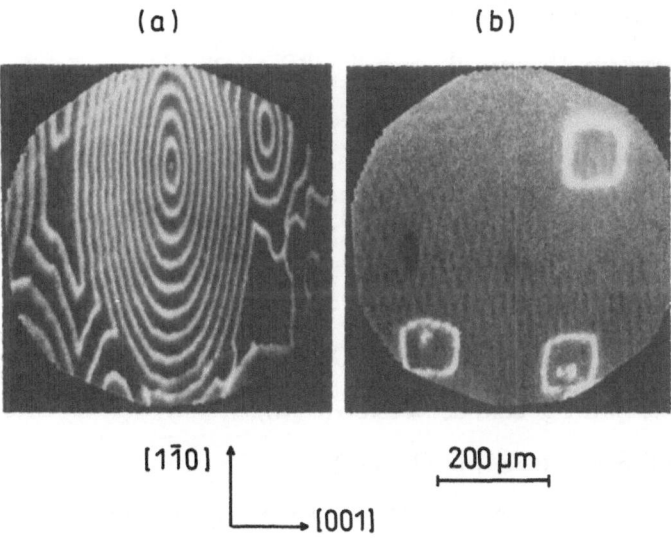

$[1\bar{1}0]$ 200 µm

[001]

Fig. 4: PEEM images showing different geometries of target patterns in the $NO+H_2$ reaction on Rh(110). From (21).
(a) Elliptical target pattern. Experimental conditions:
    T=540 K, $p_{NO}=1.6x10^{-6}$ mbar, $p_{H_2}=1.8x10^{-5}$ mbar.
(b) Nucleation of a square-shaped target pattern. Experimental conditions:
    T=595 K, $p_{NO}=1.6x10^{-6}$ mbar, $p_{H_2}=5x10^{-6}$ mbar.

4. Concluding remarks

It has been demonstrated that single crystal studies represent a suitable tool for investigating the mechanism leading to rate oscillations and spatiotemporal pattern formation. The systems are conceptually simple enough to formulate tractable mathematical models. Since practically all phenomena of non-linear dynamics can be found in oscillatory surface reactions, one can expect that these studies will contribute strongly to further progress in the field of non-linear dynamics.

References

(1)   G. Nicolis and I. Prigogine, *Self-Organization in Nonequilibrium Systems,* Wiley, New York (1977)

(2)   H. Beusch, P. Fieguth and E. Wicke, Chem. Ing. Techn. **44**, 445 (1972).

(3)   The high-pressure studies (p>1 Torr) of oscillatory surface reactions are reviewed in Ref. 4, while the single crystal investigations are referenced in the two review papers of Refs. 5 and 6.

(4)   F. Schüth, B. E. Henry and L. D. Schmidt, Adv. Catal. **39**, 51 (1993).

(5)   G. Ertl, Adv. Catal. **37**, 213 (1990).

(6)   R. Imbihl, Prog. Surf. Sci. **44** 185 (1993).

(7)   G. Ertl, Science **254**, 1750 (1991).

(8)   R. Imbihl, M. P. Cox and G. Ertl, J. Chem. Phys. 84, 3519 (1986).

[9]   S. Ladas, R. Imbihl and G. Ertl, Surf. Sci. **198**, 42 (1988).

(10) M. Sander, R. Imbihl and G. Ertl, J. Chem. Phys. **95**, 6162 (1991).

(11) J. E. Turner, B. C. Sales and M. B. Maple, Surf. Sci. **109**, 591 (1981).

(12) M. Ehsasi, C. Seidel, H. Ruppender, W. Drachsel, J. H. Block and K. Christmann, Surf. Sci. **210**, L198 (1989).

(13) S. Ladas, R. Imbihl and G. Ertl, Surf. Sci. **219**, 88 (1988).

(14) M. Ehsasi, M. Berdau, T. Rebitzki, K.-P. Charlé, K. Christmann and J. H. Block, J. Chem. Phys. **98**, 9177 (1993).

(15) M. R. Bassett and R. Imbihl, J. Chem. Phys. **93**, 9177 (1993).

(16) N. Hartmann, K. Krischer and R. Imbihl, submitted to J. Chem. Phys.

(17) J. M. T. Thompson and H. B. Stewart, Nonlinear Dynamics and Chaos, Wiley, New York (1987).

(18) T. Fink, J.-P. Dath, R. Imbihl and G. Ertl, J. Chem. Phys. **95**, 2109 (1991).

(19) R. Imbihl, T. Fink and K. Krischer, J. Chem. Phys. **96**, 6236 (1992).

(20) A. M. Turing, Philos. Trans. Roy. Soc. London **B 237**, 37 (1952).

(21) F. Mertens and R. Imbihl, submitted to Nature.

(22) J. Falta, R. Imbihl and M. Henzler, Phys. Rev. Lett. **64**, 1409 (1990).

(23) R. Imbihl, A. E. Reynolds and D. Kaletta, Phys. Rev. Lett. **67**, 275 (1991).

V.

Thin Film Growth and Interfaces

Homoepitaxial Growth of Metals and the Role of Surfactants

Matthias Scheffler, Vincenzo Fiorentini,[*] and Sabrina Oppo[*]

Fritz-Haber-Institut der Max-Planck-Gesellschaft, Faradayweg 4-6,
D-14 195 Berlin-Dahlem, Germany
[*] Dipartimento di Scienze Fisiche, Università degli Studi di Cagliari, via Ospedale 72,
I-09124 Cagliari, Italy

Abstract: Various possibilities which affect the mode of homoepitaxial growth of metals
are outlined. Special attention is given to the potential energy surface of diffusing atoms
and how it determines the growth kinetics, the island density, and the critical island size.
In particular we discuss mechanisms how surfactants may work, with special attention to
Sb on Ag(111). Using density functional theory calculations it is shown that antimony is
a strongly surface segregating species, and that its stable geometry is at the substitutional
surface site. In this geometry it acts repulsively on deposited Ag adatoms, giving rise to
an increase of the Ag island density and to irregular island shapes. As a consequence, the
growth changes from the multi-layer (at room temperature) to the layer-by-layer mode.
The theoretical results are compared with recent experiments.

1   Introduction

Crystal growth is probably the oldest topic of solid state physics but the understanding
of the controlling microscopic mechanisms is still far from complete. With the advent
of molecular beam epitaxy and molecular organic chemical vapor deposition techniques
about 20 years ago, and the technological relevance of new materials and new structures
the interest in research of growth phenomena has increased tremendously. Various experimental studies employing for example field ion microscopy (e.g., Refs. [1-4]), electron
microscopy (e.g., Ref. [5]) and scanning tunneling microscopy (e.g., Refs. [6-8]) have identified several microscopic properties of surface diffusion and growth structures, and these
studies have stimulated significant theoretical efforts. In this paper we will not review
this active field and we are even unable to reference all the relevant literature. Instead
we will focus on some special but important aspects of the kinetics of crystal growth of
close-packed metal surfaces.

A coarse classification of crystal growth distinguishes two-dimensional (or layer-by-
layer) growth and multi-layer growth (see Fig. 1). The first one implies that deposited
atoms wander around on the surface with ease and that each layer will be practically

Figure 1: Schematic plots of layer-by-layer growth by islands (left) and multi-layer growth (right). In both examples 3.6 ML (ML = monolayers) have been deposited on a flat substrate (hatched).

completed before the next layer starts to grow. This results in a flat, typically well ordered surface. There are several different ways to achieve this result. The simplest one is to increase the temperature such that the deposited atoms will be sufficiently mobile to reach a lower step or kink site of an intrinsic surface step where they are bound most favorably (see the discussion of Fig. 2 below). Crystal growth then proceeds by the "step-flow" mechanism. In fact, because the formation energy of steps is only a fraction of the surface formation energy [9], surfaces created by cleavage will have a rather high density of steps. Another source of steps arises from an intentional or unintentional miscut giving rise to a vicinal surface with parallel aligned steps. Dislocations which reach the surface will contribute to the step density as well. However, when the intrinsic step density is low, and when atoms are deposited on the surface at low temperature they are no longer sufficiently mobile to reach an intrinsic lower step edge. They will then form islands, thus creating new steps. At very low coverage the island density will increase with coverage up to a saturation value. When the island density (and resulting step density) is sufficiently high that additionally deposited adatoms can reach these step edges, the islands will no longer increase in number but will grow in size. At first they grow parallel to the surface, because at low coverage the probability that additionally deposited atoms land *on* the islands is very small. Thus, atoms land on the lower terrace, wander around, and will be caught at the island edges. A frequently found result of growth by islands is that, when some critical coverage Θ_c is exceeded, the growth mode changes from the two-dimensional island growth to multi-layer growth (e.g., Ref. [12]). Thus, deposited atoms which land on top of an island apparently stay there, join with other atoms, and grow an island on top of the island and so on (see Fig. 1, right). Such surfaces are very rough with "mountains" as high as 10-20 or more atomic layers. Beautiful pictures of deeply fissured mountain sceneries have been obtained by electron microscopy [5] and scanning tunneling microscopy (STM) [8, 13]. For technological applications a surface or interface should be well ordered and flat. Obviously, the key question here is what determines the inter-layer mass transport and how can it be influenced. Figure 2 shows a schematic plot of the potential-energy surface experienced by a deposited atom at the edge of on island. It contains:

1) An energy barrier at the step edge, $E_d^\perp$.

2) Different diffusion barriers at the upper and the lower terrace, E_d^u and E_d^l.

3) A potential gradient which is superimposed to the atomic-structure actuated energy corrugation, and in the example of the figure attracts adatoms towards the step [14].

4) A most favorable adsorption site at the lower side of the step.

One aspect which is not included in the schematic plot of the figure, but which may exist for some systems, is the possible existence of precursor energy barriers close to the edge. Figure 2 was designed following the results of *ab-initio* total energy calculations of self-diffusion at steps on aluminum (111) (see Ref. [10]). The qualitative behavior of the figure can be understood as a result that the energy of each atom scales with the square root of its coordination number [15]. Although this approach is clearly insufficient for a *quantitative* description of adsorption and surface diffusion [16] it can be used for a rough and qualitative estimate: On the flat parts of the surface an atom jumps from one stable adsorption position to an equivalent one in the neighboring surface unit cell. The energy in between these two positions is less favorable, and defines the diffusion energy barrier, $E_d^{l,u}$. In order to cross a step edge of an island on a low-index surface the coordination is typically reduced substantially, giving rise to a significant step-edge energy barrier, $E_d^\perp$. At the lower side of the step the coordination is high. Thus, here the adparticle is

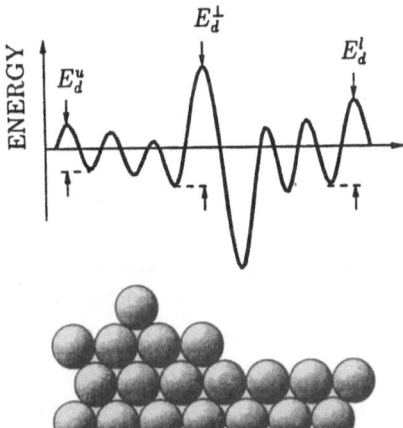

Figure 2: Schematic plot of the potential-energy surface of an adatom (assumed to be the same chemical species as the substrate) at a step. Note that the energy barrier on the lower terrace is chosen here to be larger than that on the upper terrace which may be a result of surfactants (see Section 3). There is a step-edge barrier hindering an adparticle on the upper terrace to move down, and a particularly favorable binding site at the lower side of the step. Superimposed to the energy corrugation is an energy gradient attracting adparticles on the lower and the upper terrace towards the step.

bound particularly tight. In fact, the adsorption energy of an adatom on a fcc (111) metal surface is about 60-70 % of the cohesive energy, its energy at a close-packed step is about 80-90 %, and at a kink site it equals exactly the cohesive energy. Thus, kink sites help to establish thermal equilibrium (or *local* thermal equilibrium) for which the atom's chemical potential is the bulk cohesive energy. Under growth conditions the chemical potential will be enhanced which may give rise to interesting structural changes at the surface [17]. For example, the enhanced atom chemical potential makes it easier to "create" adatom and more costly to create surface vacancies.

We emphasize that Fig. 2 only shows the potential-energy surface relevant for a movement perpendicular to the step edge. For many questions it is also important to know the energy barriers of an adatom bound at a lower step site and trying to move parallel to the edge. This determines the island shape (see for example Ref. [10] and references therein).

2 How Can Layer-by-Layer Growth Be Achieved?

It is now assumed that temperature and deposition rate are set as such that the crystal grows via islands rather than by the step flow mechanism. The definition of layer-by-layer growth mentioned at the beginning of Section 1 can be expressed in the following way: There is a critical coverage Θ_c for which, when it is exceeded, islands will grow on top of already existing islands. Obviously, when $\Theta_c = 1$ there is layer-by-layer growth. Alternatively, this trivially translates into the following condition for layer-by-layer growth:

$$A_c^{is} \geq A^s / N^{is} \quad . \tag{1}$$

Here N^{is}/A^s is the island density (N^{is} = number of islands and A^s = sample area), and A_c^{is} is the critical island size which is required that a nucleation occurs on top of an already existing island. Experimental studies, e.g. of the growth of clean silver [5, 18] as well as semi-empirical simulations of growth (e.g., Refs. [19-21]) show that these quantities are well defined. For $A_c^{is} \cdot (N^{is}/A^s) < 1$ it follows that $\Theta_c = A_c^{is} \cdot (N^{is}/A^s)$ and otherwise we set $\Theta_c = 1$. If equ. (1) is fulfilled, the islands will coalesce before a second layer has started to grow on top of islands. The island density, N^{is}/A^s, and the critical island size, A_c^{is}, are determined by the growth conditions (deposition rate F and temperature T) as well as by the different energy barriers and interactions of the deposited adatoms on the surface and by the minimal size of an island nucleus. A_c^{is} is determined by the probability that N^{nuc} adatoms land within a reasonable time interval on the same island, and that they form a nucleation center. The minimum number of atoms in such a nucleation center is denoted by N^{nuc}. It appears that for many systems N^{nuc} is as small as two or three atoms. Quite obviously, if there were no step-edge barrier then the adatoms which land on an island will move from the upper to the lower terrace to bind at the favorable sites. Thus, the formation of an island on top of the island becomes highly unlikely, and layer-by-layer growth is to be expected.

However, if a step-edge barrier exists, the probability that a nucleus is formed on top of an island is significantly increased. Once such a nucleus is formed, this will efficiently grow into a bigger and bigger island because at first the nucleus and then the lower step sites of the new island represent energetically most favorable positions which can easily be reached by adatoms which land on the same island. Thus, when a noticeable step-edge barrier exists it is indeed likely that equ. (1) is not fulfilled and three-dimensional growth will result.

When $D^l = D_0^l \, e^{-E_d^l/kT}$ is the diffusion constant of an adatom on the lower terrace, the dependence of N^{is}/A^s on temperature and deposition rate is

$$\frac{N^{is}}{A^s} \sim \left(\frac{F}{D^l} \right)^\chi \quad , \tag{2}$$

where χ typically has a value between 1/3 and 1/2 (see for example Refs. [12, 22, 23, 20, 24, 25]). Thus, the island density increases when the mobility of the atoms is reduced, e.g. by decreasing the temperature. The critical island size, A_c^{is}, has a similar dependence on the deposition rate as N^{is}/A^s but its temperature dependence is somewhat different. This is due to the fact that A_c^{is} is determined not only by the diffusion on the lower terrace but also by the diffusion energy barriers on the islands, perpendicular, and parallel to the island edges, and the corresponding pre-exponential factors. This difference in the dependence of N^{is}/A^s and A_c^{is} on the different energy barriers and diffusion mechanisms is important because it can be utilized to modify one of these quantities while leaving the other nearly uneffected (see Section 3 below). From this discussion it follows that growth can be changed from the multi-layer to the layer-by-layer mode by the following approaches:

1) Comsa *et al.* [26] have shown that on Pt(111) at very low temperatures the island shape is fractal [27]. This implies that the island perimeter is particularly long, that the islands have rough edges, and many rather thin branches. As a consequence, the step-edge barrier might be reduced substantially. Furthermore, an adatom which lands on such an island will visit the edge most frequently. This makes it likely that a possibly existing energy barrier can be overcome. We also note that the island density N^{is}/A^s is high at low temperature which reduces the probability that more than one atom land in a reasonable time interval on the same island. Altogether these properties give rise to layer-by-layer growth.

2) Also a second possibility was demonstrated by Comsa's group. Rosenfeld *et al.* [28] have shown that increasing the island density, N^{is}/A^s, is in fact sufficient to achieve layer-by-layer growth. A high island density can be obtained by lowering the

temperature or increasing the deposition rate F (see equ. 2) in the very beginning of the growth. Once the island density is increased, the growth is continued at normal (higher T and lower F) conditions. Thus, the island density N^{is}/A^s is set large (as determined by low T and high F parameters) but A_c^{is} is not reduced and assumes the value determined by the "normal" T and F growth parameters. Indeed, three-dimensional growth did not start before the layer was completed.

This approach requires that whenever a new layer is completed, the growth parameters have to be changed, in order to create again the desired high island density.

3) We also mention the possibility to enhance the mobility of deposited adatoms by photo-stimulation. However, we will not elaborate on this mechanism.

4) An interesting way to achieve layer-by-layer growth is by making use of surface contaminants (see for example Refs. [30, 29]), so called surfactants. While for liquid surfaces by definition a surfactant reduces the surface formation energy the word is used in a more general sense in the context of crystal growth. Here it labels a contaminant which, when deposited once, promotes a smoother growth. There is one necessary condition these species should fulfill: They should stay on the surface, i.e., they should not become buried in the bulk during the growth process. Thus, the total energy of a surfactant should be lower when the atom is at the surface *and* it should be kinetically easy to keep the surfactant at the surface during growth. While a good probability of surface segregation is necessary it alone would not affect the growth mode. There are the following possible mechanisms which, when active, provide that a surface segregating contaminant increases the inter-layer mass transport:

i) The simplest idea which comes to mind is that the surfactant decorates edges of steps and islands and reduces the step-edge barrier. Figure 3 demonstrates one possibility how this could be achieved.

ii) It is also possible that surface impurities induce a potential energy gradient which attracts deposited atoms towards the step; for deposited atoms which land on an island the number of visits at the edge is then increased, and, as a consequence, the probability to move down is increased as well.

iii) A third possibility is that surfactants act as nucleation centers by attracting deposited adatoms, thus increasing the island density N^{is}/A^s. This will induce layer-by-layer growth, provided that the probability that atoms which land on an island and move to the lower terrace is reduced only little (or not at all). Thus, the critical island size A_c^{is} should remain about unchanged (obviously, an increase would be optimum). This mechanism was recently discussed by Zhang and Lagally [31].

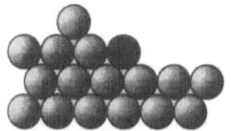  

Figure 3: The motion of an atom from the upper terrace down by the exchange mechanism. The atom at the edge is marked in black because it may be a contaminant atom: exchange can be one mechanism how surfactant atoms, if they decorate steps, increase the inter-layer mass transport. However, exchange is also found for some systems at clean steps (see for example Ref. 10 and references therein).

iv) A forth possibility was recently discussed in the context of the surfactant action of Sb on Ag(111). The basic mechanism here is that the surfactant atoms are practically *immobile* on the surface and act *repulsively* to deposited diffusing Ag atoms. This mechanism will be discussed in Section 3.

Different surfactants work by one or the other mechanism, or by a combination. For oxygen for the growth of Pt(111) Esch *et al.* [8] have recently argued that the mechanism no. *i*) is the important one. The mechanism of oxygen for the growth of Cu on Ru(0001) is still somewhat unclear [32, 33]. In Section 3 we will discuss the (at present) chief example for surfactant action on metal growth, namely Sb an Ag(111). Also for this system the mechanism no. *i*) was originally favored [34], whereas Rosenfeld *et al.* [28] and Tersoff *et al.* [18] invoked the island-density effect in terms of mechanism no. *iii*), and we suggested [35, 36] that mechanism *iv*) is active, which also gives rise to an increased island density.

3 Density Functional Theory Results for Ag and Sb on Ag(111) and Interpretation

At room temperature Ag(111) grows by the multi-layer mode (see the schematic plot of Fig. 1, right). This has been observed by electron microscopy [5], x-ray reflection [34] and scanning tunnelling microscopy [13]. It is interesting to note that the more open (100) surface of silver grows by the layer-by-layer mode [37] which indicates that for the (100) surface $E_d^{u,l}$ and $E_d^{\perp}$ are of comparable size while for clean Ag(111) $E_d^{\perp} > E_d^{u,l}$. Van der Vegt *et al.* [34] recently showed that the multi-layer growth mode of Ag(111) can be changed by deposition of Sb at the beginning of the growth process. The required coverage

Θ_{Sb} was between 0.05 and 0.2 ML. The layer-by-layer growth mode of Sb precovered Ag (111) is sustained for more than 10 or 20 deposited silver layers.

Oppo *et al.* [35, 36] recently carried out density functional theory (DFT) calculations for Ag on clean Ag (111) surfaces and on silver surfaces with preadsorbed Sb. The theoretical study was performed by using the local-density approximation (LDA) for the exchange-correlation functional and the supercell approach. The calculations [35, 36] revealed several unexpected results. It was found that Sb adsorbs on Ag (111) but that the "normal" on-surface site is only metastable. The stable geometry for Sb on Ag (111) is the surface substitutional site. This geometry has a significantly (about 1 eV) more favorable energy than the on-surface site. The surface substitutional energy is also energetically greatly favored over positions where Sb is in buried below the surface. The surface segregation energy is about 1 eV, which shows that Sb is a strongly segregating species (the necessary condition for a surfactant). While Sb is somewhat too big to be solved in the bulk, it fills nicely the surface substitutional site with only a small outward relaxation. We note that surface segregation of Sb during crystal growth is connected with a diffusion energy barrier. Thus, although strong segregation is to be expected, it will most likely not be complete.

How can the substitutional geometry be achieved? We mentioned that on-surface adsorption is metastable. This is, because the substitutional geometry can only be reached when each Sb atom kicks out a Ag surface atom. When this Ag atom has reached a kink site at the surface, the ground state energy is obtained. Thus, the substitutional geometry can be reached by annealing or it will be reached during growth after a coverage $\Theta_{Ag} = 1 - \Theta_{Sb}$ of Ag is deposited.

For high deposition rates the concentration of Ag adatom determines the local thermal equilibrium and kink sites become less important. Then it becomes possible that Ag adatoms kick out substitutial Sb, thus creating on-surface Sb. We expect that this site exchange will be hindered by an energy barrier and that it will be not very important for the growth properties. We note in passing that on-surface Sb is clearly stronger bound to the silver substrate than a silver adatom [35]. Because diffusion energy barriers are roughly proportional to the adsorption energy it follows that on-surface Sb is less mobile on Ag (111) than a Ag adatom. Obviously, substitutional Sb can be considered to be practically immobile at room temperature.

For low Sb coverage the calculations [35] predict a weak Sb-Sb repulsion, which is small for the on surface geometry (0.11 eV) and larger for the surface-substitutional site (0.39 eV) – comparing a coverage of $\Theta_{Sb} = 1/9$ and $1/4$ ML. Thus we expect that there is no island formation [on clean and defect free Ag (111)] and that the Sb will be distributed in a homogeneous and disordered way over the surface, at least when Sb has assumed the surface substitutional geometry.

To explain the effects of Sb adsorption on the growth mode of Ag, we studied Ag adsorption on clean and Sb-covered Ag (111). Only the substitutional adsorption for Sb is considered, because only this one is relevant during growth, even though on an unannealed sample we note that the first layer should grow differently. Figure 4 shows the results for an Ag adatom on the Sb free parts of the surface and close to substitutional Sb. The results are obtained with a 2×2 surface super cell. For the surface with adsorbed Sb (we used $\Theta_{Sb} = 0.25$ ML) two different Ag positions where considered: the fcc hollow site nearest and second nearest to the Sb. Due to the super-cell approach, the latter is in fact equally far away from three Sb adatoms. The results clearly show that substitutional Sb repels Ag adatoms. Thus, the diffusion barrier for Ag on Ag (111), which is usually smaller than 0.1 eV, increases to a value of about 0.4 eV when an Ag adatom has to pass close to an Sb center. As a consequence, Sb reduces the mobility of Ag adatoms. In view of the discussion in Section 2 it follows that substitutional Sb increases the islands density N^{is}/A^s.

Based on the theoretical results that substitutional Sb centers are highly immobile and repel Ag adatoms the following scenario might be expected [35, 36]:

1) The reduced mobility of Ag adatoms on the lower terrace gives rise to an increase of the island density, i.e., to the formation of many small rather than few large islands.

2) The islands will grow in directions where they will avoid to meet Sb centers. This together with the fact that Ag diffusion parallel to the island edge will be hindered by the presence of the Sb centers close to the island should give rise to an irregular island shape.

3) As a further consequence of the manner islands will grow, we expect that there are less Sb centers incorporated in the island (as long as they are small) than there are in the lower terrace.

The combination of the higher island density and the irregular island shape, which gives the Ag adatoms a higher frequency to visit the edge and to run against the step-edge

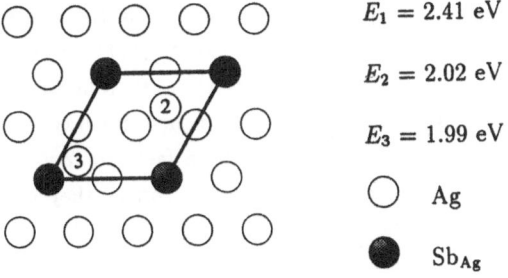

$E_1 = 2.41$ eV

$E_2 = 2.02$ eV

$E_3 = 1.99$ eV

○   Ag

●   Sb$_{Ag}$

Figure 4: Adsorption energies (eV/atom) for a Ag adatom on Ag(111), 1) at clean parts of the surface, 2) in a second-nearest neighbor distance, and 3) in a nearest neighbor distance to substitutionally adsorbed Sb. The two fcc hollow sites are marked as 2) and 3).

barrier, drives the growth mode towards the layer-by-layer growth. A reduction of the step-edge barrier E_d^+ would help to keep the critical island size, A_c^{is}, "large" and to bring the inequality (1) to work. However, for Ag(111) we have not performed calculations for diffusion at step edges and can therefore not discuss the importance of the step-edge barrier.

It is interesting to note that all this results because Sb is a *repulsive* and *highly immobile* nucleation center. Obviously, an attractive nucleation center would also reduce the Ag adatom mobility but such a center would be incorporated into the growing Ag islands particularly fast. There it would then act again as a nucleation center and thus it would reduce the critical island size A_c^{is}. Thus, a *repulsive* nucleation center requires less increase of island density N^{is}/A^s than an attractive one.

4 Comparison with Recent Experiments

In recent He-scattering experiments, Rosenfeld *et al.* [28] showed that three-dimensional growth is suppressed when the density of Ag islands is increased. We had noted this study already above in Section 2 under item no. 2). It was suggested [28] that the role of Sb in surfactant-assisted layer-by-layer growth is that of increasing N^{is}/A^s without changing very much the critical island size A_c^{is}. Everything in this study is consistent with the independently obtained theoretical analysis [35, 36].

The experimental work by Tersoff *et al.* [18] showed that for clean Ag(111) at room temperature and the used (but unfortunately not reported) deposition rate the critical island size is $A_c^{is} \approx 3 \cdot 10^6 \, \text{Å}^2$ and the inverse island density is about $6 \cdot 10^6 \, \text{Å}^2$ (deduced from Fig. 2 of Ref. [18]). This gives a critical coverage of $\Theta_c \approx 0.5 \, \text{ML}$. From their Fig. 2 we also guess that Sb increases the island density by more than a factor of 100. Thus, their results (as well their interpretations) are in agreement with our predictions.

A more detailed comparison is possible with a recent STM study of Vrijmoeth *et al.* [13]. Similarly to the results just discussed these authors find an increase in the island density due to Sb. Under their growth conditions (room temperature and not specified deposition rate) they obtained for the clean Ag(111) surface $A_c^{is} \approx 9 \cdot 10^6 \, \text{Å}^2$ and $A^s/N^{is} \approx 18 \cdot 10^6 \, \text{Å}^2$ (as deduced from their Fig. 1). Both values are a factor of 3 larger than those of the study of Ref. [18] and thus the same critical coverage of $\Theta_c \approx 0.5 \, \text{ML}$ results. Only an increase of the island density by a factor of 2 were needed (when A_c^{is} remained constant) to induce layer-by-layer growth. Indeed they report a much higher increase of the island density of the Sb covered and annealed surface of a factor of 6. Their results fully support the other studies and analyses [28, 18] as well as our calculations [35, 36].

It is interesting to compare some quantitative data of the STM work [13] with the results of the DFT-LDA calculations [35, 36]:

- Vrijmoeth *et al.* confirm that for an unannealed surface the first layer grows differently than all the subsequent ones.

- They argue (in agreement with the theoretical prediction) that the on-surface adsorption site is metastable, and that in thermal equilibrium Sb occupies the substitutional geometry with a disordered arrangement.

- Vrijmoeth *et al.* determine that Sb reduces the mobility of Ag adatoms and express this in terms of an increase of the diffusion barrier to be 0.5 eV (the theoretical studies had predicted a Ag-Sb repulsion energy of 0.4−0.5 eV).

Although the agreement between the theoretical results and the STM data is very good, Vrijmoeth *et al.* argue that the main effect which induced the change in the growth mode is not the increase in island density, but an increase in A_c^{is} which follows because the additional step-edge barrier $E_d^\perp - E_d^u$ is reduced. This argument is conceptionally unclear, and it is in fact not supported by the experimental data. The measurements show that the increase of the island density due to deposited Sb is substantial, so that the critical island size, A_c^{is}, cannot be reached. Thus, the experiments do not allow an analysis whether the Sb deposition has increased A_c^{is} over its value for the clean silver surface or whether A_c^{is} is changed only slightly (probably reduced) as it was suggested in the theoretical study [35, 36].

5 Summary

In the above Sections we discussed the various possibilities which may be used to change the growth of a metal from the multi-layer to the layer-by-layer mode. In particular we concentrated on the system Sb on Ag (111). We summarized the results of the DFT-LDA calculations and the suggested scenario for the Sb induced change of the growth mode. A detailed comparison with recently performed experimental studies seems to confirm all the (initially puzzling and unexpected) theoretical predictions. These were in particular the result that the stable adsorption site of Sb is the substitutional geometry, making the Sb very immobile, and the prediction that substitutional Sb and deposited Ag repel each other. The first prediction seems to be confirmed by STM. With respect to the second prediction we have to note that it is (also) consistent with all reported experimental data but at this point we cannot argue that it is confirmed. It would be important to continue the experimental studies and to identify the nature of the interaction between substitutional Sb and Ag adatoms.

We thank J. Meyer for discussions.

References

[1] G. Ehrlich and F. G. Hudda, J. Chem. Phys. **44**, 1039 (1966).

[2] S. C. Wang and G. Ehrlich, Phys. Rev. Lett. **71**, 4174 (1993).

[3] T. T. Tsong, *Atom-Probe Field Ion Microscopy*. Cambridge University Press, Cambridge 1990.

[4] G. L. Kellog, Phys. Rev. Lett. **72**, 1662 (1994).

[5] K. Meinel, M. Klaua, and H. Bethge, phys. stat. sol. **110**, 189 (1988); and J. Cryst. Growth **89**, 447 (1988); H. Bethge, in: *Kinetics of Ordering and Growth at Surfaces*, Ed. M. G. Lagally. Plenum Press, New York 1990, 125.

[6] Y. W. Mo, J. Kleiner, M. B. Webb, and M. G. Lagally, Phys. Rev. Lett. **66**, 1998 (1991).

[7] M. Horn-von Hoegen, to be published in Appl. Phys. A.

[8] S. Esch, M. Hohage, T. Michely, and G. Comsa, Phys. Rev. Lett. **72**, 518 (1994).

[9] The step formation energy of metals is typically a fraction of the surface energy. For close-packed steps on Al(111) the calculations predict 0.24 eV per step-edge atom (see Ref. 10). For 4d transition metals see Ref. 11.

[10] R. Stumpf and M. Scheffler, Phys. Rev. Lett. **72**, 254 (1994); and submitted to Phys. Rev. B.

[11] M. Scheffler, J. Neugebauer, and R. Stumpf, J. Phys.: Condens. Matter **5**, A91 (1993).

[12] S. Stoyanov and D. Kashchiev, in *Current Topics in Materials Science*, Ed. E. Kaldis. North-Holland, Amsterdam 1981, Vol. 7, 69.

[13] J. Vrijmoeth, H. A. van der Vegt, J. A. Meyer, E. Vlieg, and R. J. Behm, to be published.

[14] For Al(111) there is indeed an energy gradient superimposed to the atomic-structure corrugation, which gives rise to an attraction of about 0.15 eV between a position of the adatom just at the step and one far away (see Ref. 10). This attraction results not from the electrostatic fields of the adatom and step dipoles which is, in fact, slightly repulsive, but is mediated by adatom- and step-induced surface states.

[15] Typically, coordination number models scale the energy of each atom with the square root of its local coordination. Often also a small linear term is added. The square root behavior takes the bond saturation into account which makes this approach very similar to the embedded-atom and effective-medium methods [see for example I. J. Robertson *et al.*, Europhs. Lett. **15**, 301 (1991), Phys. Rev. Lett. **70**, 1944 (1993), and M. Methfessel *et al.*, Appl. Phys. A **55**, 442 (1992)]. All these methods are often labeled as "glue-type models".

[16] Glue-type models (see Ref. 15) give an energy barrier for self-diffusion of Al on Al(111) which is by more than a factor of ten higher than what is found in the DFT-LDA calculations. This might be an extrem case of the inaccuarcy of glue-type models, because for Al(111) the DFT-LDA barrier is particularly small (0.04 eV) due to the partly covalent nature of Al.

[17] M. Bott, M. Hohage, T. Michely, and G. Comsa, Phys. Rev. Lett. **70**, 1489 (1993).

[18] J. Tersoff, A. W. Denier van der Gon, and R. M. Tromp, Phys. Rev. Lett. **72**, 266 (1994).

[19] S. Stoyanov and I. Markov, Surf. Sci. **116**, 313 (1982).

[20] J. A. Venables, G. D. T. Spiller, and M. Hanbücken, Rep. Prog. Phys. **47**, 399 (1984).

[21] G. S. Bales and D. C. Chrzan, Phys. Rev. B, in press (1994).

[22] J. Villain, A. Pimpinelli, and D. E. Wolf, Comments Cond. Mat. Phys. **16**, 1 (1992); J. Villain, A. Pimpinelli, L. Tang, and D. E. Wolf, J. Physique I **2**, 2107 (1992).

[23] J. G. Amar, F. Family, and P.-M. Lam, in *Mechanisms of Thin Film Evolution*, Eds. S. M. Yalisove, C. V. Thompson, and D. J. Eaglesham, Materials Research Society, to be published.

[24] M. C. Bartelt and J. W. Evans, Phys. Rev. B **46**, 12675 (1992).

[25] E. S. Hood, B. H. Toby, and W. H. Weinberg, Phys. Rev. Lett. **55**, 2437 (1985).

[26] R. Kunkel, B. Poelsema, L. K. Verheij, and G. Comsa, Phys. Rev. Lett. **65**, 733 (1990); M. Bott, T. Michely, and G. Comsa, Surf. Sci. **272**, 161 (1992).

[27] Island with dendritic shape have been observerd for Au and Ag (111) (see Ref. 5), for Au on Ru (1000) by R. Q. Hwang, J. Schröder, R. Günther, and R. J. Behm, Phys. Rev. Lett. **67**, 3279 (1991), and for Pt (111) (see Ref. 26).

[28] G. Rosenfeld, R. Servaty, C. Teichert, B. Poelsema, and G. Comsa, Phys. Rev. Lett. **71**, 895 (1993).

[29] W. F. Egelhoff, Jr. and D. A. Steigerwald, J. Vac. Sci. Technol. **A7**, 2167 (1989).

[30] H. L. Gaigher, N. G. van der Berg, and J. B. Malherbe, Thin Solid Films **137**, 337 (1986).

[31] Z. Zhang and M. G. Lagally, Phys. Rev. Lett. **72**, 693 (1994).

[32] H. Wolter, M. Schmidt, and K. Wandelt, Surf. Sci. **298**, 173 (1993).

[33] H. Wolter, M. Schmidt, M. Nohlen and K. Wandelt, this volume, p. 232

[34] H. A. van der Vegt, H. M. van Pinxteren, M. Lohmeier, E. Vlieg, and J. M. C. Thornton, Phys. Rev. Lett. **68**, 3335 (1992).

[35] S. Oppo, V. Fiorentini, and M. Scheffler, Phys. Rev. Lett. **71**, 2437 (1993).

[36] S. Oppo, V. Fiorentini, and M. Scheffler, MRS Proc. **314**, 111 (1994).

[37] Y. Suzuki, H. Kikuchi, and N. Koshizuka, Jap. J. Appl. Phys. **27**, L 1175 (1988).

Metal-on-Metal Heteroepitaxy and the Influence of a Surfactant: Cu/O/Ru(0001)

H. Wolter, M. Schmidt, M. Nohlen and K. Wandelt
Institut für Physikalische und Theoretische Chemie, Universität Bonn,
Wegelerstr. 12, D-53115 Bonn, Germany

Abstract. Cu-on-Ru(0001) has been a suitable model system for metal-on-metal heteroepitaxy for many years because of the immiscibility of the two components. In contrast to most *ex-situ* studies the present work emphasizes dynamical work function change ($\Delta\varnothing$) measurements using a special KELVIN-capacitor probe as being a sensitive method to monitor the film morphology *in-situ*, that is *during* film growth. In particular, it is demonstrated that preadsorbed oxygen on the Ru(0001) substrate permanently "floats out" to the Cu surface and acts as a *surfactant* for layer-by-layer growth at a temperature of 400 K, at which in the absence of oxygen non-ergodic growth leads to the formation of pyramidal clusters and, hence, a very rough film. The mechanistic influence of the *surfactant* oxygen is discussed in detail and may certainly be carried over to other metal-on-metal systems.

1 Introduction

The characterization and, even better, the control of the growth morphology of thin films is of great practical importance. The fabrication and performance of, for instance, electronic devices generally relies on films of uniform thickness as well as on sharp interfaces between them. It is therefore very important to understand and to control the influence of those parameters which ultimately determine the structure of a thin film. In the case of simple physical vapor deposition (PVD) these are mainly the substrate temperature and the deposition rate, i.e. the number of impinging particles per second and unit area. Depending on the choice of these two quantities the growth of *hetero*epitaxial films may be determined either by kinetic or by energetic parameters of the respective metal-on-metal system. Kinetic effects originate from activation barriers for surface diffusion. Energetic driving forces arise from differences in surface free energy (bond energies) and surface strain (e.g. due to lattice mismatch).

In order to form a film of uniform thickness, atoms which during the growth process happen to impinge on an existing island in a given layer, need to migrate into lower, still incomplete layers as indicated in Fig. 1a. The occurrence of this interlayer mass transport may be kinetically impeded by the existence of an extra step-edge barrier ϵ_s^* for step-down migration [1,2] and will in any case be controlled by the total energy difference $\Delta E_{1,2}$ of the system with the atom after and before overcoming this barrier. $\Delta E_{1,2}$ depends on the change in effective coordination number of the atom in the initial and the final position as well as on the energy per bond, that is on the underlying material and the geometrical structure at both locations. Transition of the step-edge barrier ϵ_s^* requires thermal activation either during film growth or by post-annealing the deposited film. (Notice that there is a limitation to the tolerable deposition or annealing temperature in order to avoid interdiffusion of the deposited and the substrate metal and to maintain a sharp interface as emphasized above). However, even in the case of equilibration $\Delta E_{1,2}$ determines whether the film will assume a uniform thickness with all atomic layers completely filled or whether the film will remain rough.

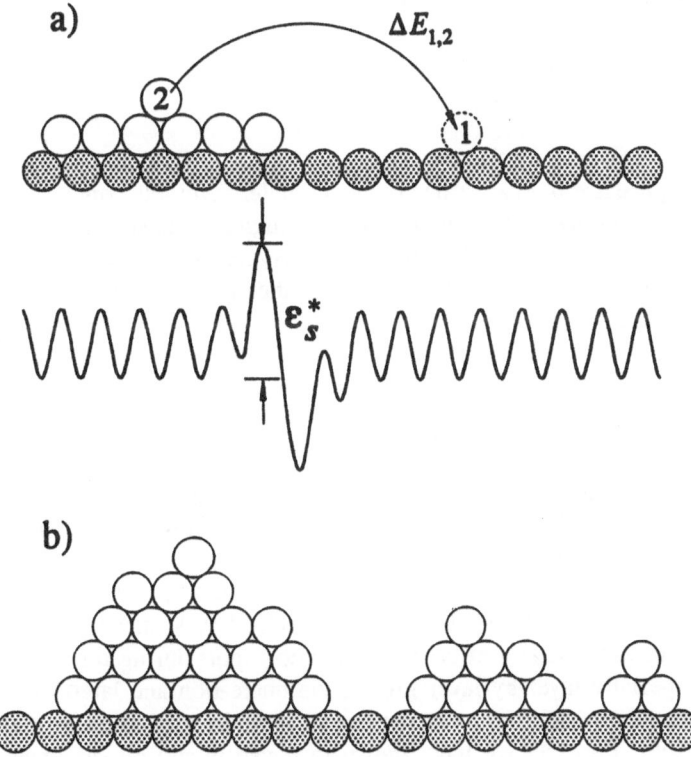

Fig. 1 Schematic representation of a) interlayer mass transport as a prerequisit for layer-by-layer growth, and b) 3D cluster growth. The potential energy curve in a) illustrates the sinusoidal potential barriers for surface migration across flat surface regions as well as the extra barrier ϵ_s^* for step crossing from the higher to the lower surface terrace. $\Delta E_{1,2}$ is the difference in total energy of the system with the adatom in the first (1) and second (2) layer.

If $\Delta E_{1,2} < O$ atoms will tend to complete lower layers, if $\Delta E_{1,2} > O$ atoms will remain in higher layers leading to the formation of three dimensional pyramidal clusters and an overall rough surface morphology (Fig. 1b). $\Delta E_{1,2}$ determines whether the equilibrium growth mode for a given metal-on-metal system is of the Franck - v.d. Merwe (FM) -, Volmer - Weber (VW) -, or Stranski - Krastanov (SK) - type [3,4].

We realize that not only the non-ergodic growth process but also the VW- and the SK-mode of an ergodic-growth process result in non-uniform, rough films.

In view of the fact that adsorbates may lead to a considerable structural reconstruction of surfaces it is not surprising at all that coadsorbed species may also influence the mode of thin film growth. It was, in fact, discovered recently [5-8] that certain adsorbed species, socalled surfactants, may alter the growth mode of a given metal-on-metal system from a rough structure (without surfactant) to a layer-by-layer growth (in the presence of the surfactant) under otherwise unchanged conditions as far as substrate temperature and deposition rate are concerned. Reported examples of surfactants for *homo*epitaxial systems are antimony for Ag-on-Ag(111) [7] and oxygen for Pt-on-Pt(111) [8].

The present work reviews the growth of Cu films on a Ru(0001) substrate with and without the presence of coadsorbed oxygen. Under favorable conditions the oxygen acts as a surfactant and turns a non-uniform, rough growth structure into a uniform layer-by-layer growth mode. This role of the oxygen is discussed in terms of its influence on the nucleation behavior and on the interlayer mass-transport of the growing Cu film [9-11].

2 Methodical

A distinction between the two growth structures illustrated in Fig. 1 is obviously possible by their difference in surface roughness. The most appropriate methods for investigating the growth of thin metal films are in-situ methods since they provide information about the growth mode *during* deposition. This is commonly done with diffraction techniques like RHEED, SPA-LEED, X-Ray Scattering or TEAS and the observation of the intensity of certain diffraction spots during the epitaxial growth [12-15]. Especially layer-by-layer growth via nucleation and lateral island growth per layer manifests itself in the occurrence of periodic oscillations of the diffracted beam intensity. Recently we have shown that also in-situ measurements of work function changes ($\Delta\varnothing$) by means of a specially designed KELVIN-probe are a suitable tool to monitor the film growth [9-11].

This KELVIN-probe is basically a combination of those described by Hölzl et al. [16] and Besocke and Berger [17], and is sketched in Fig. 2. The KELVIN capacitor consists of a reference electrode (R) made of gold and the substrate crystal surface. In contrast to ordinary KELVIN-probes the periodic variation of the capacity

$$\frac{dC}{dt} = \frac{d}{dt}\left(A\frac{\sin(2\omega t)}{D}\right)$$

is effected by vibrating the reference electrode *parallel* to the sample surface at a constant separation D. This leads to a periodic variation of the "overlap-area" A between both electrodes and a resultant alternating current

$$I = \frac{dQ}{dt} = \Delta\phi_{CPD}\frac{dC}{dt}$$

in the external circuit, which depends on the contact potential difference $\Delta\phi_{CPD}$ between R and the crystal. Hence, changes of $\Delta\phi_{CPD}$ due to adsorption on the sample surface will cause changes of I. I can be continuously suppressed by an electronic feedback-loop which in every instant applies a compensation voltage U = - $\Delta\phi_{CPD}$, so that U becomes a (dynamical) measure of work function changes $\Delta\phi$

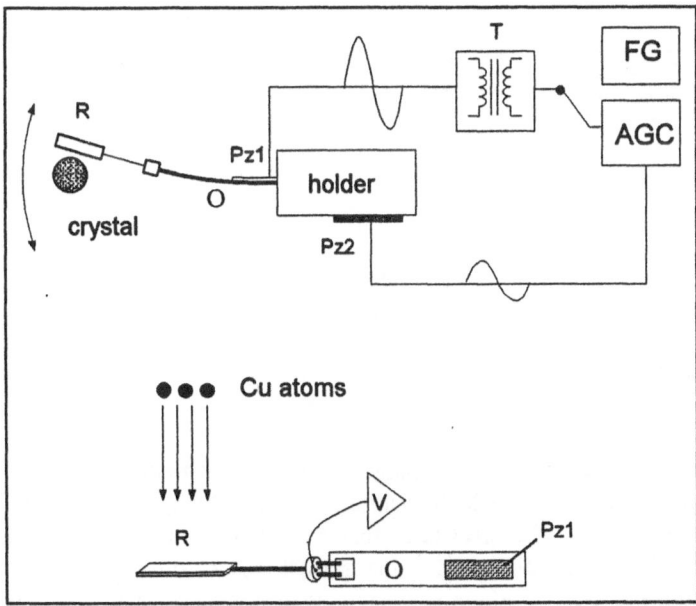

Fig. 2 Principle of the KELVIN-probe used in the present study. R = reference gold electrode vibrating *parallel* to the crystal surface. Pz1 and Pz2 = piezo ceramics, T = transformer, FG = frequency generator, AGC = auto-gain amplifier, V = preamplifier, O = oscillator arm (see text).

of the sample surface. The vibration of the reference electrode is driven by applying an alternating voltage to a piezoceramics (Pz 1) glued onto the electrode arm (see O Fig. 2). This can be done either by using a frequency generator (FG) or by the amplified signal of a second piezoceramics (Pz 2) which is attached to the holder of the vibrating electrode. In the latter case the system always stabilizes itself at the mechanical resonance frequency with a constant amplitude. The parallel vibration mode enables the deposition of metal atoms from an evaporation source located perpendicular to the sample surface *behind* the reference electrode, and the accuracy achieved with this setup allows to monitor accompanying work function *changes* with a noise level of ≤ 2 meV.

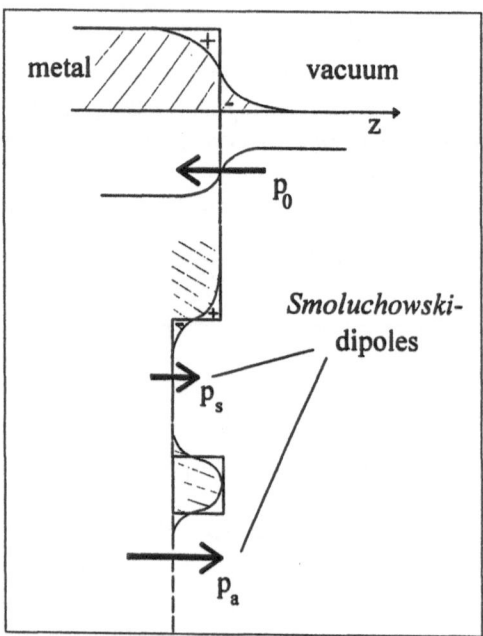

Fig. 3 Schematic representation of the correlation between work function and surface roughness due to the Smoluchowski-electron-smoothing-effect.

The sensitivity of work function change measurements towards surface roughness is based on the so-called Smoluchowski-effect [18]. Owing to this effect a redistribution (smoothing) of the electronic surface charge at step-edges leads to local dipoles (p_s) which are antiparallel to the normal surface dipole of flat surfaces (p_o) as depicted in Fig. 3. This effect is even more pronounced for lower coordinated or single adatoms whose dipole moment (p_a) is even larger. Hence, step defects (and adatoms) cause a lowering of the work function and earlier works with regularly stepped vicinal surfaces [19,20] have suggested a linear relationship between the step density and the resultant work function decrease. Using Photoemission of Adsorbed Xenon (PAX) as a local work function probe it is even possible to determine the *local* work function decrease at individual step-sites [20,21]. For instance, values of -700 meV and -150 meV were found at step-sites on a Ru(0001)- and a Cu(111)-surface, respectively [20]. Dynamic work function change measurements are therefore at every moment a very sensitive monitor of the total step-edge length at the surface of a continuously growing film and, thus, of the roughness of the film. This will be taken advantage of for the investigation of the growth of Cu films on Ru(0001) with and without the presence of coadsorbed oxygen as described in the following sections.

For details of the experimental procedure the reader is referred to earlier publications [9-11]. Here we mention only that Cu was evaporated onto the Ru(0001) surface (diameter: 6 mm) from a small resistively heated tungsten basket. For a given set of experiments the constancy of the evaporation rate was checked periodically (see below) and was found to be better than 97 %. The evaporation time could be precisely controlled by a mechanical shutter in front of the Cu source. Oxygen was dosed via a leak valve. After completion of each deposition experiment a clean Ru(0001) surface could be restored by heating the sample to ~1600 K. Under these conditions both oxygen (if present) and copper desorb [22] and leave a clean Ru surface behind for the next experiment. The area under the Cu desorption peak serves to determine the relative Cu coverage (see below).

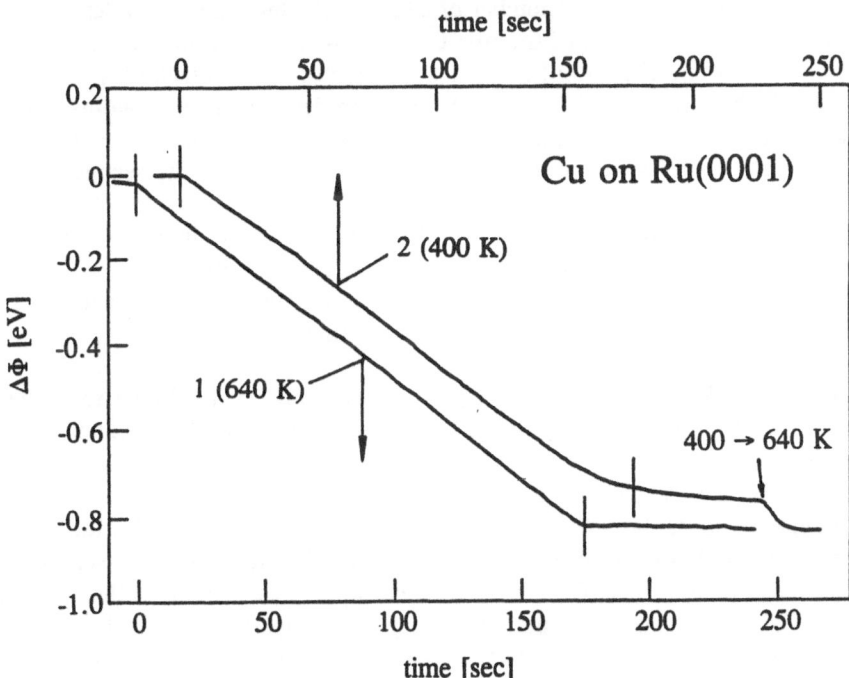

Fig. 4 Work function change (ΔØ) curves as a function of Cu deposition time onto a Ru(0001) substrate at 640 K (1) and 400 K (2), respectively. Both curves are registered during Cu evaporation with the same deposition rate. Near 230 sec during registration of curve 2 the temperature was raised from 400 K to 640 K. Note the different time axes for both curves.

3 Cu growth on clean Ru(0001)

The system Cu/Ru has received considerable attention in the literature [e.g. 23-30] and can be regarded as a model system for the growth of heteroepitaxial thin metal films. The Ru(0001) surface has a close-packed hexagonal structure and may be prepared with a low step density [30]. Cu is immiscible with Ru [27] and has been shown to grow epitaxially *on* the clean Ru (0001) surface forming a pseudomorphic film only in the first layer [25,26,29,30]. Submonolayer coverages of Cu lead to the formation of twodimensional (2D) pseudomorphic islands of monoatomic thickness which was concluded among others [27,30] from dynamic $\Delta\emptyset$-measurements as well as thermal desorption (TD) curves of Cu as displayed in Fig. 4 and 5.

Fig. 4 shows two work function change curves, 1 and 2, for Cu deposition on Ru(0001) at substrate temperatures of 640 K and 400 K, respectively, as a function of deposition time [31]. At 640 K the work function decreases strictly linearly until a sharp knee is reached after ∼170 sec. The linear decrease is indicative of a constant dipole moment and, in view of Fig. 3, of a similar environment of each deposited Cu atom. This is easily reconciled with 2D island growth; at 640 K substrate temperature the Cu atoms are sufficiently mobile to reach and form large islands with a relatively low number of edge (step) sites. The knee after ∼170 sec corresponds to completion of the first Cu layer as follows from a comparison with the Cu TD-spectra shown in Fig. 5 as a function of Cu coverage [32]. The evolution of the high temperature peak corresponds to desorption of Cu atoms in immediate contact with the Ru substrate, i.e. to atoms within the first Cu layer. The low temperature curves arise from desorption of Cu atoms from the second Cu layer [23,32]. Saturation of the high temperature TD-peak (maximum at 1300 K) coincides with the appearance of the knee in the 640 K $\Delta\emptyset$-curve in Fig. 4, both, thus, being indicators for completion of the first Cu layer on Ru(0001). $\Delta\emptyset$-measurements for Cu deposition at 640 K were therefore utilized to calibrate and check the evaporation rate of the Cu-source from time to time.

Besides the relative coverage two further conclusions can be drawn from the Cu TD-spectra in Fig. 5. Firstly, the higher desorption temperature of Cu atoms from the first layer compared to that of Cu atoms from the second layer indicates that the Cu-Ru bond is stronger than the Cu-Cu bond. This results in $\Delta E_{1,2} < 0$ in Fig. 1a and supports initial layerwise growth of Cu on Ru(0001). Secondly, the common and exponential rise of all TD-curves under the first layer and the second layer peak, respectively, resembles zeroth order desorption kinetics and is thus consistent with 2D-island formation at partial coverage of either layer (at sufficiently high temperature). In fact, a more detailed and quantitative analysis of these Cu TD-spectra using Monte-Carlo simulation methods yielded an attractive interaction between Cu atoms in the first layer on Ru(0001) [33].

Returning to Fig. 4 the $\Delta\emptyset$-curve at 400 K (registered with the same deposition rate as the 640 K curve) lacks a clear knee after deposition of one monolayer, that is after 170 sec. Instead, this curve is more rounded and levels off at a slightly higher

239

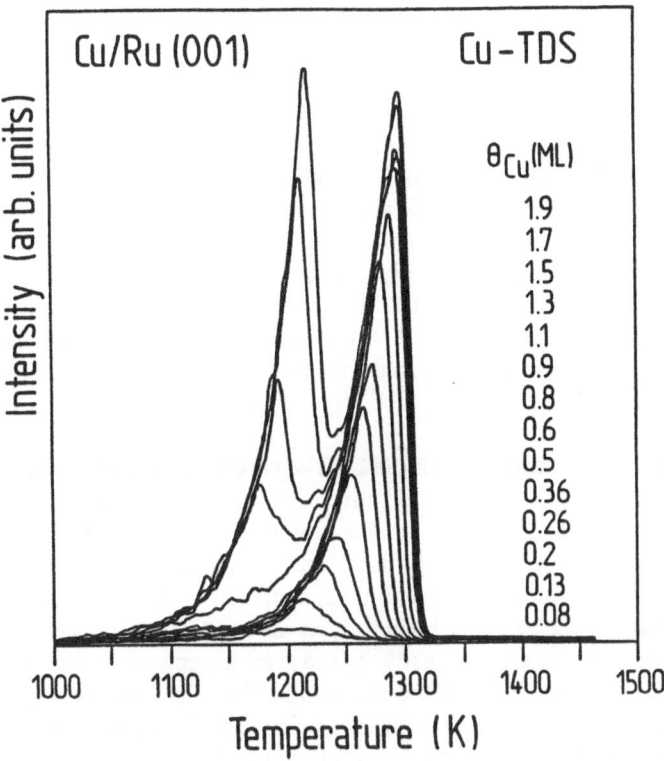

Fig. 5 Thermal desorption (TD) curves for various Cu precoverages Θ_{Cu} in monolayers (ML) on Ru(0001). The two evolving peaks correspond to population of the first ($T_{max} \sim 1300$ K) and the second ($T_{max} \sim 1200$ K) Cu layer on Ru(0001).

Ø-value than the 640 K curve. This behavior is consistent with an imperfect growth of the first layer; a certain fraction of the deposited Cu atoms remains in the second layer leaving the Ru-surface (of higher work function) partially uncovered. Subsequent heating of this imperfect structure to 640 K (see Fig. 4) heals the defects and causes the work function to drop to the same level as the 640 K curve, thereby supporting the above interpretation of the 400 K curve. Fig. 4, thus, reinforces the high structure sensitivity of ΔØ-measurements.

Since the interatomic distance within a Cu(111) plane is 5,5% smaller than that within the Ru(0001) surface, the first pseudomorphic Cu layer on Ru(0001) is expanded and has an ~11% reduced atomic density. This is in agreement with a relatively shallow dispersion of the twodimensional 3d-valence bands measured with angular resolved photoemission for 1 ML Cu/Ru(0001) (as well as for submonolayer 2D islands) [24,32].

The second Cu layer on Ru(0001) is contracted (within one of three possible domains) in one direction practically to the interatomic distance of the Cu(111) surface, while in the perpendicular direction the Cu rows are still in registry with the Ru substrate [30]. Only in higher Cu layers is the lattice constant and, hence, the atomic density of the Cu(111) plane reached [34]. Deposition at room temperature followed by annealing results in a layer-by-layer completion up to about 5 ML as determined by Park et al. [25] and Kalki et al. [22,32]. For higher coverages (> 5 ML) at room temperature STM images show an epitaxial film with several incomplete layers giving rise to a pyramidal roughness [29] and an *increasing* number of incompleted layers during the film growth. Even at 400 K no layer-by-layer growth of pure Cu films on Ru(0001) occurs as indicated again by the rounded $\Delta\emptyset$-curve 1 in Fig. 6 which (compared to Fig. 4) extends up to an equivalent of about 8 deposited layers.

Summarizing this section, deposited Cu does not form surface alloys with a bare Ru(0001) substrate at any temperature. Under ergodic conditions submonolayer coverages form 2D islands of monoatomic thickness which grow until completion of the first monolayer. Both the islands and the monolayer are pseudomorphic, that is the Cu-Cu interatomic distance is expanded by ~5,5 % compared to that in a Cu(111) plane. Successively higher layers in equilibrated thicker Cu films assume various compression structures [34] in order to accommodate the misfit between the first pseudomorphic Cu layer and the regular Cu bulk structure. Up to 5 ML of Cu have been shown to grow in a layerwise fashion. At room temperature and up to 400 K the growth is non-ergodic resulting in rough surfaces with pyramidal clusters and an increasing number of incompleted atomic layers. The latter growth mode manifests itself in a rounded $\Delta\emptyset$-curve as displayed in Fig. 6, curve 1.

4 Cu growth on oxygen precovered Ru(0001)

Fig. 6 shows two work function change curves both registered at a substrate temperature of 400 K and with the same Cu deposition rate. Curve 1 passes through a rounded minimum and reincreases towards a final value which nearly corresponds to the difference in work function between Ru(0001) and Cu(111). As described in the previous section this curve is characteristic of the growth of a rough Cu film on *bare* Ru(0001); with ongoing Cu deposition the number of incomplete Cu layers and, hence, the total step-edge length increases until a steady state is reached. No features are evident which mark the completion of individual layers. Curve 2 looks strikingly different and corresponds to Cu deposition onto an oxygen precovered Ru(0001) surface. This curve shows an initial work function increase by ~0.9 eV due to the preadsorption of 0.4 ML oxygen [9] on the clean substrate. The work function then remains constant until Cu evaporation is started (indicated by the arrow). This Cu deposition first causes a rapid decrease of the work function by about 1,4 eV, followed by the appearance of eye-catching work function oscillations with a constant period. Under favourable conditions up to 80 periods have been observed [9]. In the inset of Fig. 6 an enlarged section of curve 2 is plotted against

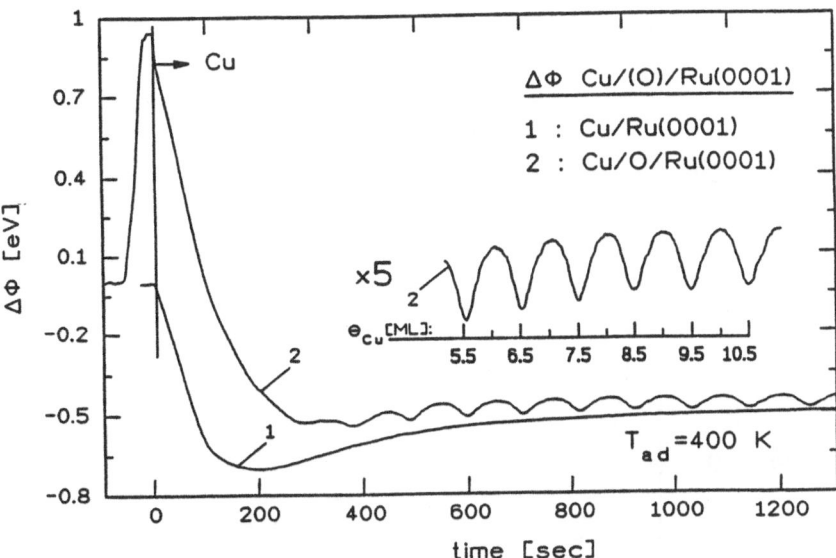

Fig. 6 Dynamically monitored work function change (ΔØ) curves as a function of Cu deposition time onto a clean (curve 1) and an oxygen precovered (curve 2) Ru(0001) surface at 400 K. The initial rise of curve 2 corresponds to the oxygen preadsorption. Cu deposition starts at the arrow (t = 0 sec) and the deposition rate is the same in both cases. The inset shows an amplification of the oscillatory structure of curve 2 as a function of Cu coverage in terms of Cu(111) monolayers (see text and ref. 35). The origin of this coverage axis coincides with t = 0 sec.

the coverage in terms of Cu(111) monolayers [35]. The origin of this scale corresponds to the start of the Cu evaporation. As can be seen, above ~4-5 ML one period corresponds to a deposited amount equivalent to one Cu(111) layer. The strict periodicity suggests a layer-by-layer growth mode and the number of oscillations becomes a count of the deposited Cu(111) layers. Furthermore a minimum corresponds to about a half filled layer while a maximum is always reached at completion of the respective layer. In Fig. 6 the amplitude of the oscillations is about 50 meV, but this value was found to depend strongly on the exact experimental conditions, in particular on the oxygen precoverage as well as on the substrate temperature [9,10], and amplitudes of up to 300 meV could be found under certain conditions [9]. 0.4 ML as used in Fig. 6 were found to be the optimal oxygen precoverage [10]. The influence of the substrate temperature is demonstrated in Fig. 7. No oscillations develop at T ≤ 370 K and T ≥ 450 K. Within this rather narrow temperature range ~400 K happens to be the optimal temperature for the occurrence of many oscillations. No such oscillations were every detected for Cu adsorption on the oxygen free Ru(0001) substrate between 300 K and 800 K. Oxygen preadsorption has obviously a dramatic influence on the properties of the growing Cu film, even up to the 50th or 80th layer! Since it is hard to imagine that oxygen at the *Cu/Ru-interface* can still exercise influence on the growth mechanism in the

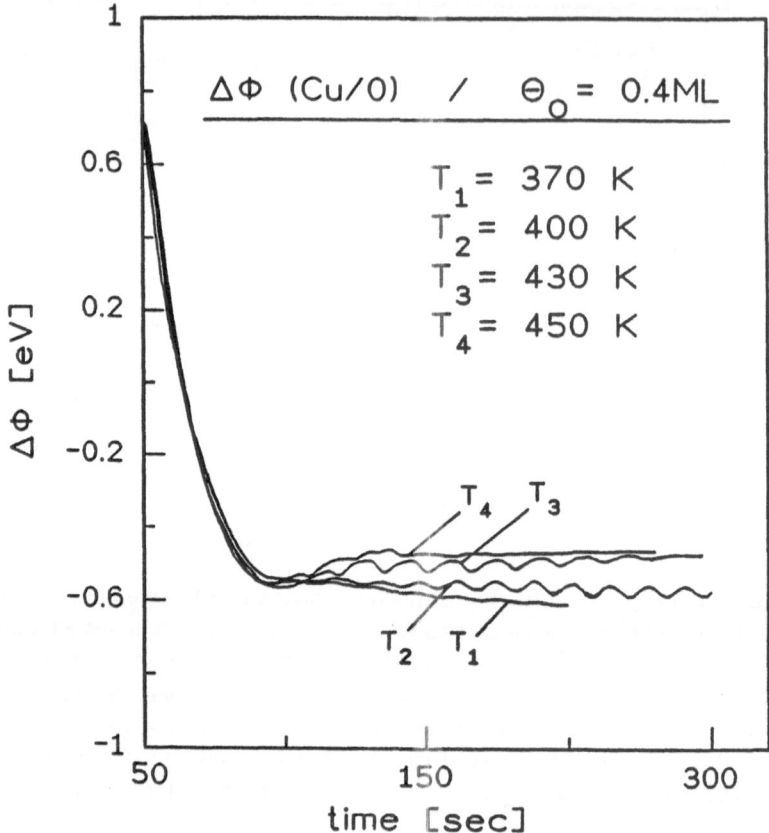

Fig. 7 Temperature dependence of the amplitude of the ΔØ oscillations observed for Cu deposition on an oxygen precovered Ru(0001) surface.

50th or 80th layer, one is lead to conclude that the oxygen "moves" with the outer surface of the growing film. This was, indeed, verified by Auger-spectroscopic measurement which showed about ~85% of the initial oxygen precoverage permanently "floating" to the Cu film surface [22,32].

Summarizing this section, we find that in the presence of oxygen Cu grows in a layer-by-layer fashion on the Ru(0001) substrat at 400 K, in total contrast to the growth behavior without oxygen. The oxygen, thus, acts as a *surfactant*. The layerwise growth manifests itself in the form of periodic work function oscillations. Each period corresponds to the deposition of one Cu(111) layer. Since the oxygen permanently "floats" to the surface while the film is thickening it supports the layer-by-layer growth up to very high layers. This surfactant-effect, however, is restricted to a rather narrow range of oxygen precoverages (between 0.3 and 0.42 ML of oxygen) and substrate temperatures (between 370 K and ~450 K) [10].

5 Discussion of growth mechanisms

Based on the correlation between the work function and the roughness (i.e. step density) of a surface as visualized in Fig. 3 the work function change curves shown in Figs. 4, 6 and 7 suggest three different growth mechanisms of Cu on Ru(0001).

Curve 1 in Fig. 4 is characteristic of 2D equilibrium [36] within the growing Cu film on *bare* Ru(0001). From the analysis of the TD-spectra shown in Fig. 5 we recall that the Cu-Ru bond is stronger than the Cu-Cu bond and that Cu atoms on Ru(0001) attract each other. At a temperature of 640 K the Cu atoms are mobil enough to condense into 2D islands within the first layer as well as to overcome the step-edge barrier ϵ_s^* (see Fig. 1a) if they happened to impinge onto an existing island. The step-down migration is also energetically favourable because $\Delta E_{1,2} < 0$. As a consequence each impinging Cu atom reaches an equilibrium site, incorporated in a 2D island within the first layer, with the same coordination and, thus, with the same dipole moment. This explains the linear decrease of curve 1 in Fig. 4 up to monolayer completion at the sharp knee. By the way, the same scenario holds for the growth of the second Cu layer on bare Ru(0001) at 640 K because it is still $\Delta E_{2,3} < 0$ [32]; the work function change is again linear up to a second knee (not shown here) [9].

As mentioned earlier curve 2 in Fig. 4 and curve 1 in Fig. 6, both registered at a substrate temperature of 400 K, are indicative of non-equilibrium within the growing Cu film on *bare* Ru(0001). At this lower temperature deposited Cu atoms are not able to overcome the step-edge barrier ϵ_s^* (Fig. 1a) in order to reach energetically more favourable sites in the lower layer. As a consequence islands grow on islands as sketched in Fig. 1b, and both the roughness and the total step-edge length increase up to a steady state. The resultant $\Delta\emptyset$-curves are rounded, and show no marked features of completion of individual layers. This growth mode leads to a pyramidal roughness as verified with STM at room temperature [29].

Obviously, curve 2 in Fig. 6 represents the most interesting case. Even though the deposition temperature was again only 400 K the periodic $\Delta\emptyset$ oscillations suggest that in this experiment the permanent presence of the oxygen at the surface causes the Cu to grow layer-by-layer. This must be concluded from the fact that each period corresponds to the deposition time of just one Cu(111) layer. An interpretation of this *periodic* behavior of the work function change in terms of a microscopic growth mechanism has to consider also the characteristic phase of the oscillations, namely the observation of minima at half-filled layers and of maxima at completed layers (see inset Fig. 6).

Before entering into the detailed discussion of the oxygen induced layer-by-layer growth mechanism we like to emphasize again that the preadsorbed oxygen is always (up to ~80 layers) found at the surface of the growing film [32] and that this "floating" oxygen layer is a prerequisite for the occurrence of the $\Delta\emptyset$ oscillations.

244

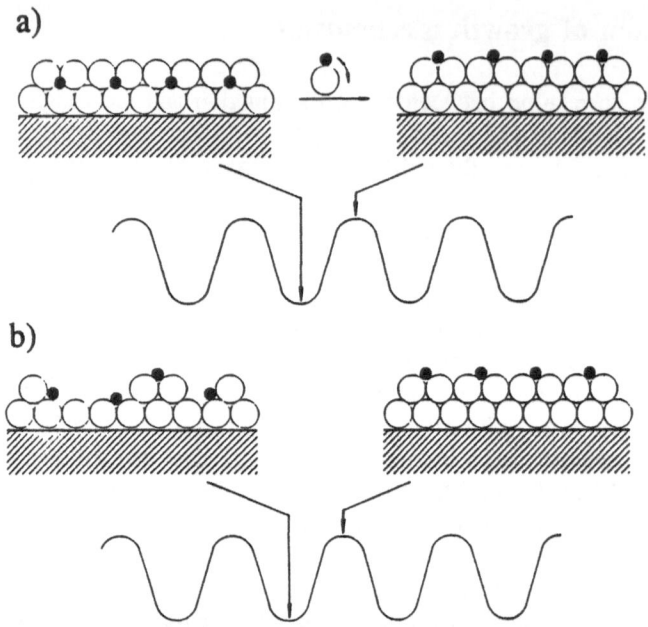

Fig. 8 Schematic representation of two possible mechanisms explaining the ΔØ oscillations observed for Cu deposition onto an oxygen precovered Ru(0001) surface. Hatched area = substrate; white circles = Cu atoms; black dots = oxygen atoms. For details see text.

Neither the intensity of the Cu(60)- nor that of the O(523)-Auger signal shows periodic variations during Cu deposition [10]. Therefore we exclude any model based on a periodic exchange between subsurface oxygen (burried during deposition) and a copper overlayer as sketched in Fig. 8a. We do not see reasonable arguments why this should happen with monolayer periodicity. The *permanent* presence of the oxygen at the surface seems to exclude the oxygen dipole moment as being the immediate origin for the work function *oscillations*. We rather suggest that during the oxygen induced *layer-by-layer* growth the roughness, namely the total step-edge length, varies periodically for each growing layer. Each new layer starts with the adsorption of adatoms followed by the formation of nuclei which grow into islands. The islands expand, coalesce and eventually fill the whole layer, before the whole sequence starts again. Obviously the total step-edge length along island boundaries first increases during nucleation and formation of (many small, see below) islands, passes through a maximum due to island coalescence and finally vanishes at completion of each layer while the oxygen is always on the surface as sketched in Fig. 8b. This minimum-maximum-minimum variation of the total step-edge length converts into a maximum-minimum-maximum variation of the work function during the growth of each new layer and, thus, explains the observed ΔØ-oscillations.

This interpretation in terms of a periodically varying surface roughness is supported by the temperature dependence of the $\Delta\emptyset$ oscillations as displayed in Fig. 7. Layer-by-layer growth can only occur if the adsorbed Cu atoms are mobile enough to overcome the step-edge barrier ϵ_s^* (Fig. 1a). Obviously at 400 K this barrier is too high without the oxygen, and sufficiently low in the presence of the oxygen in order to enable step-down diffusion (see below). At even lower temperature, namely 370 K, the atoms are again unable to surmount this (lower) activation barrier ϵ_s^* and, as a consequence, the corresponding curve in Fig. 7 shows no oscillations but rather keeps on decreasing because the roughness continues to increase. Conversely, at higher temperature namely 450 K, the mobility of the Cu atoms is eventually so high (in the presence of the oxygen) that all atoms reach and condense at *existing* steps on the surface so that the step-edge length and, hence, the work function practically does not change ("step-flow mechanism"). This situation is, after some time, realized in the T_4-curve in Fig. 7. The absolut work function at 450 K becomes higher than at 370 K because the surface is flatter at the higher temperature. The behavior of both these limiting cases in terms of surface roughness indirectly support the interpretation of the oscillations at 400 K in terms of a nucleation and 2D growth mechanism.

It remains to be explained, however, how the oxygen influences the microscopic growth mechanism, in particular, how it modifies the step-edge barrier for interlayer mass transport. As a starting point for this explanation we consider the basic rate limiting process for the interlayer mass transport

$$Cu_{up} \xrightarrow{k_i} Cu_{low}$$

namely that a Cu adatom on the *upper* terrace has to jump down on the *lower* terrace. The rate constant k_i can be expressed by an Arrhenius ansatz:

$$k_i = \nu \cdot e^{-\epsilon_s^*/RT}$$

with ν being the frequency factor (attempt frequence for step crossing) and ϵ_s^* is again the step-edge barrier (Fig. 1a). For given T (i.e. 400 K in the present case, see Fig. 6) there are obviously two possibilities to increase k_i, namely by increasing ν or by decreasing ϵ_s^*. There is experimental evidence to suggest that the presence of the surface oxygen gives rise to both effects.

Firstly, as demonstrated by STM measurements [8] the presence of surface oxygen may result in a much higher nucleation density and, as a consequence, in the formation of many small islands in contrast to the formation of few larger islands in the absence of the oxygen. A similar effect by Sb on the homoepitaxial growth of Ag on Ag(111) [7] is treated in detail in the contribution by Scheffler et al. in this book. Cu atoms which in the present case happen to impinge and diffuse about a *small* island hit much more often upon the step-edge barrier at the island boundary. The attempt frequency ν increases with decreasing island dimensions. Secondly, using a Cu vicinal surface it has been demonstrated by means of angular resolved X-ray photoelectron spectroscopy [37], that oxygen atoms prefer to occupy the high-coordination sites at the bottom of Cu surface steps as illustrated in Fig. 9. This oxygen will certain modify the electronic properties of the Cu step atoms which, in turn, may very well be expected to change (possibly reduce) the barrier ϵ_s^* for step crossing.

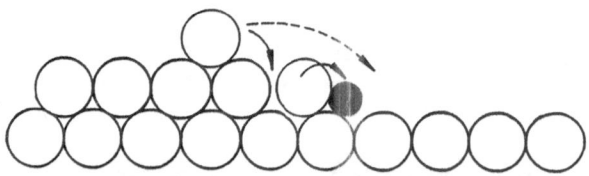

Fig. 9 Illustration of two possible mechanisms for step crossing of an adatom from an upper to a lower terrace site. The dashed path requires a transition over a step-edge barrier as shown in Fig. 1a. The incorporation process indicated by the two full arrows, however, follows a path of lower activation barrier and is expected to be even facilitated by the presence of oxygen (black atom) at the bottom of the step (see text).

Yet another process may be even more likely to facilitate interlayer mass transport. Instead of actually crossing the step-edge barrier ϵ_s^* (as indicated by the dashed arrow in Fig. 9) an atom on the island may be incorporated in the position between the outer step atom and the inner of the island pushing the step atom one position outward (see full arrows in Fig. 9). This incorporation process follows a path with an overall lower activation barrier because the effective coordination of the penetrating atom remains always higher than in the case of a simple step crossing. The adjacent oxygen atom may now be expected to assist the incorporation process in that its interaction with the outer step atom will cause a bond weakening between this step atom and the inner island Cu atoms, that is it will cause an outward relaxation of the step atom as indicated in Fig. 9 by the slightly enlarged interatomic separation. This oxygen induced outward relaxation of the (onedimensional) step-edge is expected in analogy to the observed outward relaxation of a whole (twodimensional) surface due to the presence of adsorbates [e.g. 38].

All in all there are several possible ways in which the *surfactant* oxygen may influence the kinetics of interlayer mass transport on the atomic scale. While the reduced island size due to oxygen has been shown with STM an experimental verification of the oxygen induced outward relaxation of the step-edge is not known to us. The relevance of the incorporation process is inferred from corresponding field-ion-microscopic observations [39]. As a consequence the presence of the *surfactant* oxygen enables the growth of Cu films of (more) uniform thickness on Ru(0001) at a relatively lower temperature than this would be possible without the surfactant. As emphasized in the introduction this is an important characteristics of a surfactant in order to obtain sharp interfaces for miscible metal-on-metal systems. Of course, another important condition for a coadsorbate to qualify as a surfactant is its *im*miscibility in the growing material, i.e. its permanent "floating-out", in order to obtain pure metal films. The solubility of oxygen in copper is < 0.001% at ~400 K [40].

6 Summary

Originally motivated by its catalytic properties the bimetallic system Cu/Ru has been an ideal model system for the investigation of metal-on-metal heteroepitaxy for many years. In particular the immiscibility of the Cu/Ru(0001)-system has simplified both the execution of the experiments as well as the interpretation of the results. The application of a large number of surface techniques, from early LEED/AES-measurements up to recent STM-investigations, have led to a rather complete picture of the growth mechanism of Cu on bare Ru(0001). Hereafter Cu grows clearly kinetically controlled at temperatures below ~400 K resulting in the formation of pyramidal clusters and a pronounced roughness. The equilibrium growth structure at temperatures $\geq$ 650 K is of the Stranski-Krastanov type with at least 5 complete layers. All these results where obtained by ex-situ methods, that is by measurements *after* the respective film preparation.

In the present work we have demonstrated that dynamic work function change ($\Delta\emptyset$) measurements with a specially designed KELVIN-probe are very sensitive to the (average) roughness of a surface and, most importantly, are best suited to monitor this surface roughness in-situ, that is *during* the film deposition. Using this approach we also demonstrated here that preadsorbed oxygen on the Ru(0001) substrate is an effective *surfactant* for the Cu-on-Ru(0001) system. The presence of the oxygen which permanently "floats out" to the outer Cu surface enhances the inter-layer mass transport so that at a temperature as low as 400 K good layer-by-layer growth is observed. This manifests itself in the occurrence of periodic $\Delta\emptyset$-oscillations which are interpreted in terms of a periodic variation of the total step-edge length during nucleation and 2D island growth per layer. The microscopic influence of the oxygen on the step-edge barrier which prevents spontanous interlayer diffusion is discussed in terms of a reduction of the step-edge barrier ϵ_s^* due to electronic effects as well as an improved incorporation (rather than step crossing) mechanism.

References

[1]   G. Ehrlich and F.G. Hudda, J. Chem. Phys. **44**, 1039(1966)
[2]   R.L. Schwoebel and E.J. Shipsey, J. Appl. Phys. **37**, 3682(1966)
[3]   E. Bauer, Appl. Surf. Sci., **11/12**, 479(1982)
[4]   R. Kern, G. Lelay and J.J. Metois: *Basic Mechanisms in the Early Stages of Epitaxy,* Chapter 3, in *Current Topics in Material Science*, Vol. 3 (North-Holland, Amsterdam, 1979)
[5]   W.F. Egelhoff, Jr. and D.A. Steigerwald, J. Vac. Sci. Technol. **A7**, 2167(1989)
[6]   M. Copel, M.C. Reuter, E. Kaxiras and R.M. Tromp, Phys. Rev. Lett. **63**, 632(1989)
[7]   J. Vrijmoeth, H.A. van der Vegt, J.A. Meyer, E. Vlieg and R.J. Behm, Phys. Rev. Lett. **72**, 3842(1994)
[8]   S. Esch, M. Hohage, T. Michely and G. Comsa, Phys. Rev. Lett. **72**, 518(1994)
[9]   H. Wolter, M. Schmidt and K. Wandelt, Surf. Sci. **298**, 173(1993)
[10]   M. Schmidt, H. Wolter and K. Wandelt, Surf. Sci. **307-309**, 507(1994)
[11]   M. Schmidt, H. Wolter, M. Nohlen and K. Wandelt, J. Vac. Sci., Technol. **A12**, 1818(1994)
[12]   M. Henzler, Prog. Surf. Sci. **42**, 297(1993)
[13]   J.M. Van Hove, C.S. Lent, P.R. Pukite and P.I. Cohen, J. Vac. Sci. Technol. **B1**, 741(1983)
[14]   J.E. Parmeter, R. Kunkel, B. Poelsema, L.K. Verheij and G. Comsa, Vacuum **41**, 467(1990)
[15]   H.A. van der Vegt, J.M.C. Thornton, H.M. van Pinxteren, M. Lohmeier and E. Vlieg, Phys. Rev. Lett. **68**, 3335(1992)
[16]   J. Hölzl, G. Porsch and P. Schrammen, Surf. Sci. **97**, 529(1980)
[17]   K. Besocke and S. Berger, Rev. Sci. Instrum. **47**, 840(1976)
[18]   R. Smoluchowski, Phys. Rev. **60**, 661(1941)
[19]   B. Krahl-Urban, E.A. Niekisch and H. Wagner, Surf. Sci. **64**, 52(1977)
[20]   K. Wandelt, in *Chemistry and Physics of Solid Surfaces VIII*, Vol. 22 of *Springer Series in Surface Science*, Eds. R. Vanselow and R. Howe (Springer, Heidelberg, 1990) p. 314
[21]   K. Wandelt, in *Thin Metal Films and Gas Chemisorption*, Ed. P. Wissmann, Vol. 32 of *Studies in Surface Science and Catalysis* (Elsevier, Amsterdam, 1987) p. 280
[22]   K. Kalki, M. Schick, G. Ceballos and K. Wandelt, Thin Solid Films **228**, 36(1993)
[23]   K. Christmann, G. Ertl and H. Shimizu, J. Catal. **61**, 397(1980)
[24]   J.C. Vickermann, K. Christmann, G. Ertl. P. Heimann, F.J. Himpsel and D.E. Eastman, Surf. Sci. **134**, 367(1983)
[25]   C. Park, E. Bauer and H. Poppa. Surf. Sci. **187**, 86(1987)
[26]   J.E. Houston, C.H.F. Peden, D.S. Blair and D.W. Goodman, Surf. Sci. **167**, 427(1986)

[27]   K.S. Kim, J.H. Sinfelt, S. Eder, K. Markert and K. Wandelt, J. Phys. Chem. **91**, 2337(1987)

[28]   H. Tochihara, G. Rocker, M. Martin and T. Yates, Surf. Sci. **203**, 44(1988)

[29]   G. Pötschke, J. Schröder, C. Günther, R.Q. Hwang and R.J. Behm, Surf. Sci. **251/252**, 592(1991)

[30]   G. Pötschke and R.J. Behm, Phys. Rev. **B44**, 1442(1991)

[31]   M. Schmidt, Diplom-thesis, University Bonn, 1992

[32]   K. Kalki, PhD-thesis, University Bonn, 1992

[33]   J. Schäfer, P. Reinhardt, H. Hoffschulz and K. Wandelt, Surf. Sci. **313**, 83(1994)

[34]   C. Günther and R.J. Behm, private communication

[35]   The deposition time per period of the oscillations is, indeed, 11% longer than the deposition time to reach the first knee in a $\Delta\emptyset$-curve at 640 K on bare Ru [9]. This is in perfect agreement with the reduced atomic density of the first pseudomorphic layer compared to a perfect Cu(111) plane.

[36]   There is, of course, no equilibrium between the surface and the 3D vapor phase. The fact that the film grows is a consequence of non-equilibrium, namely of a supersaturation of the Cu vapor phase at the given substrate temperature. However, if reevaporation of Cu atoms into the gas phase is negligible one can still speak of equilibrium or non-equilibrium conditions *within* the 2D adsorbed layer only

[37]   C.S. Fadley, S. Kono, J.T. Lloyd and K.A. Thompson, Proc. 4th ICSS & 3rd ECOSS (Cannes, 1980) Suppl. Le Vide, Les Couches Minces, No. 201, Vol. 1, p. 665

[38]   M.A. Van Hove in: *Nature of the Surface Chemical Bond"*, Eds. T. Rhodin, G. Ertl (North-Holland, Amsterdam, 1979)

[39]   G. Ehrlich, Surf. Sci. **331-333**, 865(1995)

[40]   Landolt-Börnstein, *Zahlenwerte und Funktionen*, 6. ed., Vol. 4, Part 2b, p. 685

Energetic Condensation as a Means of Inducing the Growth of Films Containing High Pressure Phases

D.R. McKenzie, N.A. Marks, P. Guan, B.A. Pailthorpe, W.D. McFall, Y. Yin
School of Physics, University of Sydney, NSW, 2006, Australia

Abstract. Molecular dynamics simulations are used to show how compressive stress in thin films depends on the energy of the incoming species used in the growth process. For all other conditions held constant, there is an impact energy which gives a local maximum in the compressive stress. The Gibbs free energy is used to show how preferred orientation effects arise in anisotropic materials under biaxial compressive stress fields and to derive a stress value for the transition from the hexagonal phase of carbon and boron nitride to the cubic phase. Predictions are made concerning the microstructures of carbon and boron nitride films grown under compressive stress conditions and comparison made where possible with experiment.

1. Introduction

Film growth by physical vapour deposition (PVD) is carried out, in its simplest form, by directing a beam of a condensable atomic or ionic species onto a substrate. The value of the impact energy has a strong effect on the microstructure of the resulting film. There is a consistent pattern of behaviour displayed by a wide range of materials as the impact energy is increased [1-4]. At low energies there is a tendency for the film to show a columnar structure containing voids and to show a tensile stress due to attraction across the void spaces. At higher impact energies, of a value greater than the bond energy, there is an increasing tendency for the density to increase as the void spaces are eliminated. The stress passes from tensile to compressive. The increase in density has been ascribed to a 'subplantation' process [5] in which the incoming atoms bury themselves below the surface of the growing film. At high impact energies, mobility induced by the impact allows compressive stress to be relieved. The high impact energies also result in implantation into the substrate which can cause intermixing of film and substrate atoms. Such an intermixing may result in improved adhesion between the film and substrate. These ideas are summarised in the diagram of Fig. 1.

This paper is concerned with the effects on microstructure resulting from high compressive stress conditions which occur in certain energy ranges. These conditions can lead to preferred orientation effects and if the stress is sufficiently high, microstructures which are normally stable only at high pressures may be stabilised.

As an example, boron nitride thin films deposited by ion plating techniques will be discussed in this paper. To assist in understanding the behaviour of real systems, useful insights can be obtained from molecular dynamics simulations.

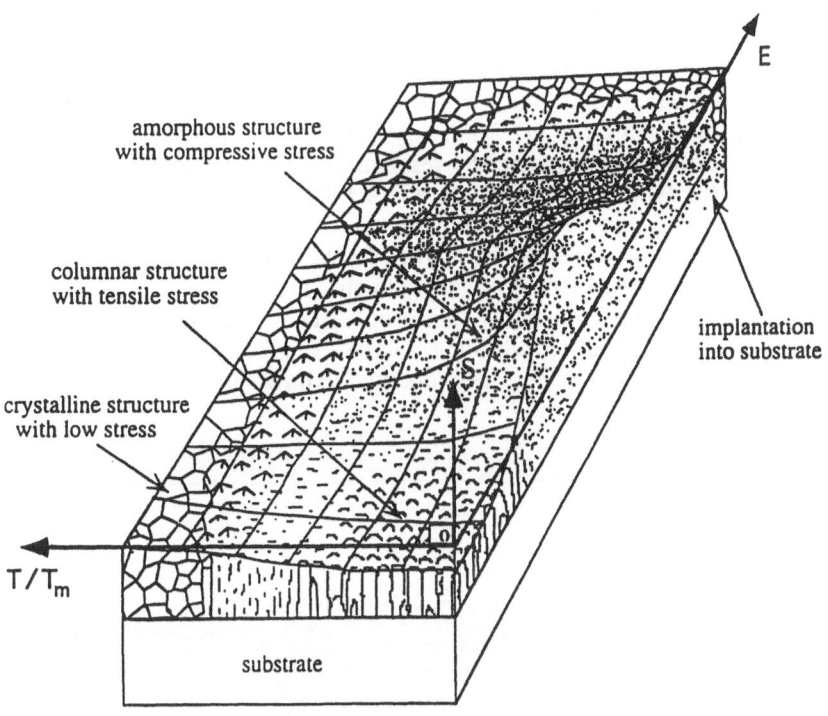

amorphous structure
with compressive stress

columnar structure
with tensile stress

crystalline structure
with low stress

implantation
into substrate

E

T/T_m

substrate

Fig. 1: A schematic diagram illustrating the effects on film microstructure and stress levels of the two variables E, the impact energy, and T/T_m, the ratio of growth temperature to melting point. The film surface height is used to represent the stress level in the film. Surface heights less than zero refer to tensile stress, above zero to compressive stress.

2. Molecular Dynamics Simulations

In order to gain information about the details of processes occurring under energetic impacts, molecular dynamics simulations have been used to examine the behaviour of model systems at very small time steps of 0.1 ps. Experimental studies on real systems with this time resolution and atomic spatial resolution are unattainable at present.

Simulations of metals have been made with interactions via central potentials in both two and three dimensions. This work is an extension of that of Muller [6, 7] who carried out two dimensional simulations of metal like systems. Covalent systems have also been described in two dimensions using a Stillinger-Weber type potential and such a system has been used to study stress generation in films.

3. Lennard-Jones Systems

A Lennard-Jones 6-12 potential appropriate for nickel was used to study the phenomena occuring below the surface of a crystal at room temperature after energetic impact. An array of 25 nm x 11 nm (5200 atoms) was used. Energy was removed from the boundaries by rescaling of velocities. Focussed collision sequences of the type originally proposed by Silsbee [8] were observed to occur along close packed directions and propagated at a velocity greater than that of sound in the medium, generating wedges of sound waves rather like the sound waves generated by a supersonic projectile in air. The energy is removed from the collision sequence or 'focuson' by this process and by damping caused by the electron cloud, providing a limit to the range of the focussed collision sequence [9]. An example showing these phenomena is shown in Fig. 2 for an incident energy of 50 eV.

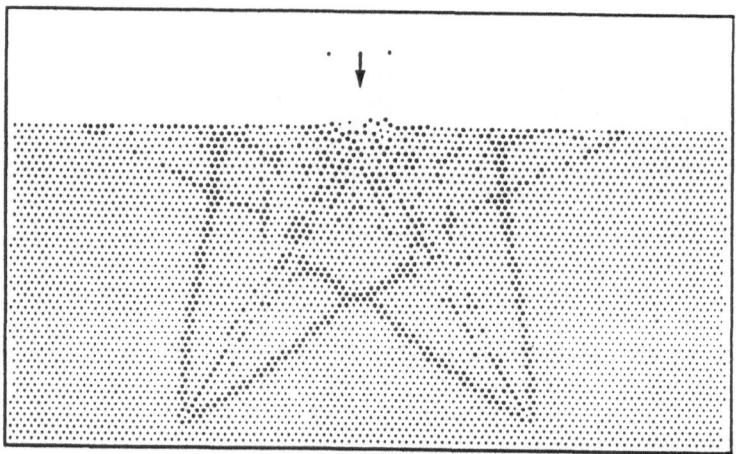

Fig. 2: The phenomena occurring below the surface of a 2D array of atoms interacting via a Lennard-Jones potential 0.2 ps after impact. Atoms with energies greater than 0.2 eV are shown as larger discs. Focussed collision sequences emanating from the impact site and the sound waves generated by them are clearly visible.

In addition, a localised region of roughly semicircular shape appears beneath the impact site. This region may be termed a 'thermal spike' as it represents a region with an initially high temperature which rapidly cools to the ambient temperature. This region is best visualised with a parameter which measures the local distortion of the array. Such a parameter can defined as the deviation from the equilibrium bond angle (60°) between nearest neighbours. Figure 3 shows an example of such a thermal spike.

A three dimensional Lennard-Jones simulation was also carried out on a 5.7 nm x 5.7 nm x 2.9 nm (9216 atoms) array of a face centred cubic lattice. In general, focussed collision sequences were less readily initiated than in 2D but could be generated easily by impacts directed along the close packed (110) directions. The focussed collision sequence also generated wedges of sound waves as in 2D, but these were confined to the close packed (111) planes. Figure 4 shows two sound wedges produced in the two (111) planes which intersect along the (110) direction of the focussed collision sequence.

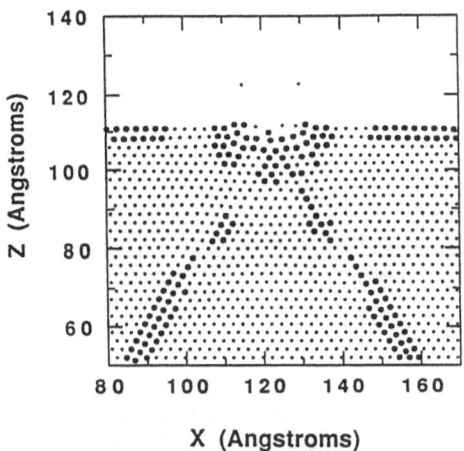

Fig. 3: The semi-circular thermal spike region occurring below the impact site 0.5 ps after impact in the 2D array of Fig. 2. The large discs represent sites of large local distortion.

4. Covalent Systems

A two dimensional covalent system was modelled by an array of atoms in which each atom had three bonds to its neighbours producing a structure equivalent to a single sheet of graphite. A Stillinger-Weber potential was used to describe the interactions. Impact studies showed that focussed collision sequences did not occur in this system but a region analogous to the thermal spike of the previous section was observed.

Film growth was simulated in this system by adding atoms at random positions onto the surface of a substrate within an interval. This simulated a monoenergetic incident beam of atoms. Four different energies were used to grow films: 1 eV, 10 eV, 30 eV and 50 eV. The films were all amorphous. The lateral stress in the film was calculated by evaluating the force across a vertical line drawn in the film and averaging the results for many such lines. The stress showed a transition from tensile to compressive as the ion energy increased as shown in Fig. 5, in agreement with experimental observations [1-3].

For energies above 50 eV, the sputter yield became excessive so that the computer time required to grow a film became prohibitive. To obtain results for the stress at 75 eV and 100 eV, 60 atoms were used to bombard the film grown at 50 eV.

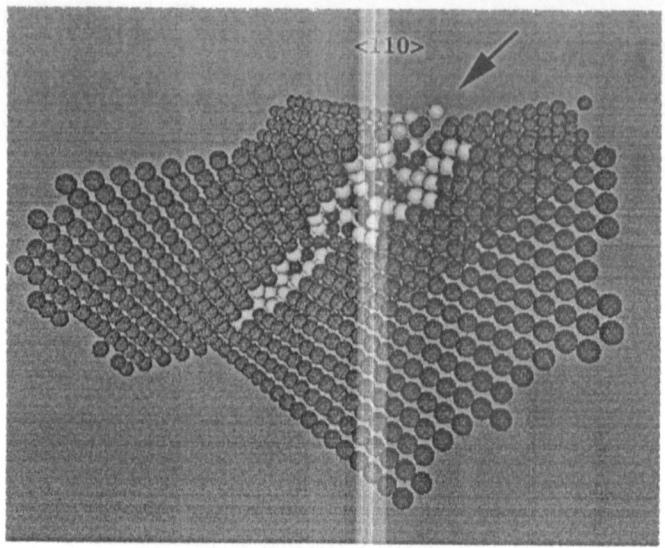

Fig. 4: A 3D view of a focuson propagating along a <110> direction and generating two wedges of sound waves in the two (111) planes which intersect along the <110> direction containing the focuson. White atoms have an energy greater than 0.20 eV.

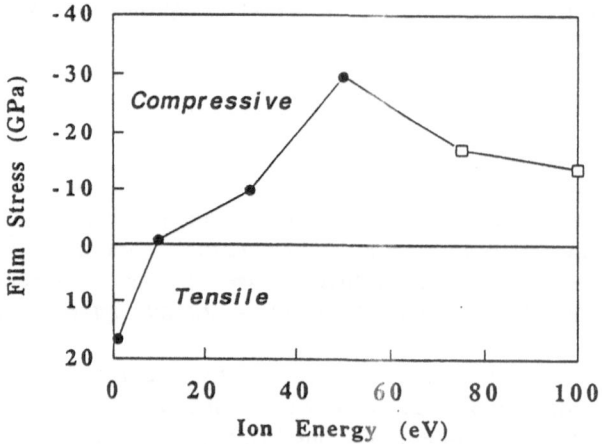

Fig. 5: The stress in films grown in a molecular dynamics simulation from monoenergetic beams of energy E for atoms interacting by a Stillinger-Weber potential. Hollow symbols indicate that the film was produced by bombarding the film grown at 50 eV with ions of the energy specified.

The films grown at 50 eV showed some degree of mixing of the deposited atoms with the substrate atoms. This process is expected to become, at higher impact energies, an implantation effect in which the mixing layer extends deep into the substrate.

5.   Thermodynamics of Compressive Stress

The processes occurring during the film growth from an energetic beam are nonequilibrium processes. Mobility of the atoms taking part in the thermal spike and the other phenomena is only available for a short time until the energy is extracted into the surrounding material giving a rapid chilling effect. This process does not give enough time for equilibrium structures to evolve and so, at least for covalent systems, amorphous or microcrystalline rather than fully crystalline structures will result. An equilibrium thermodynamic analysis should therefore be considered only a first step in the understanding of energetic film growth phenomena.   The importance of elastic strain energy in the development of preferred orientation in thin films has been known for some time [10-12]. The treatment presented here is a generalisation of previous ideas to include strains resulting from structural phase transformations. However, in this treatment we consider bulk strain energy only and neglect the contribution from the surface energy.   This approximation becomes more accurate as the film thickness increases.

The Gibbs free energy function is useful in determining which of various competing structures will be the most stable and given conditions of stress and temperature. For a general stress field the Gibbs free energy is given by [13]:

$$G = U - TS - \sum_{ij} \varepsilon_{ij} \sigma_{ij} \tag{1}$$

where U, S, ε_{ij} and σ_{ij} are the internal energy, entropy, strain tensor and stress tensor respectively.

As an example, G will be evaluated for two different orientations of an anisotropic material such as hexagonal graphite or hexagonal boron nitride. For a thin film on a substrate, the stress tensor σ_{ij} is biaxial, having the form:

$$\sigma_{ij} = \begin{bmatrix} \sigma & & \\ & \sigma & \\ & & 0 \end{bmatrix} \tag{2}$$

for x and y axes chosen within the plane of the film. Comparing G for orientations of the crystallographic c axis lying normal to the stress plane and within the stress plane, we obtain:

$$G_h^{\perp}(T_o, \sigma) = G_h(T_o, 0) - \frac{\sigma^2}{2}(s_{11}^h + s_{12}^h) \tag{3}$$

$$G_h^{||}(T_o, \sigma) = G_h(T_o, 0) - \frac{\sigma^2}{2}(s_{11}^h + 2s_{13}^h + s_{33}^h) \qquad (4)$$

where σ is the magnitude of the stress and the s_{ij}^h are the elastic compliances of the hexagonal material. A reference temperature of T_o is chosen. The expressions (3) and (4) may be evaluated using published values of the elastic compliances for graphite [14]. The results in Fig 6 show that $G_h^{|}(T_o, \sigma)$ is always lower than $G_h^{\perp}$ (T_o, σ) at non zero stress levels. This will give rise to a special kind of preferred orientation in which the c axis is constrained to lie in the stress plane, but its direction in that plane is not constrained. The crystallographic, a and b axes may rotate around the c axis without changing the Gibbs free energy. The microstructure shown in Fig. 7 is therefore expected to occur in graphite and hexagonal boron nitride films under compressive stress conditions. The graphite sheets are normal to the surface and form ribbons which meander over the surface. The bonding pattern within the ribbons, however, is not constrained in orientation and will vary randomly from place to place. The electron diffraction pattern resulting from such a microstructure is shown schematically in Fig. 8 for an electron beam incident on the side of the film.

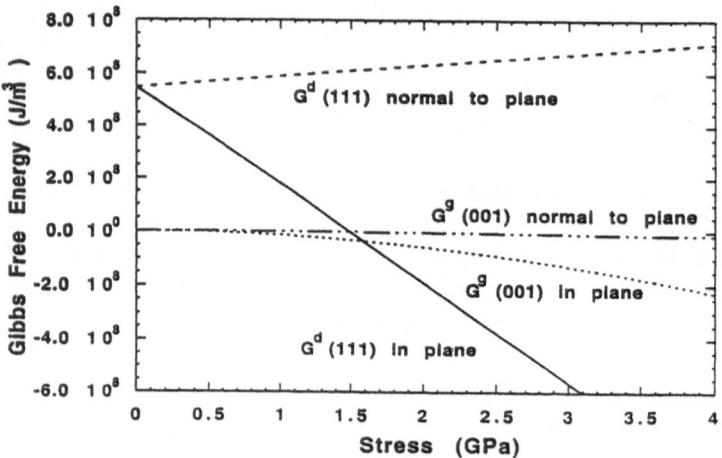

Fig. 6: The Gibbs free energy of cubic diamond G_d and hexagonal graphite G_g as a function of biaxial compressive stress. In each case two orientations are considered.

In order to see how the diffraction pattern of Fig. 8 arises, the following operations are carried out on the reciprocal lattice of graphite.
(a) Spin the reciprocal lattice around the c axis ((001) direction) so that all angles are taken with equal probability. The c axis is drawn horizontally in Fig. 8.
(b) Rotate the result of (a) around a vertical line so that all angles are taken with equal probability.

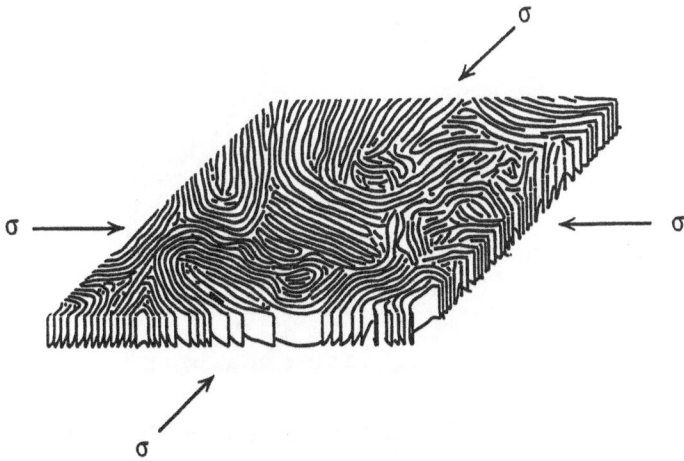

Fig. 7: A schematic diagram of the expected microstructure of hexagonal boron nitride or graphite under conditions of biaxial compressive stress. The crystallographic c axis is normal to the ribbons and the a and b axes are randomly oriented in the plane of the ribbons.

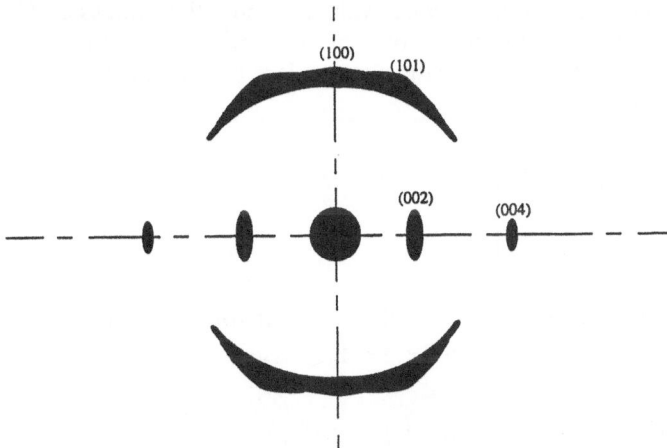

Fig. 8: A schematic diagram of the electron diffraction pattern which would be obtained from the microstructure of Fig. 7, viewed from the side.

(c) Take the intersection of the result of (b) with the plane of the page, which represents the zero order have zone of the diffraction pattern. The principal features of the resulting diffraction pattern are:
   (i)   A series of $(00l)$ spots lying along a horizontal line, the brightest originating from the (002) and (004) reciprocal lattice sites.
   (ii) Two arcs of scattering originating from the (100) reciprocal lattice sites. These arcs have maximum intensity at two points on the vertical axis.
   (iii) A set of arcs originating from the (101) reflection with maximum intensity either side of the vertical axis and close to it.

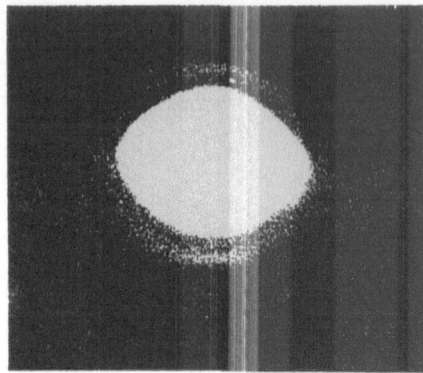

Fig. 9: The observed electron diffraction pattern for a h-BN film viewed from the side.

The observed diffraction pattern for a film of h-BN under biaxial compressive stress is shown in Fig. 9 and shows good agreement with the prediction of Fig. 8.

For a diamond phase in thermal equilibrium with graphite under a biaxial stress field the Gibbs free energy may be evaluated using the expression (1). The strain of diamond relative to graphite may be evaluated by assuming the graphite to diamond transition to occur via the intermediary of a rhombohedral graphite phase. The strain tensor is then:

$$\varepsilon_{ij} = \begin{bmatrix} -\delta & & \\ & -\delta & \\ & & -\gamma \end{bmatrix}. \tag{5}$$

The Gibbs free energy of cubic diamond for the (111) diamond direction perpendicular to the plane of the stress is

$$G_c^{\perp}(T_0, \sigma) = \Delta G\,(T_0, 0) - \sigma\,(2\delta) - \frac{2\sigma^2}{3}\,(s_{11}^c + s_{12}^c + s_{44}^c) \tag{6}$$

and for the (111) direction lying in the plane of the stress:

$$G_c^{\|}(T_0, \sigma) = \Delta G\,(T_0, 0) - \sigma\,(\gamma + \delta) - \frac{\sigma^2}{2}\,(\tfrac{3}{2}\,s_{11}^c + s_{12}^c + \tfrac{5}{4}\,s_{44}^c). \tag{7}$$

Using published values of the elastic compliances of diamond, γ, δ [15] and the free energy difference between graphite and diamond $\Delta G\,(298, 0)$ [16], the curves shown in Fig. 6 are obtained.

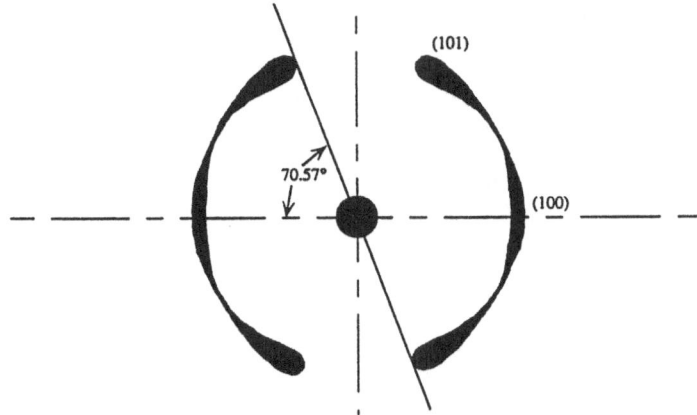

Fig. 10: A schematic diagram of the electron diffraction pattern for a c-BN film under high biaxial compressive stress, viewed from the side. The (111) diffraction ring is broken into segments as shown, with the width indicating approximate intensity.

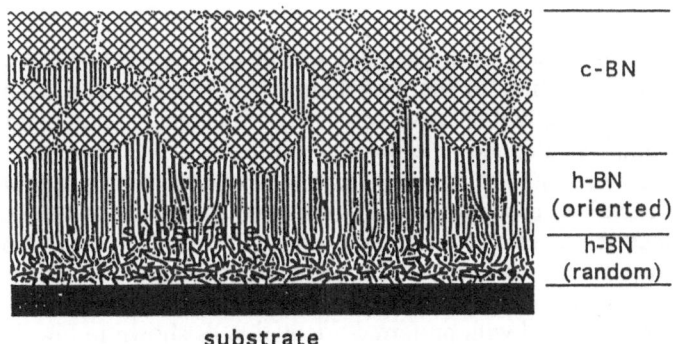

Fig. 11: The predicted sequence of microstructures in a boron nitride film deposited by ion plating. The sequence is formed as the compressive stress progressively increases.

These curves predict that, as the biaxial compressive stress is increased from zero, the most stable phase is initially graphite with its c-axis in the plane of the stress. At a stress level of approximately 1.6 GPa, the diamond phase becomes the stable phase, with its (111) axis lying in the plane of the stress. Since the elastic properties of hexagonal boron nitride (h-BN) are very similar to those of graphite and the elastic properties of cubic boron nitride (c-BN) are very similar to those of diamond, these results may be used for the boron nitride case also.

The preferred orientation of the cubic phase under compressive stress is of interest. The preferred orientation is such that a (111) direction lies in the stress plane. The Gibbs free energies not affected by rotations around this (111) axis, so all orientations of the other three (111) axes will be equally likely. Similarly, the Gibbs free energy is not affected by the direction of the (111) axis in the stress

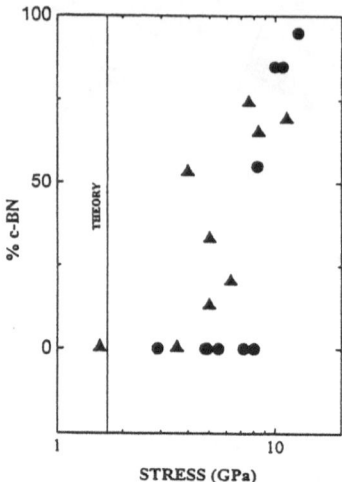

Fig. 12: The fraction of c-BN in percent determined by infrared absorption spectroscopy as a function of the compressive stress for BN films prepared by ion playing.

plane, so all directions in the plane will be equally likely. The diffraction pattern of a diamond film in thermodynamic equilibrium with graphite oriented as described above will be as shown in Fig 10. The (111) reflection is the strongest reflection for diamond or c-BN and will have a distribution as shown, with two bright zones lying horizontally and four bright arcs and a missing segment. This result is derived using procedures similar to those used for the graphite case above.

Since the level of compressive stress progressively rises from zero as a thin film grows from the substrate [17], it is expected that BN will show a progression from h-BN with no preferred orientation, to h-BN with preferred orientation followed by c-BN with preferred orientation as shown in Fig. 11. This sequence of structures has been observed experimentally [18]. The only remaining prediction which has not yet been clearly observed is the expected preferred orientation of the c-BN.

The fraction of c-BN in films deposited by ion plating is shown in Fig. 12 as a function of the measured compressive stress level. There is a clear transition from zero c-BN content to high levels at a critical stress level. This level is somewhat above the predicted value of 1.6 GPa. The reasons for the discrepancy could lie in the use of graphite and diamond property data in the theoretical formulae.

The data of Fig. 12 were obtained using a boron evaporation source, a r.f. excited plasma and an r.f. biased substrate holder heated to 300°C during deposition [19]. Details of the stress measurement technique can be found in an earlier paper [17].

The formation of a diamond structure phase under high compressive stress conditions in films deposited from carbon beams produced by cathodic arc plasma sources has already been demonstrated [20]. In contrast to the BN case, the material is amorphous but shows a highly tetrahedral character and a density 85% of that of crystalline diamond [20]. A very similar critical stress level is required for the formation of this material, referred to as tetrahedral amorphous carbon [20].

6. Conclusions

The use of energetic condensation to create high compressive stress conditions has been illustrated using molecular dynamics simulations. The existence of an energy region for the formation of high compressive stress has been shown, in agreement with experiment. Equilibrium thermodynamics has been used as a first step in showing how high compressive stress conditions can induce microstructural effects. Preferred orientation is produced as a consequence of the minimisation of free energy. High pressure phases of materials such as cubic boron nitride and diamond are formed when the compressive stress is sufficiently high.

7. Acknowledgment

The authors acknowledge the financial support of the Australian Research Council.

8. References

1. J. A. Thornton and D. W. Hoffman, J. Vac. Sci. Technol., **14**, 164 (1977).
2. P. J. Martin, R. P. Netterfield, T. J. Kinder and L. Descotes, Surf. Coat. Technol., **49**, 239 (1991).
3. P. M. Fabis, R. A. Cooke and S. McDonough, J. Vac. Sci. Technol., **A8**, 3809 (1990).
4. J. A. Thornton, Ann. Rev. Mater. Sci., **7**, 239 (1977).
5. Y. Lifshitz, S. R. Kasi and J. W. Rabalais, Phys. Rev. Lett., **62**, 1290 (1989).
6. K-H. Muller, Appl. Phys., **A40**, 209 (1986).
7. K-H. Muller, J. Appl. Phys., **62**, 1796 (1987).
8. R. H. Silsbee, J. Appl. Phys., **28**, 1246 (1957).
9. N. A. Marks, P. Guan, D. R. McKenzie and B. A. Pailthorpe, Proceedings of Materials Research Society Conference, Boston, December 1993.
10. F. Witt and R. W. Vook, J. Appl. Phys., **39**, 2773 (1968).
11. T. C. Huang, G. Lim, F. Parmigiani and E. Kay, J. Vac. Sci. Technol., **A3**, 2161 (1985).
12. T. Nakano, S. Baba, A. Kobayashi, A. Kinbara, T. Kajiwara and K. Watanabe, J. Vac. Sci. Technol., **A9**, 547 (1991).
13. J. F. Nye, *Physical Properties of Crystals*, Clarendon Press, Oxford (1960), p. 175.
14. B. T. Kelly, Ed., *Physics of Graphite*, Applied Science, London, (1981) p. 74.
15. J. E. Field, Ed., *The Properties of Diamond*, Academic, London (1979) p. 647.
16. R. Berman and F. Simon, Z. Elektrochem., **59**, 333 (1955).

17. D. R. McKenzie, W. D. McFall, W. G. Sainty, C. A. Davis and R. E. Collins, Diamond and Related Materials, **2**, 970 (1993).
18. D. J. Kester, K. S. Ailey, R. F. Davis and K. L. More, J. Mater. Res., **8**, 1213 (1993).
19. D. R. McKenzie, W. D. McFall, S. Reisch, B.W. James, I. S. Falconer, R. W. Boswell, H. Persing, A. J. Perry and A. Durandet, Thin Solid Films (submitted).
20. D. R. McKenzie, D. A. Muller and B. A. Pailthorpe, Phys. Rev. Lett., **67**, 773 (1991).

Atomically Abrupt Interfaces of Compound Semiconductor Heterostructures: The AlAs/GaAs Case

D. Bimberg, J. Christen* and W. Wittke
Institut für Festkörperphysik, Technische Universität Berlin,
Hardenbergstraße 36, 10623 Berlin, Germany

D. Gerthsen** and D. Stenkamp
Institut für Festkörperforschung, Forschungszentrum Jülich, Germany

D.E. Mars and J.N. Miller
Hewlett-Packard Laboratories, Palo Alto, California 94303

present address:

*   Otto-von-Guericke Universität, 30916 Magdeburg, Germany

**   Laboratorium für Elektronenmikroskopie der Universität, Kaiserstr. 12, 76128 Karlsruhe, Germany

Abstract: The structure of AlAs/GaAs/AlAs quantum well interfaces grown with 100 s interruption at each of the interfaces is investigated. Cathodoluminescence (CL) peak energy histograms, wavelength images of the excitonic recombination and high-resolution transmission electron microscopy pictures yield partly identical and partly complementary atomic-scale information in the spatial frequency range from 1 mm^{-1} to 3 nm^{-1}. Large smooth islands extending more than 2000 nm at the AlAs on GaAs interface are unambiguously observed by CL. TEM-pictures show interfaces without any steps within the whole area of observation being 150 nm, thus corroborating the CL results. In contrast to this the island extension at the GaAs on AlAs interface is 2-4 nm. All interfaces are found to be graded in growth direction due to cation exchange.

1. Introduction

Optical as well as transport properties of low-dimensional semiconductor structures depend to a large extent on the crystallographic-chemical properties of the hetero-interfaces confining the carriers:

-   The usually dominant broadening mechanism of luminescence from quantum wells /1/ or wires /2,3/ at any temperature /4/ is inhomogeneous, caused by roughness of the interfaces at a subexcitonic length scale. Variations of the mean interface position(s) across distances larger than the exciton Bohr-diameter induce in-plane localization of the 2D- (1D)-excitons leading to violation of the k-selection rule /4,5/.

    Formation of interface islands of monoatomic height extending over distances d_I much **larger than the Bohr-diameter ($d_I > a_B$) and the ambipolar diffusion length ($d_I > l_D$)** lead to an observable decomposition of quantum wells into quantum columns and subsequent splitting of the QW luminescence line into a multiplet showing **n o thermalization** of the components even at lowest temperature /6/.

-   The mobility of two-dimensional (2D) carriers in thin quantum wells or at modulation-doped inverted GaAs/AlGaAs heterostructures depends on interface roughness scattering /7,8/ similar to the case of 2D electrons in Si MOSFET inversion layers.

-   The peak to valley ratio and the voltage swing in tunneling structures like double barrier diodes depends on interface roughness.

Variations in the parameters of the growth, like growth rate, flux ratio, growth interruption time /9,10/ leading to modifications of its kinetics result in changes of the interface roughness. In order to take full advantage of the potential provided by modern growth technologies to engineer surfaces and interfaces resulting in profound changes of electronic properties, we must possess means to characterize them microscopically on an atomic scale and we must determine precisely what the limits of such engineering efforts are. In particular it is essential to know to which extent the limit of an **ideally smooth** interface showing no in-plane variations of the chemical composition for a given fixed lattice plane can be approached in reality.

A number of authors have recently tried to address this question for the $Al_xGa_{1-x}As/GaAs/Al_xGa_{1-x}As$ model system, using mostly luminescence techniques, including rather sophisticated cathodoluminescence wavelength imaging (CLWI) /6,11/ and to a lesser extent equally sophisticated high-resolution transmission electron microscopy (HRTEM) /12-14/.

These results and their interpretation are, however, subject of large controversy. In particular the existence of large monolayer islands is questioned /17,18/. Luminescence results from different samples or even the same samples seem to contradict each other /e.g. 15-18/ and conclusions drawn from HRTEM and luminescence on the same sample were found to be difficult to reconcile. Only recently for a particular case for the first time at least qualitative agreement between conclusions based on these mutually complementary methods was obtained /19/. Detailed reviews of the present state of understanding of the morphology of the GaAs/AlGaAs-interface is given in /10/.

With this paper we intend to resolve the controversy by presenting results of comprehensive CL, CLWI and HRTEM study of the AlAs/GaAs type I QW model system grown under particular conditions by molecular beam epitaxy together with a very careful data evaluation. We give unambiguously evidence for the heavily disputed existence of monoatomic AlAs on GaAs interface islands **extending over several hundred nm.** Information obtained from luminescence and HRTEM experiments is shown to be in quantitative agreement and to be partly complementary leading to a complete description of the structure of **both** interfaces of the QW at spatial frequencies from 1 mm^{-1} to 3 nm^{-1}. Histograms of CL peak positions taken at appropriate temperatures /10/ and monochromatic CL images (CLI) /6/ and CWLI /10/ of the columnar structure of the QW are found to be particularly important to avoid possible misinterpretations of the information on interfaces contained in luminescence spectra.

2. Samples and Experimental Methods

The samples investigated here are grown on exactly oriented (001) semi-insulating GaAs substrates (misorientation $\pm$ 0.1°) covered by a 500 nm GaAs buffer. They contain from top to bottom a 80 nm $Al_{0.4}Ga_{0.6}As$ cap layer followed by three single GaAs QWs of 2, 5 and 8 nm width sandwiched between 18 nm AlAs barriers. The growth temperature was 620°C and the growth rates of the GaAs and AlAs layers were 1 μm/h and 0.67 μm/h, respectively. The samples were rotated with 25 rpm to reduce lateral inhomogeneities. The growth was either interrupted for 100 s at both interfaces or uninterrupted.

We have chosen barriers of AlAs instead of ternary AlGaAs to facilitate the HRTEM experiments and to avoid ambiguities in their interpretation. In the alloy the two cations Al and Ga show (at least) statistical composition fluctuations /21/.

stronger inhomogeneous broadening, and the doublet is unresolved, the peak energy of the convolution of the two components varies across the surface even if the energy of the free exciton peak is not varying. Keeping the local position of excitation constant and varying its intensity I_{exc}, the convoluted peak energy varies by several meV due to differences in the I_{exc}-dependence of the free and the bound exciton recombination. This effect might be one source of misinterpretations in literature.

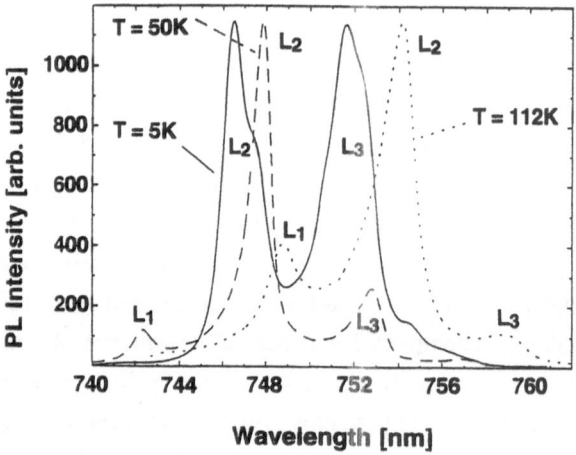

Fig. 1:   Cathodoluminescence spectra of a nominally 5 nm thick quantum well at three different temperatures.

Fig. 1 shows three CL-spectra taken at 5 K, 50 K and 112 K. 3 lines L_1-L_3 are visible, showing a pronounced dependence of their intensities on sample temperature. At 5 K the spectrum is dominated by L_2 and L_3 being of equal intensity, each exhibiting the doublet splitting just discussed. Line L_1 is visible only on a semilogarithmic scale at this temperature. At temperatures above 30 K the bound excitons become thermally activated and the respective luminescence disappears. Between 5 K and 30 K the free exciton peak does not move in energy. At 50 K the high energy line L_1 is now clearly visible and L_3 has dramatically dropped in intensity. Increasing the temperature to 112 K the intensities of L_1 (L_3) show a further increase (decrease). Increasing the temperature any further does not lead to further changes of the relative intensities.

Fig. 2 shows a histogram of the peak wavelengths of L_2 and L_3 at T = 5K. The number of pixels $N(\lambda)$ showing a certain peak wavelength is plotted versus

Composition fluctuations of the ternary barriers (or the ternary well in systems like InGaAs/InP) at the interface lead to the same type of intermixing of atoms in a given lattice plane as structural roughness does. A structurally flat ternary interface would appear exhibiting atomic scale roughness. Thus, a precise identification of the interface position and determination of roughness on an atomic scale is not possible in such structures. The AlAs barriers have the additional technical advantage of showing a larger contrast than AlGaAs in chemical lattice imaging.

Similarly, interpretation of luminescence experiments is facilitated if only binary barriers (and wells) are used. E.g. alloy fluctuations in ternary barriers at the interface leads to atomic scale barrier height and well width fluctuations inducing inhomogeneous broadening of the excitonic recombination.

The cathodoluminescence set-up is described in detail elsewhere /6,11/. It´s key feature for the investigations reported here the extremely high yield in collecting the CL from the sample allowing to work at low acceleration voltages of a few keV and beam currents << 1 nA for maintaining a narrow electron beam diameter. Thus the lateral resolution as given by the maximum of the two quantities electron beam diameter and Bethe-range is improved to well below 100 nm. A detailed description of the way the HRTEM experiments are performed was given recently /14/. The pattern recognition method as described in /12/ was applied to image quantitatively the chemical composition across and laterally along the interfaces.

3. Results

Here we concentrate on typical results obtained for the 5 nm QW of a sample grown with 100 s growth interruption at both interfaces. Upon growth interruption at the present substrate temperature the GaAs surface becomes substantially smoother /20/ showing an island-like structure. A strong reduction of the inhomogeneous broadening of the luminescence results. As a consequence a doublet splitting into a lower energy exciton bound to an interface acceptor /14,20/, and a higher energy free exciton component of each recombination line is clearly observable at low lattice temperatures without refering to lineshape deconvolution procedures. A proper distinction between these two recombination processes is of largest importance. We observe generally, not only for this sample, a variation of acceptor concentration across the sample surface leading to strong variations of the **relative** intensities of the two components. As a consequence for samples where luminescence lines show

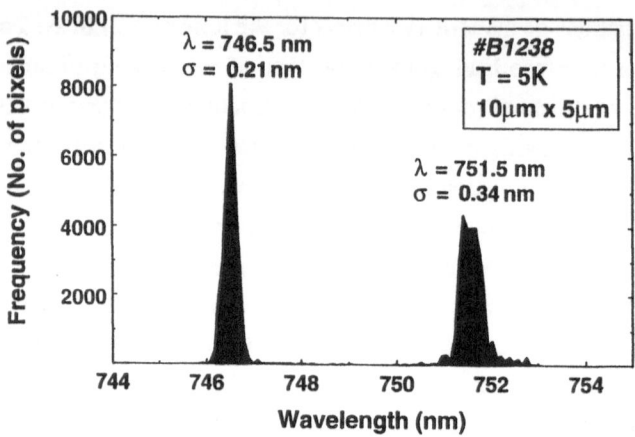

Fig. 2: Peak wavelength histograms of the 2 components L_2, L_3 of the spectrum of Fig. 1 taken in an area of 10 μm x 5 μm.

wavelength λ. The histogram is composed of 51200 spectra taken across an area of 10 μm x 5 μm. Histograms taken at various parts of the sample and various magnifications yield almost identical results. The standard deviation σ_E of lines L_2 and L_3 is as small as 0.47 meV equivalent to 0.04 monolayers. This value presents an upper limit of the "true" standard deviation. The spectra are measured with an intensified Si-diode array, having an interpixel distance of 25 μm. The halfwidth of each of the three histogram L_1-L_3 is equivalent to only 3-4 pixels of the diode array.

A detailed comparison of the experimentally observed temperature dependence of L_1-L_3 with theoretical predictions /22/ using $E_g (T) = E_g (0) - aT^2/(T+B)$ and an investigation of the I_{exc}-dependence of L_1-L_3 at different temperatures leads to the conclusion that lines L_1-L_3 are indeed caused by excitonic recombination.

An inspection of the luminescence lines L_1-L_3 taken at T > 30 K on a semilogarithmic scale shows a predominantly Lorentzian shape similarly to what we observed earlier for AlGaAs/GaAs QW´s grown with interruption of the growth /4,6/. No Maxwell-Boltzmann tail on the high energy side is observed at any temperature /14/. Such a tail is indicative of momentum nonconserving recombination of excitons localised in potential minima of a depth larger kT induced by fluctuations of the mean QW width across distances larger than the exciton diameter and smaller than its diffusion length /4,5/. Based on this observation and on the almost vanishing width of the histogram lines in Fig. 2 we conclude that such mean well width variations do not exist in the present sample on the length scale investigated here, in agreement with HRTEM results shown below.

The mean width of the QW is now calculated directly from the spectral positions of the L_1-L_3 peaks after adding the exciton binding energy /23/ using the envelope function approximation /24/, and the same band structure parameters as in /6/. We find L_z = 15.9 ML, 16.8 ML and 17.7 ML for lines L_1...L_3, respectively. Given the uncertainties of the input parameters (e.g. ΔE_c = 0.62 E_g is used here) the difference of thickness between the various "columns" of the well is close to 1 ML, the value theoretically expected. We fully attribute this a/2-splitting to smoothing of the GaAs surface in agreement with previous work of some of us, where we investigated separately the effects of growth interruption at GaAs and AlGaAs surfaces and observed a triplet upon GRI at the GaAs surface but no splitting upon GRI at the AlGaAs surface which remains rough on a subexcitonic scale /20/. A model of the structure of the QW is shown schematically in Fig. 6.

The information contained in the spectra and the histograms is used now to determine the interface structure quantitatively. Observing /20/ that the roughness spectrum of the lower GaAs on AlAs interface causes the remaining inhomogeneous line broadening σ_E after GRI we calculate the mean lateral extension d of a growth island at the lower interface. The surface of a growth island at such an interface can be defined by the condition that the Al-concentration below the surface is larger than 50% and above the surface below 50%. The standard deviation of the histograms in Fig. 2 is identical to the standard deviation of the inhomogeneous broadening of our luminescence spectra /10/. However, since Lorentzian broadening is dominating the shape of the exciton lines in this situation a determination of the inhomogeneous contribution directly from a single lineshape would be subject to a large systematic error.

Using /4/

$$d = \frac{\sigma_E}{\delta \sqrt{p(1-p)}} \frac{2 a_B(L_z)}{\left|\frac{\partial E_g(L_z)}{\partial L_z}\right|_{L_z}} \tag{1}$$

where $a_B(L_z)$ = 7.52 nm is the two-dimensional Bohr-radius /23/ for L_z = 5 nm, δ = 0.283 nm is the height of a monolayer step, $\partial E_g (L_z)/\partial L_z$ = 1.34 meV/nm and a relative coverage p = 0.5 by monolayer islands we obtain a most probable island extension d $\approx$ 21 Å. The value we used here for p does introduce some error. For the range p = 0.2 - 0.8 d ranges from 26 Å (p = 0.2) to 21 Å (p = 0.5) to 26 Å (p = 0.8). We have implicitly assumed here the existence of a one step structure at the lower interface. Much smaller "islands" might exist. It is obvious from Equ. (1) that such irregularities would not contribute to any spectral broadening.

We will discuss now the temperature dependence of the luminescence spectra shown in Fig. 1. The observation of three discrete excitonic lines L_1, L_2 and L_3, which vary strongly in intensity upon an increase in temperature, signifies that the lateral extension of the respective quantum columns is much larger than the exciton diameter but smaller than the diffusion length as far as L_1 and L_3 is concerned and somewhat larger than the diffusion length as far as L_2 is concerned. L_2 dominates the spectrum at any position of the electron beam and at any temperature larger than 30 K up to 300 K. The highest energy line L_1 is not visible at 5 K since all excitons created in the L_1 columns thermalize into columns L_2 which have a lower energy. If the extension of the L_2 columns would be also smaller than the diffusion length, similarly L_2 would be not visible at 5 K, since all excitons would thermalize into the lowest energy columns L_3. Time-resolved experiments of Tu et al. /25/ show directly such ambipolar inter-column relaxation. Hillmer et al. /26/ investigated the mobility of 5 nm GaAlAs/GaAs QWs grown under similar conditions as ours (e.g. 2 min growth interruption) and found a constant mobility of ≈ 4200 cm^2/Vs in the temperature range 30 K - 100 K. Assuming a lifetime of 0.3 ns /27/ and an effective exciton temperature of 20 K /28/ at a lattice temperature of 5 K we find $l_D \approx 0.46$ μm using the relation $l_D = (kT t \mu /e)^{1/2}$. Our spectra show that at temperatures larger than 80 K the lines L_1-L_3 are in thermal equilibrium with each other. Here $l_D \approx 0.92$ μm. The extension of the L_2 columns is thus larger than 0.9 μm, twice the exciton diffusion length at 20 K, and smaller than 1.8 μm, twice the exciton diffusion length at 80 K. From an extrapolation of the relative intensity ratios of L_1-L_3 for T -> ∞ we estimate more than 60% of the QW to be covered by columns L_2. Still the smallest L_1 columns are expected to extend much beyond 15 nm.

Low temperature CLWI and CLI confirm these results. The lateral intensity distribution of L_2 in Fig. 3 indicates that indeed connected areas of the order of $\approx$2-3 μm exist at the AlAs on GaAs interface caused by the L_2 columns. The lateral L_3-intensity distribution displayed in Fig. 3 is not showing the correct size of the respective islands but an upper limit. The images are broadened by twice the diffusion length.

Fig. 4 shows a <100> high-resolution image of the nominally 8 nm thick QW with GRI. The sample thickness of the observed area is between 25 and 35 nm. It is, therefore, not suited for a quantitative composition analysis. However, as for the thinner samples, the appearance of the GaAs is dominated by the {220} and the appearance of the AlAs by the {200} lattice fringes. The HRTEM image pattern depends on the dynamic interaction of the various beams and, at sample thicknesses

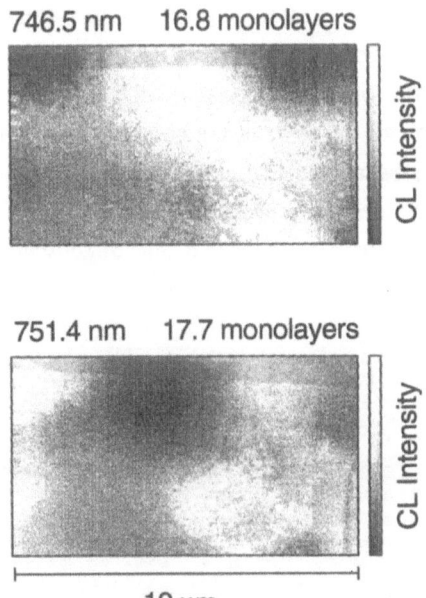

746.5 nm   16.8 monolayers

751.4 nm   17.7 monolayers

10 μm

Fig. 3:   Monochromatic cathodoluminescence images of lines L_2 (746.5 nm) and L_3 (751.4 nm) of Fig. 1.

Fig. 4:   <100> high resolution TEM image of the 8 nm QW of the sample with 100 s interruption of growth at both interfaces.

above 25 nm, significantly on absorption which determines the average image intensity. Due to the stronger absorption, the GaAs appears darker than the AlAs. The average is not grossly affected by noise. The demarcation line at the top interface between GaAs and AlAs does not change its position along the whole interface section. It extends over more than 100 nm in perfect agreement with the predictions from our luminescence experiments. For the bottom interface, visual inspection by eye already indicates that the interface is rough in lateral direction on a scale which is distincly below the exciton diameter. An island like structure with extensions of the order of 3 nm as predicted by the analysis of the histograms, is observed. A comparison of both interfaces also proves that the blurring of the picture which occurs because the steps are predominantly oriented towards <110> and <1$\bar{1}$0> /13,29,30/ does not prevent their observation. The roughness difference between the top and the bottom interface is thus directly visualized by Fig. 4.

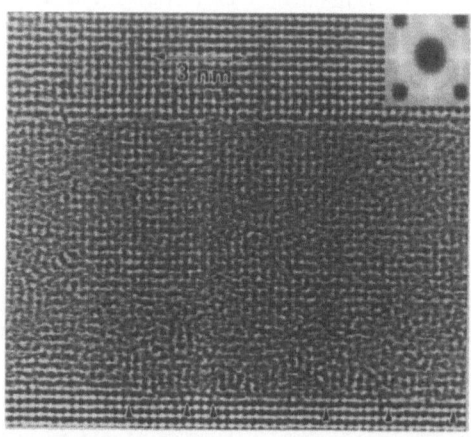

Fig. 5:   Fourier filtered section of the 8 nm QW grown with interruption of growth with the used mask inserted (rectangular distortions are due to the printer). The bottom interface shows roughness in contrast to the top interface.

Fig. 5 shows a magnified section of Fig. 4 which was digitized by a CCD camera, Fourier transformed and filtered using a mask (inserted in Fig. 5, the rectangular distortion is caused by the printer), which blocks out the {220} reflections and the small spatial frequencies. The GaAs shows faint {200} fringes because of the specimen thickness. The large differences between the bottom and the top interface are again clearly visible in perfect agreement with the discussion above and the work of Köhrbrück /20/. This observation is in agreement with the idea that the mobility of gallium atoms on an arsenic surface is much larger than that of aluminum atoms. Arrows mark tentative edges of islands. The extension of the islands vary in the 1-4 nm range. The arrows in Fig. 5 mark positions where the composition between two adjacent atomic columns drastically changes. However, a simple interpretation in terms of island edges is not possible due to the thickness of the sample (≈25 nm).

Only the average composition for each atomic column is observed in the HRTEM image. To explaun the observed small-scale roughness (5 nm scale) it can be assumed that one atomic column parallel to the electron beam cuts several small islands. If one assumes the scale of the islands to be 10 nm or less, drastic compositional changes may indeed occur in adjacent atomic columns as observed in Fig. 5.

Chemical lattice images /12/ indicate that a vertical composition grading extends over 2-4 ML across both interfaces being superimposed to the interface position as defined by Fig. 3 /14/. The images also clearly show that the z-position of a line dividing areas with Al-concentrations larger than and smaller than 50% is almost invariant across the whole image for the upper but not for the lower interface in agreement with Fig. 3. For the upper interface rarely 1 ML high extrusions up to 2 ML wide are found. Such a vertical grading of concentration probably caused by interdiffusion of Al and Ga does not contribute to broadening in luminescence nor does it contribute to the widths of the histograms shown in Fig. 2.

4. Conclusion

In conclusion we have presented here the essential results of a study of the structure of the interfaces of the model QW system AlAs/GaAs. Only in a binary/binary system crystallographic structure which can be modeled to some extent by varying growth parameters like growth rate or interruption time /10/ can be easily distinguished from effects resulting from statistical or nonstatistical distributions of the elements on their respective lattice sites in ternary or multernary materials being on one or both sides of the heterostructure. Thus only in such a system the question how smooth an interface can really be made can be answered with reliability. We have compared the results of excitonic luminescence peak position histograms obtained over areas of varying extension, yielding quantitative information on the GaAs on AlAs interface, of cathodoluminescence wavelength images of the columnar structure of the QW, yielding information on large monolayer islands at the AlAs on GaAs interface with HRTEM pictures yielding information on both interfaces. These methods give results on spatial frequencies from the 1 mm^{-1} to 3 nm^{-1} regime and are partly complementary. There is, however, sufficient overlap of the spatial frequency ranges of the optical and structural methods to compare quantitatively the respective results. We find close to perfect agreement. In particular the existence of large almost ideally flat growth

274

islands at AlAs on GaAs interfaces extending over several hundred nm to a few μm is unambiguously proved. The sizes of islands at the GaAs/AlAs interface of extension much less than the exciton Bohr-radius derived from luminescences and TEM also agrees with each other and are of the order 2-4 nm. Thus, the roughness spectrum in Fourier space is indeed at least bimodal. Our understanding what smooth means can be redefined in light of the observation that the actual structure of a growth surface is not simply frozen in upon overgrowth by a material of different chemical composition but that exchange of atoms in growth direction occurs, depending on the actual growth temperature and other parameters. Growth interruption or a change of growth rate will not affect this exchange. This omnipresent atomic scale roughness which is constant across the interface plane does not displace the interface position sampled across several lattice constants and does not introduce any observable broadening in luminescence or absorption spectra, nor does it show up in high-resolution TEM images like the one in Fig. 3. Chemical lattice images, however, make it visible. We therefore have also a means to

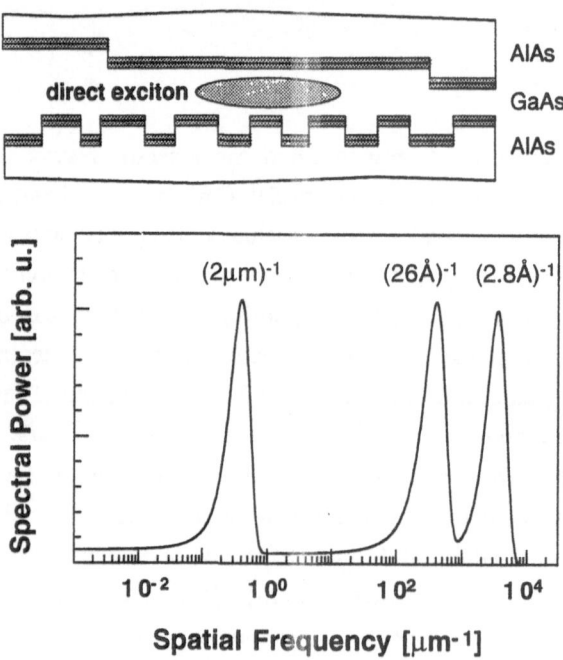

Fig. 6:   Qualitative plot of the power spectrum of the QW interface roughness as compiled from a combination of TEM, CL and PL analysis and visualization of the quantum well interface structure.

distinguish the crystallographic smoothness of an interface from the superimposed vertical grading. We suggest to use for a **quantitative description** of the "smoothness" of an interface the joined information we obtain from a **numerical evaluation** of the inhomogeneous broadening of optical spectra, peak energy histograms, line splittings and high resolution TEM images.

Fig. 6 qualitatively visualizes the overall "interface roughness" of the quantum wells in our sample. The power spectrum of the roughness is plotted over a range of spectral frequencies from 10^{-3} μm^{-1} up to 10^4 μm^{-1} compiled from all three experimental methods described above. Due to the three isolated and identified different disordering mechanisms (i.e. $D \approx (2\mu m)^{-1}$ at the AlAs on GaAs, $D \approx (26\text{Å})^{-1}$ at the GaAs on AlAs, and $D \approx (2.8\text{Å})^{-1}$ due to atom exchange at both interfaces), three distinct maxima show up resulting in a multimodal power spectrum of interface roughness. In this sense "atomically smooth interface islands" extending over many hundred nm up to several μm were shown here to exist at the AlAs on GaAs interface.

Conclusions on the general "nonexistence" of such **smooth** islands drawn recently by some authors who claim to have lacked their observation, either due to the actual nonexistence in the limited number of "high" and "highest" quality samples investigated grown under conditions which are actually not optimal for this purpose and/or due to inadequate experimental procedures and evaluations (e.g. disregard of carrier diffusion effects /16/) unfortunately confuse "(non)observation" with "(non)existence".

Acknowledgements

One of the authors is very indebted to A.C. Gossard and P. Petroff for helpful discussions during a preliminary stage of this work. The TUB part of the work is funded by DFG in the frame of SFB1424.

References:

/1/   C. Weisbuch, R.C. Miller, R. Dingle, A.C. Gossard and W. Wiegmann; Solid State Commun. **37**, 219 (1981).

/2/   see e.g. M.S. Miller, H. Weman, C.E. Pryor, M. Krishnamurty, P.M. Petroff, H. Krömer, and J.L. Merz; Phys. Rev. Lett. **68**, 3464 (1992).

/3/   J. Christen, E. Kapon, M. Grundmann, D.M. Hwang, M. Joschko, and D. Bimberg; Phys. stat. sol (b) **173**, 307 (1992) and Appl. Phys. Lett. **61**, 67 (1992).

/4/   J. Christen and D. Bimberg, Phys. Rev. **B42**, 7213 (1990).

/5/   R.F. Schnabel, R. Zimmermann, D. Bimberg, H. Nickel, R. Lösch and W. Schlapp; Phys. Rev. **B46**, 9873 (1992).

/6/   D. Bimberg, J. Christen, R.K. Bauer, D. Oertel, D.E. Mars, and J.N. Miller; J. Vac. Sci. Technol. **B4**, 1014 (1986).

/7/   H. Sakaki, T. Nada, K. Hirakawa, M. Tanaka, and T. Matsusue; Appl. Phys. Lett. **51**, 1934 (1987), and Appl. Phys. Lett. **57**, 1651 (1990).

/8/   N.M. Cho, S.B. Ogale, and A. Madhukar; Appl. Phys. Lett. **51**, 1016 (1987).

/9/   J.H. Neave, B.A. Joyce, P.J. Dobson, and N. Norton; Appl. Phys. **A31**, 631 (1983).

/10/   M. H. Herman, D. Bimberg, and J. Christen, J. Appl. Phys. **70**, R1 (1991) and J. Christen, Advances in Solid State Physics 30, (U. Rösler ed.) Vieweg, Braunschweig 1990, p. 239.

/11/   J. Christen, M. Grundmann, and D. Bimberg; J. Vac. Sci. Technol. **B9**, 2358 (1991).

/12/   A. Ourmazd, F.H. Baumann, M. Bode, and Y. Kim; Ultramicroscopy **34**, 237 (1990).
A. Ourmazd, J. Cryst. Growth **98**, 72 (1989).
M. Bode, A. Ourmazd, J. Cunningham, and M. Hong; Phys. Rev. Lett. **67**, 843 (1991).

/13/   N. Ikarashi, M. Tanaka, H. Sakaki, and K. Ishida; Appl. Phys. Lett. **60**, 1360 (1992).

/14/   D. Bimberg, F. Heinrichsdorff, R.K. Bauer, D. Gerthsen, D. Stenkamp, D.E. Mars, and J.N. Miller; J. Vac. Sci. Technol. **B10**, 1793 (1992) and ibid **B11**, 126 (1993).

/15/   R.F. Kopf, E.F. Schubert, T.D. Harris, R.S. Becker; Appl. Phys. Lett. **58**, 631 (1991).

/16/   C.A. Warwick, W.Y. Jan, A. Ourmazd, and T.D. Harris; Appl. Phys. Lett. **56**, 2666 (1990).

/17/   D. Gammon, B.V. Shanabrock, and D.S. Katzer; Appl. Phys. Lett. **57**, 2710 (1990).
D. Gammon, B.V. Shanabrock, and D.S. Katzer, Phys. Rev. Lett. **67**, 1547 (1991).

/18/   C.A. Warwick and R.F. Kopf; Appl. Phys. Lett. **60**, 386 (1992).

/19/   U. Morlock, J. Christen, D. Bimberg, E. Bauser, H.J. Queisser, and A. Ourmazd; Phys. Rev. **B44**, 8792 (1991).

/20/   R. Köhrbrück, S. Munnix, D. Bimberg, D.E. Mars, and J.N. Miller; J. Vac. Sci. Technol. **B8**, 789 (1990), and Appl. Phys. Lett. **57**, 1025 (1990).

/21/   P.M. Petroff, A.Y. Cho, F.K. Reinhart, A.C. Gossard, and W. Wiegmann; Phys. Rev. Lett. **48**, 170 (1982).

/22/   M.B. Panish, H.C. Casey; J. Appl. Phys. **40**, 163 (1968).

/23/   D.B. Tran Thoai, R. Zimmermann, M. Grundmann, and D. Bimberg; Phys. Rev. **B41**, 5906 (1990).

/24/   G. Bastard and J.A. Brum; IEEE J. Quant Electron., **QE22**, 1625 (1980).

/25/   C.W. Tu, R.C. Miller, P.M. Petroff, R.F. Kopf, B. Deveaud, T.C. Damen, and J. Shah; J. Vac. Sci. Technol. **B6**, 610 (1988).

/26/   H. Hillmer, A. Forchel, R. Sauer, C.W. Tu; Phys. Rev. **B42**, 3220 (1990).

/27/   J. Christen, D. Bimberg, A. Steckenborn, and G. Weimann; Appl. Phys. Lett. **44**, 84 (1984), and D. Bimberg, J. Christen, A. Steckenborn, G. Weimann and W. Schlapp, J. Lumin. **30**, 562 (1985).

/28/   C. Colvard, D. Bimberg, R.K. Bauer, D.E. Mars, and J.N. Miller; Phys. Rev. **B39**, 3419 (1989).

/29/   S. Permogorov, A. Naumov, G. Gourdon, and P. Lavallard; Solid State Commun. **74**, 1057 (1990).

/30/   M.S. Skolnick, T.D. Harris, C.W. Tu, T.M. Brennan, and M.D. Sturge; Appl. Phys. Lett. **46**, 427 (1985).

Deposition and Characterization of Thin Polyimide Films

P.Y. Timbrell and R.N. Lamb

Surface Science and Technology, School of Chemistry, University of New South Wales, Kensington, NSW 2033, Australia.

G. Comelli

Dipartimento di Fisica, Universita di Trieste, 34127 Trieste, Italia.

Abstract. Polyimide thin films (~ 40Å) can be produced via the vapour deposition polymerisation (VDP) of the two monomers pyromellitic diimide (PMDI) and 4,4'-oxydianiline (ODA). Angle dependent X-ray photoelectron spectroscopy (XPS) was used to investigate the interfacial bonding of the PMDI monomer to polycrystalline Ag substrates. The interfacial N 1s and C 1s XPS components indicate that PMDI bonds in an upright configuration, with minimal molecular fragmentation, to the Ag substrate via the imide nitrogen. This differs to the bonding of the more commonly used monomer, pyromellitic dianhydride (PMDA), where near-edge X-ray absorption fine structure (NEXAFS) results from Ag(110) substrates, suggest a tilted bonding configuration. The potential advantages of PMDI over PMDA, in terms of polymer film adhesion and sequential organic film growth are also discussed.

1. Introduction

Polyimides are an important class of polymer that find wide application in the aerospace and microelectronics industries as thermally stable coatings or as a device packaging material [1]. The adhesion, and durability, of such polymer films is of practical importance and has resulted in the investigation new methods for producing such films as well as the examination of film-to-substrate interfacial bonding. Thick polyimide (PI) films (> 1000 Å) are typically produced for planar coatings and films using spin coating methods with the polymeric precursor, polyamic acid (PAA), in a polar solvent solution (eg. NMP). Thinner polyimide films can be produced in a monolayer-by-monolayer fashion using Langmuir Blodgett techniques [2] using the ammonium salt derivatives of the polymeric precursor polyamic acid, again in solvent. In this work thin films (< 50Å thickness) of polyimide were produced using the vapour deposition polymerisation (VDP) method. In VDP the monomers are directly co-deposited onto the substrate in high vacuum (effectively solventless) to form the polymeric precursor (PAA) prior to thermal curing (~ 250°C) to produce the resultant polyimide film. The two monomers, 4,4'-oxydianiline (ODA) and pyromellitic dianhydride (PMDA) are commonly used to form thin

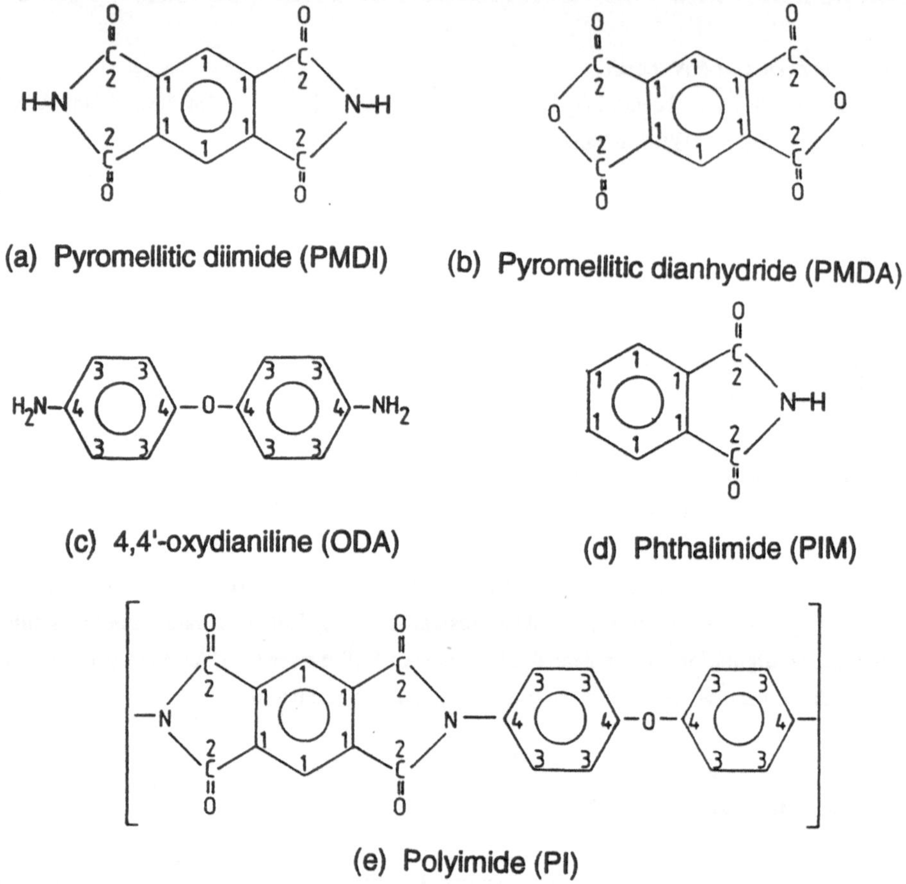

(a) Pyromellitic diimide (PMDI)

(b) Pyromellitic dianhydride (PMDA)

(c) 4,4'-oxydianiline (ODA)

(d) Phthalimide (PIM)

(e) Polyimide (PI)

Fig.1 Structural formulae of the monomers (a) pyromellitic diimide (PMDI), (b) pyromellitic dianhydride (PMDA) and (c) 4,4'-oxydianiline (ODA); (d) the model compound phthalimide (PIM); (e) the repeat unit of polyimide (PI).

polyimide films using the VDP method [3 - 8]. Upon deposition at room temperature a polymeric precursor film of polyamic acid (PAA) is initially formed which then imidises with heating to produce the final polyimide film.

In this study we explore an alternative monomer to pyromellitic dianhydride (PMDA), namely pyromellitic diimide (PMDI). The structural formulae of polyimide (PI), and the various monomer (PMDA, PMDI, ODA) and model compound molecules (PIM) discussed in this paper are illustrated in Fig.1. In PMDI the anhydride oxygen is replaced with an imide nitrogen. Although PMDI has served as polyimide model compound [9 - 11] it has only recently been utilised to produce VDP polyimide films [12,13].

X-ray photoelectron spectroscopy (XPS) was used to examine the ODA-PMDI polymerisation reaction and PMDI-to-silver substrate interfacial bonding. In order to further probe the monomer-to-substrate molecular orientation near-edge X-ray absorption fine structure (NEXAFS) data were obtained for PMDA deposited on Ag(110) substrates. The interfacial bonding and molecular orientation should directly affect the initial film growth and macroscopic polymer film adhesion.

2. Experimental Details

Polycrystalline silver (99.99%) substrates were polished and Ar ion sputter cleaned prior to the film deposition. The latter was performed in a high vacuum preparation chamber (~ 1 X 10^{-7} mbar) attached to main XPS analysis chamber. The co-deposition was achieved using two resistively heated copper Knudsen cells containing ODA (Aldrich, zone refined, 99+%) and PMDI (Aldrich, > 97%) which was further refined prior to deposition. The monomers were co-deposited on room temperature (~ 20°C) Ag substrates and the resultant film progressively heated to ~ 450 °C.

The XPS spectra were recorded with a Kratos XSAM/AXIS 800 spectrometer using unmonochromatized Mg K_α radiation operating at 150 W power. The hemispherical analyser pass energy was set to 20 eV and the overall instrumental energy resolution was 0.9 eV. For the angle dependent XPS studies the sample was tilted from the horizontal (normal emission) to 60°. In other words the XPS spectra were recorded at photoelectron emission angles (measured from the sample surface normal) of 0° and 60°, respectively. The quoted average film thicknesses were estimated from the attenuation of the Ag $3d_{5/2}$ photoelectron substrate signal ($E_K = 885.3$ eV) using mean-free-path of $\lambda \sim 12$ Å [14].

The near-edge X-ray absorption fine structure (NEXAFS) data above the carbon K-edge for PMDA deposited on Ag(110) substrates were obtained using the grasshopper monochromator (1200 lines/mm holographic grating) on beam line I-1 at the Stanford Synchrotron Radiation Laboratory. The Ag(110) crystal was cleaned by cycles of Ar ion bombardment and flash desorption after exposure to O_2 at room temperature. The NEXAFS spectra at normal (90°) and grazing (20°) X-ray incidence (measured from the sample surface plane) were recorded by total-electron-yield detection [15]. The spectra were normalized by the signal from a metal grid reference monitor which eliminated possible time-dependent instabilities of the X-ray flux from the storage ring and by the independently recorded signal form a clean Si wafer which was used to eliminate energy dependent structures in the monochromator transmission function.

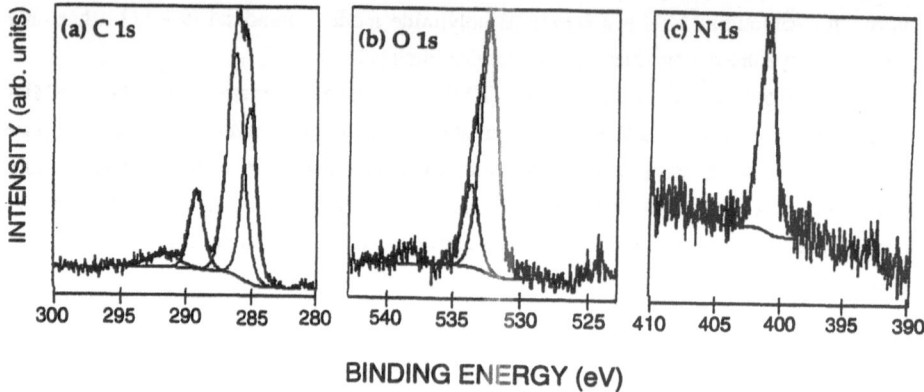

Fig.2 XPS spectra of the (a) C 1s, (b) O 1s and (c) N 1s regions of a PMDI-ODA co-deposited on room temperature Ag and then subsequently heated to ~ 400°C to produce a polyimide film of 39 Å thickness. The polyimide polymer film was thermally stable to 480°C.

3. Polyimide via the Vapour Deposition Polymerisation of ODA and PMDI

Fig. 2 shows the C 1s, O 1s and N 1s XPS region spectra from co-deposited PMDI and ODA on room temperature polycrystalline Ag substrates after heating to 400°C. The fractions of ODA and PMDI in the deposited precursor film (64 Å thickness) were estimated from the carbonyl to phenyl C 1s peak areas and an approximately equal mix was obtained during the initial deposition. Broadened XPS lineshapes were observed for the initial co-deposited film and are attributed to the formation of a stable polydiamide (PDA) polymeric precursor film [12, 13]. Upon progressive heating to 400°C a narrowing of all the peak shapes occurred, the film reduced in thickness from 64 Å to 39 Å, and the characteristic XPS spectrum of polyimide was observed (Fig. 2). This polyimide spectrum first emerges after heating to ~ 180 °C and does not change significantly with further heating to 480 °C [12].

The polyimide C 1s region can be deconvoluted into a carbonyl and two phenyl components (Fig. 2(a)), the asymmetric O 1s region into ether and carbonyl components (Fig.2(b)), while a single imide component peak at E_B = 401.0 eV is observed for the N 1s region (Fig. 2(c)). The measured elemental C: O : N atomic percentages of 75.9 : 16.7 : 7.4 of the film are in good agreement with the stoichiometric values (75.9 : 17.2 : 6.9) for polyimide. The measured carbonyl to phenyl carbon ratio of 0.17, and carbonyl to ether oxygen ratio of 3.27, both fall short of the stoichiometric polyimide ratios of 0.22 and 4, respectively. This suggests that the polyimide film may, in fact, contain a slight excess of unreacted ODA units which would lead to an overall reduction in the relative carbonyl intensities.

The postulated PMDI-ODA polymerisation reaction is analogous to the PMDA-ODA system except that NH_3, rather than H_2O, is evolved during the imidization step and that a PDA (polydiamide), rather than PAA (polyamic acid) polymeric precursor is involved [12].

4. Interface Studies Ultra-Thin Films of PMDI on Ag

At the initial stages of film growth the monomer to substrate bonding and molecular orientation are important as they ultimately influence the polymer film formation and adhesion. Hence the motivation to investigate the interfacial bonding of alternate polyimide monomers, such as PMDI, to metallic and oxide substrates. The PMDI-Ag substrate interface region was studied by comparing the XPS spectra obtained from vapour deposited ultra-thin (< 10 Å) films, in which the interface region is probed, with thicker (~ 40 Å) PMDI films.

Fig. 3 shows the C 1s and N 1s XPS spectra obtained for ~ 40 Å and ~ 7Å thickness PMDI films deposited on polycrystalline Ag. For the 40 Å thickness film the C 1s region has a distinct phenyl and carbonyl components separated by 3.3 eV (Fig. 2(a)). This is less than the value of 3.7 eV reported for PMDA [5] as these carbonyl carbons now share an imide nitrogen instead of an anhydride oxygen. Two distinct π - π^* shake-up satellites are observed, shifted by about 5.5 eV (to higher binding energy) from their parent phenyl and carbonyl peaks. The N 1s region is characterised by a single imide nitrogen component at E_B = 400.6 eV. A shake-up region is also present and is attributed to π - π^* transitions in the phenyl ring.

The ultra-thin (~7Å) film data is shown in Figs. 3 (c) and (d). The main difference from thicker film data is that interface related photoelectron peak components are apparent in both the C 1s and N 1s regions. The C 1s interface feature is derived from the carbonyl band and is shifted by 1.2 eV (to lower binding energy) from the carbonyl peak at E_B = 288.9 eV. A phenyl π - π^* shake-up region is still observed, of comparable relative intensity to the thick film case, which suggests that the aromaticity of the PMDI is still preserved in these ultra-thin films. The apparent lack of a carbonyl $\pi - \pi^*$ shake-up (at E_B = 295 eV) in the 7Å film is attributed to a low signal to noise ratio and a curve fit was not attempted for this weak feature. The N 1s region is the most revealing in terms of an interface related component and a new peak, shifted by some 2.0 eV to lower binding energy, is clearly in evidence (Fig. 3(d)). The large peak shift of the interface nitrogen component suggests that the nitrogen is directly bonded to the silver substrate.

The main "bulk" C 1s and N 1s peaks exhibit a broadening of approximately 25%, when compared to the thick film case, and is likely due to the distribution of chemical environments in the interfacial region. The measured C : O : N atomic percentages in the ~ 7 Å PMDI film are

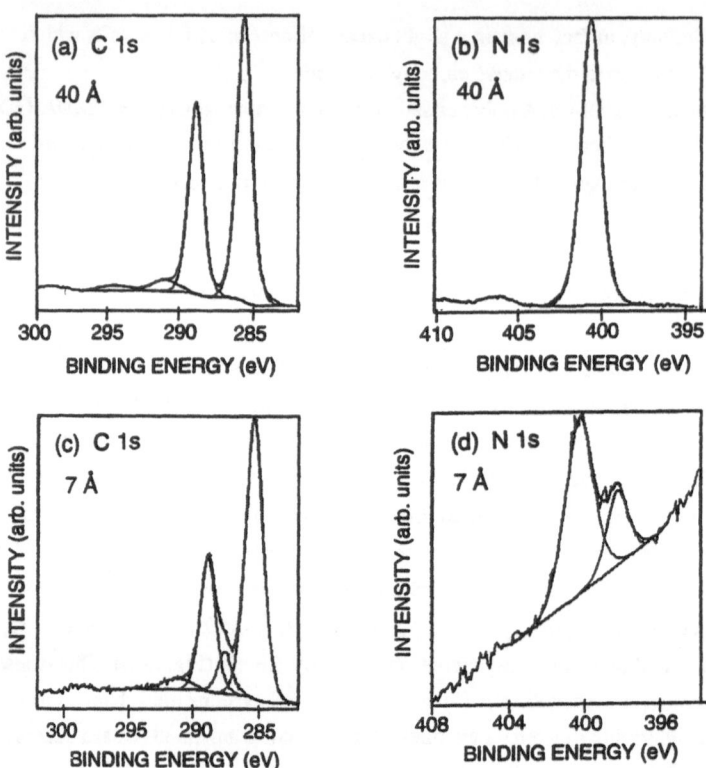

Fig.3 C 1s and N 1s XPS spectra for thin (~40Å) and ultra-thin (~7Å) films of PMDI on polycrystalline Ag substrates. Interfacial C 1s and N 1s components are observed at E_B = 287.7 eV and 398.5 eV, respectively, in the ultra-thin (~7Å) PMDI film.

59.2 : 26.6 : 14.2 (see Table 2) which, as in the thicker film case, are in reasonable agreement with the stoichiometric values. There are no obvious signs of PMDI molecular fragmentation upon bonding to the silver substrate, unlike the case of PMDA where there is a marked decrease in the carbonyl carbon intensity and overall oxygen content (consistent with a loss of a carbonyl group) upon bonding [4,5]. For the 7Å thin PMDI film the total carbonyl intensity (i.e the peak at E_B = 289 eV and the interface component at E_B = 287.7 eV) is reduced from the 40 Å film case (by approx. 15 %) which suggests that there may be some loss of carbonyl groups upon adsorption on the Ag. There is a slight charging shift of 0.2 eV (on average) to higher binding energies in the main XPS peaks in the ~ 40 Å thick film compared to the ultra-thin ~ 7 Å film.

Additional information concerning the PMDI molecular orientation/bonding can be inferred from angle dependent XPS experiments. Fig. 4 shows the N 1s region of a ~ 6Å thick PMDI film recorded at sample tilts of 0° (horizontal sample, normal emission) and 60°. The tilt angle

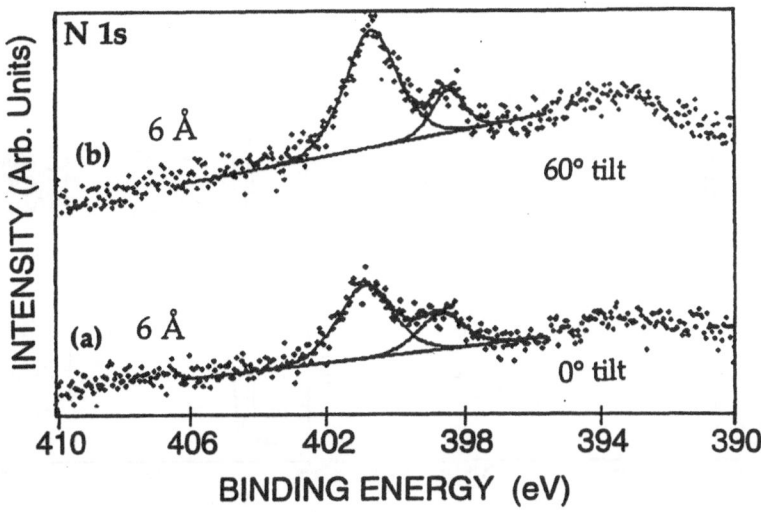

Fig. 4 N 1s spectra of a PMDI film (~ 6Å) on Ag recorded at (a) 0° and (b) 60° sample tilt (i.e. photoelectron emission angle). The lower binding component, at E_B = 398.5 eV, is attributed to the N-Ag bond is attenuate relative the "bulk" molecular imide component, at E_B = 400.5 eV, with increasing sample tilt.

of 60° leads to a factor of two decrease in the effective XPS sampling depth in the overlayer and a subsequent attenuation of the interfacial, relative to the polymer layer, photoelectron intensities. The interfacial component (E_B = 398.5 eV) decreases in intensity, with respect to the bulk N 1s (E_B = 400.6 eV) with increasing sample tilt. This confirms that the additional N 1s photoelectron feature in the ultra-thin PMDI films is indeed due to interfacial (sub-surface) nitrogen bonded to the Ag substrate. The measured binding energy difference between the bulk and interfacial N 1s components is 2.1 eV for this film, and values in the 2.0 - 2.2 eV range were measured for all the thin PMDI films studied.

Interfacial XPS studies of the model polyimide compound, phthalimide (PIM), provide insight into the interfacial bonding of PMDI on Ag. The PIM molecule is similar to PMDI except that it only has a single imide group on the phenyl ring (see Fig. 1). In the case of PIM adsorbed on copper substrates a 2.2 eV shift in the N 1s peak, a 1.2 eV shift in the C 1s carbonyl peak as well as a slight (less than 1 eV) upwards shift in the O 1s carbonyl peak have been reported [9]. The observed trends and chemical shifts in our PMDI/Ag data closely follow those observed for PIM/Cu case which suggests that they are bonding in analogous fashion to their respective metallic substrates.

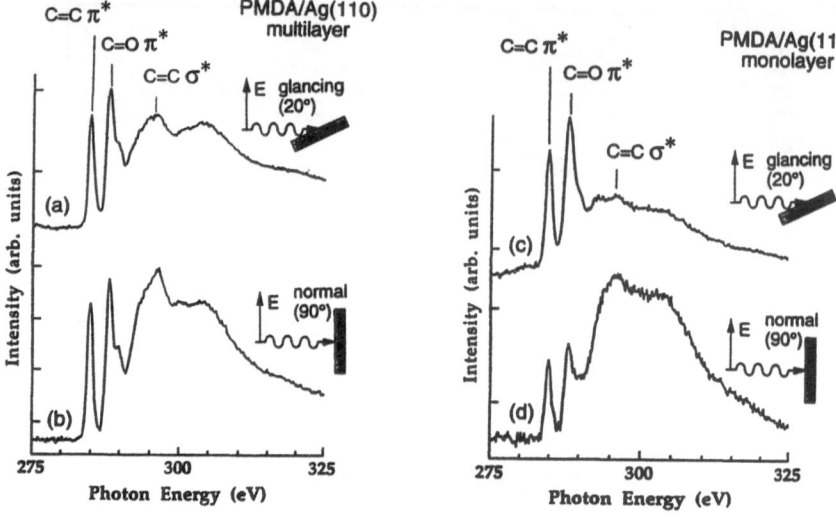

Fig. 5 Carbon K-edge NEXAFS spectra for multilayer PMDA on a Ag(110) substrate at (a) glancing (20°) and (b) normal (90°) X-ray incidence; and monolayer PMDA on Ag(110) at (c) glancing (20°) and (d) normal (90°) X-ray incidence as illustrated in the insets.

PIM bonds to copper substrates in vertical configuration (ring-plane perpendicular to the substrate) through the imide nitrogen [9]. The PIM and PMDI chemical shifts can be interpreted in terms of a bonding model whereby charge transfer occurs from the metallic substrate to molecule (via the nitrogen atom) which then further induces a secondary shift in the neighbouring carbonyl carbon atoms (the increased screening leads to observed shift to lower binding energy). In other words the interfacial PMDI carbon component is not due to change in the carbonyl C=O functionality but electronic screening effects.

5. NEXAFS studies of monomer orientation

Fig. 5 shows carbon K-edge NEXAFS data obtained from multilayer, and monolayer, PMDA films deposited on clean single crystal Ag (110) substrates. NEXAFS spectroscopy can be used to deduce molecular orientation by measuring the polarisation dependence of the various π^* and σ^* resonances of the adsorbed organic molecule [15]. In the case of multilayer PMDA there is not a significant dependence on the NEXAFS spectra with the electric field orientation (i.e. X-ray incidence angle). This is to be expected from a thick film in which the

PMDA molecules are in a more less random orientation and the influence of the Ag (110) substrate does not predominate. For PMDA films in the monolayer regime, however, there is a distinct polarisation dependence in the π^* and σ^* intensity ratios. For normal incidence (Fig.5 (d)) the σ^* resonance intensity increases as one would expect for flat lying phenyl rings. However there is still a admixture of π^* and σ^* resonance intensities which suggests that the PMDA is in a tilted bonded geometry on Ag (110) as illustrated in Fig. 6(b).

The molecular orientation of ~3 Å thick PMDI films deposited on highly-oriented pyrolitic graphite (HOPG), using Langmuir-Blodgett techniques, has been investigated using nitrogen K-edge NEXAFS [17]. The strong polarisation dependence of the nitrogen π^* resonances indicate, in contrast to PMDI on Ag, that this weakly bonded layer of PMDI molecules are flat-lying with their ring-planes parallel to the graphite substrate. The molecular orientation of PMDI on single crystal Ag surfaces has not yet been investigated by NEXAFS. On the basis of the XPS results one would expect a clear polarisation dependence of the π^* and σ^* resonances that would support the view that the PMDI ring plane are perpendicular to the Ag surface.

6. Discussion

Fig. 6(a) illustrates the proposed bonding scheme, which is consistent with the XPS data, for PMDI on Ag. The PMDI is bonded to the Ag via one of the imide nitrogens and is in an upright geometry. This model further is supported by surface enhanced Raman spectroscopy (SERS) studies, in which the interfacial region is also probed, of spin-coated PMDI on Ag island films [10,11]. The authors conclude that the PMDI molecules adsorb with their ring-planes perpendicular to the surface with one imide group in contact with the surface on the basis of symmetry considerations and the observed strong perpendicular axial vibrational modes (eg. CNC stretch mode at 1372 cm $^{-1}$). In contrast PMDA is believed to bond via the opening of one the anhydride rings and the formation of a carboxylate bond to the substrate [5, 18].

It is interesting to speculate on the possible advantages of polyimide thin films produced using PMDI-ODA, instead PMDA-ODA, from an interfacial bonding point of view. PMDA is known to fragment at the Ag surface, as evidenced by the loss of the carbonyl component upon adsorption, which may influence the interfacial bonding [5]. PMDI, however, appears to remain intact when bonded to the Ag substrate. This bonding differs from the PMDA case, where the anhydride ring opens so that bonding through the oxygens may occur, in that PMDI can anchor itself to the substrate without significant fragmentation. In addition to strong interfacial bonding an ideal monomer would also have a free reactive end in order to polymerise with its co-monomer. PMDI appears to fulfil this criterion and the angle dependent XPS results indicate that this monomer bonds intact in an upright configuration leaving a reactive imide end that is free to link to ODA units.

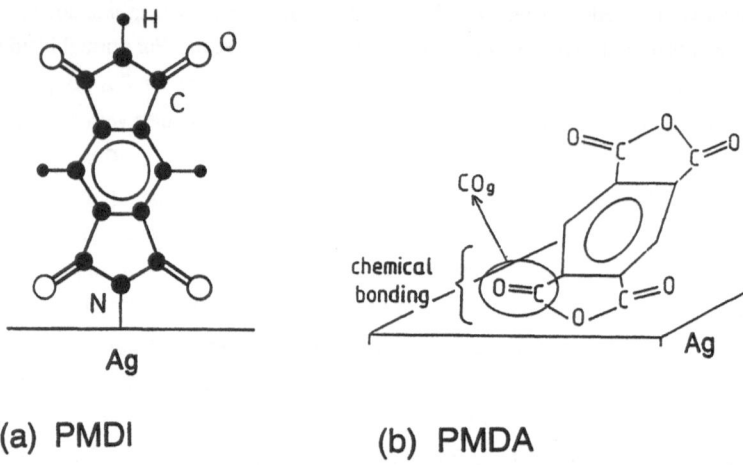

Fig. 6 Proposed bonding models for (a) PMDI and (b) PMDA chemisorbed on Ag substrates.

Frederick *et al.* have recently demonstrated that is possible to deposit organic bilayer films by the careful sequential dosing of a monolayer of PMDA, followed by aniline to terminate the sequence [18]. It would be interesting to repeat such experiments with PMDI as one of the building blocks. PMDI is a good bifunctional molecule: one end is bonded to the substrate while the other is free to react with the next layer (eg. ODA, aniline). It should prove possible to deposit ordered organic layers using PMDI given its vertical bonding configuration (high symmetry) onto Ag and other metallic (eg. Cu) substrates.

As a final note it is interesting to consider whether the nature of the PMDI-Ag interfacial bond would actually lead to an improvement in the macroscopic adhesion of PMDI-ODA, over PMDA-ODA, produced polyimide films. The determining factor in the adhesion of polyimide films appears to be cohesive failure, as opposed to pure adhesive at the interface, within the interphase region of the film [19]. It is in this interphase region were the film makes the transition from a more or less ordered interface phase to the random chains within the bulk of the polymer film. The additional ordering that one might expect at the PMDI-Ag (compared to the PMDA-Ag) interface might, in fact, weaken the cohesive strength within the interphase region of the polymer film.

7. Conclusion

PMDI has been used as an alternative to PMDA (pyromellitic dianhydride) to produce thermally stable polyimide thin films via the vapour deposition polymerisation of PMDI and ODA (4,4'-oxydianiline). XPS studies of deposited films of pyromellitic diimide (PMDI) on polycrystalline silver as a function of film thickness and sample tilt (i.e. photoelectron emission angle) exhibit C 1s and N 1s features due the interfacial PMDI-Ag bonding. The XPS results indicate that PMDI bonds strongly to the silver substrate, via the imide nitrogen, in an upright bonding configuration (ring-plane perpendicular to the surface) with minimal molecular fragmentation. NEXAFS can be used to further investigate the monomer-to-substrate molecular orientations and in the case of PMDA on Ag(110) a tilted bonding geometry is inferred.

Acknowledgments

The authors would like to thank M. Grunze and J. Stöhr for informative discussions.

References

1. D.S. Soane and Z. Martynenko, "Polymers in Microelectronics", Elsevier (1989).
2. W. Meyer, M. Grunze, R. Lamb, A. Ortega-Vilamil, W. Schrepp and W. Braun, Surface Science 273 (1992) 205.
3. J.R. Salem, F.O. Sequeda, J. Duran, W.Y. Lee and R.M. Yang, J. Vacuum Sci. Technol. A4 (1986) 369.
4. R.N. Lamb, J. Baxter, M. Grunze, C.W. Kong and W. N. Unertl, Langmuir 4 (1988) 249.
5. M. Grunze and R.N. Lamb, Surface Science 204 (1988) 183.
6. T. Strunskus, M. Grunze and S. Gnanarajan, in: "Metallization of Polymers", eds. E. Sacher, J.-J. Pireaux and S.P. Kowalczyk (American Chemical Society, 1990) p.353.
7. S.C. Perry and A. Campion, Surface Science Letters 234 (1990) L275
8. S.S. Perry and A. Campion, J. Electron Spec. and Rel. Phenom. 54/55 (1990) 933.
9. P. Bodo, K. Uvdal and W.R. Salanek, in: Metallized Plastics 2, ed. K.L. Mittal (Plenum Press, New York, 1991) p.189.
10. F.J. Boerio, P.P. Hong, H.W. Tsai and J.T. Young, Surface and Interface Analysis 17 (1991) 448.
11. W.H. Tsai, J. T. Young, F. J. Boerio and P.P. Hong, Langmuir 7 (1991) 745.

12. N. Than-Trong, P.Y. Timbrell and R.N. Lamb, Chemical Physics Letters 205 (1993) 219.

13. P.Y. Timbrell, N. Than-Trong and R.N. Lamb, Surfaces and Colloids A: Physicochemical and Engineering Aspects (1994), in press.

14. D.T. Clark, in: Chemistry and Physics of Solid Surfaces, Vol. 11, ed. R. Vanselow (CRC Press, Bocca Raton, 1979).

15. J. Stöhr, R. Jaeger, J. Feldhaus, S. Brennan, D. Norman and G. Apai, Appl. Opt. 19, (1980) 3911.

16. J. Stöhr and D.A. Outka, Phys. Rev. B36 (1987) 7891.

17. Th. Schedel-Neidrig, M. Keil, H. Sotobayashi, T. Schilling, B. Tesche and A.M. Bradshaw, Ber. Bunsenges. Phys. Chem. 95 (1991) 1385.

18. B.G. Frederick, N.V. Richardson, W.N. Unertl and A. El. Farrash, Surface and Interface Analysis 20 (1993) 434.

19. M.Grunze, G.Hähner, Ch. Woll and W. Schrepp, Surface and Interface Analysis 20 (1993) 393.

VI.

Characterization of Catalysts

The Application of Surface Physics in Catalysis Science: Understanding Heterogeneous Catalysts for Partial Oxidation

Robert Schlögl

Fritz Haber Institut der Max-Planck Gesellschaft, Faradayweg4, D - 14195 Berlin

Abstract

The relation between surface science and catalysis science is discussed emphasising the aspects of the analysis of heterogeneous catalysts and its implications for model studies. The need to replace a-priori assumptions about the catalyst nature under reaction conditions by solid experimental evidence is pointed out. Only then a surface science study using the correct model system is expected to be successful in analysing the mode of operation of a catalyst. This aspect becomes increasingly important as surface science moves towards more complicated reaction systems.

Two examples from the analysis of technical catalysts for partial oxidation reactions are used to illustrate the general statements. It is demonstrated that in cases of heteropolyacids and elemental silver as catalysts for selective dehydrogenation the bridging of pressure and material gaps cannot be achieved by linear extrapolation of reactivity and structure from conditions of model studies to practical operating conditions.

Introduction

This contribution summarises a personal view about the current state of the relation between surface physics and catalysis science. In 1991, a euro-american symposium on the subject was held[1] which arrived at the conclusion that there is a mutual interaction between both fields of science which had brought about a large body of fundamental understanding. The different levels of complexity in both fields studied today is, however, a significant barrier for a practical application of surface physics results in modern catalysis science.

In a recent correspondence in the journal „Angewandte Chemie" the same issue was controversly[2] discussed. The question was put forward about the possibility of the „design" of a catalytic process using our knowledge about the mode of operation of existing systems. The counter-hypothesis was that our solid knowledge about existing processes under realistic conditions is so scarce[3] that we have to rely in devising a catalyst on extrapolation and on chemical intuition fuelled by the results from surface physics. Such a devised system requires for practical viability still a great deal of empirical testing and development. This surface-physics-assisted mode of catalyst development may be more efficient than the traditional purely heuristic approach by a reduction of the screening candidates but the procedure is still far away from the „design" stage reached by the application of principal understanding and universal concepts.

The necessity to improve our arsenal of catalytic processes and the achievements made so far in the field are pointed out in a recent review[4] over the whole field of catalysis. In spite of the significant progress made in surface physics, solid state chemistry and chemical engineering, the fundamental ingredients in a design strategy, their individual results and methodologies are not coherent and cannot be simply correlated by linear extrapolation.

After a general analysis of the relations between different fields in catalysis science, the presentation will give two examples of how an analytical approach can be used to help to understand existing catalyst systems.

Generalities

From industrial circles we often hear the statement that surface physics was a failure as tool to speed up the development process of catalysts. The reason is commonly identified with the insufficient complexity of the study objects of surface physics which „must move to study reactions on more complex surfaces under more realistic conditions"[1]. The keyword for this context is „in-situ experiment". With this term we define an experiment which analyses a physical property of a catalytic system in operation under practical conditions with simultaneous detection of the catalytic performance[3]. Such experiments are far from trivial and require extensive design work on both the experimental end as well as on the theory required for a correct interpretation of the data. It is noted that the present efforts concentrate much on the side of the detection of the physical properties[5 6] leaving aside the problem of on-line detection of catalyst performance. A significant potential in this field has the recently commercialised method of ion-molecule reaction mass spectrometry.

The apparent shortfall of useful complexity on the physics side of catalysis science is paralleled by a shortfall in system definition on the catalyst side. Here we find a large number of a-priori assumptions which lack experimental backing. The problem of defining the active material in terms of chemical composition and bulk structure is very prominent in this field. If this part of the problem definition is, however, incorrectly assumed, all model studies performed by surface physics are no longer related to the process under study and corresponding conclusions are bound to be inapplicable. The magnitude of this problems increases with the increasing complexity of reactions and catalysts.

Using unproved a-priori assumptions on the technical level of catalyst characterisation leads to unnecessary controversies and artificial complications. Incorrect spectroscopic assignments[7] and incomplete kinetic characterisations[8] of reaction systems are frequent sources of such complications which may have obscured the view of the practitioner on the significant body of fundamental and useful information obtained so far by surface science[10,11]. Essentially all concepts in heterogeneous catalysis which are used in our everyday work were derived from surface physics. A discussion of this point can be found in the review articles [1,4].

The known modes of actions for catalytic systems are to be collected in a data base which then ultimately may serve as the rational basis for the much awaited „catalyst design" process. It is obvious that we are at present still far away from a useful completeness of such a hypothetical data base. The enormous progress in understanding catalysis made in recent times[1,4] allows, however, to be optimistic about the speed at which this information will accumulate in the future. Our experience with solved cases so far may be used to device a strategy for more efficient problem analysis.

It should be mentioned in this context that there exists no general theory of catalysis up to now and that it was questioned if such theory can be found at all[9]. It is also true[4] that we were in the past much better in finding new catalysts by trial and error than explaining why the existing systems operate. These statements were derived from attempts to solve the catalyst problem using kinetic methods. This most important aspect is not considered here as we deal with the material science - surface analytical aspects of heterogeneous catalysis. From this

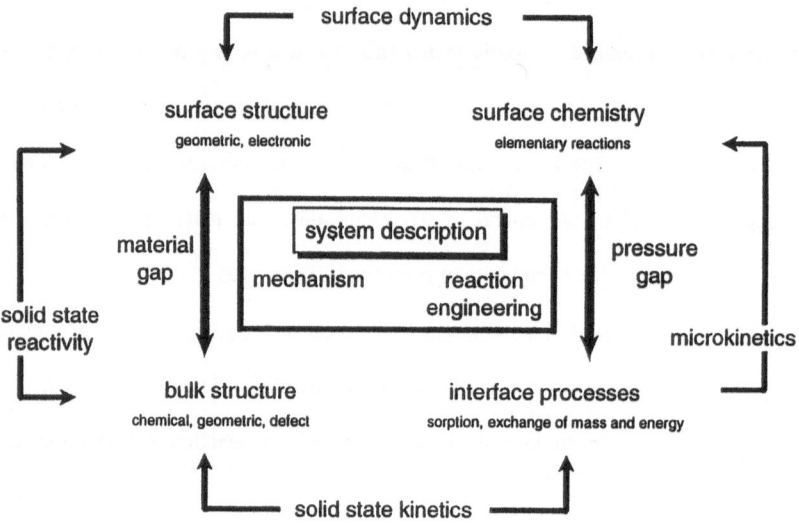

Figure 1

Relations between various disciplines of catalysis and surface science. The system description (central) is given by four fields of research (corners). Between the two fields of surface science (top) and those related to the practical world (bottom) exist two gaps which cannot be bridged by simple assumptions. The four outer research topics are targets for in-situ experiments the results of which allow to experimentally bridge the gaps.

point-of-view, however, sufficient progress can be expected to overcome the ambiguities of kinetic modelling without experimentally verified parameter sets.

The interaction of chemistry and physics is the basis of any rational approach towards the analysis of a given catalytic process. Figure 1 illustrates such relationships. Central is the *system description* of the catalytic process under study. This description is complete if we know the reaction mechanism in qualitative and quantitative aspects and the implication of the chemical engineering parameters (heat and mass flow, reactor design, feed and product conditions) on the process.

As we consider here heterogeneous processes, four major research fields are addressed in the creation of the system description. Interface processes such as adsorption of educts, desorption of products and mass and energy transport to the interface will regulate the macroscopic reaction behaviour of the system. The surface chemistry describing the elementary reactions and their kinetic parameters controls the product spectrum and the energy balance of the system. The surface structure with its geometric and simultaneous electronic details causes the surface chemistry to take place in every turnover. The material basis described by the bulk structure with its chemistry, crystal structure and defect details is the foundation of the surface structure.

These four areas are investigated in separate approaches in traditional strategies. The fields of surface chemistry and surface structure which form main topics of surface science are studied using appropriate assumptions about the solid state structure and gas phase conditions of the practical reaction. This concept is addressed in the literature[10,11] as the „single crystal approach" in catalysis science. Chemical engineering and solid state chemistry are separate fields of their own and are rarely considered in context with each other or with surface science. An exception to this is the area of macrokinetic modelling where input about the microkinetic steps is required which is taken -if available- from surface science studies[12] with, however, varying success.

One main result of the past research efforts is the information that the parameters of each of these four fields are all interdependent with difficult-to-predict relationships. One reason for this difficulty in extrapolation between different parameter sets of different types of experiment is the unclear status of the reacting system with respect to thermodynamic equilibrium. Whereas it is commonly accepted that the educt-product reaction system is often far away from thermodynamic equilibrium, it is assumed that the equilibrium surface chemistry frequently known only at room temperature and ambient atmosphere applies for the solid part of the reaction system. This problem causes a material gap to open between model systems and practical catalyst materials. This gap opens on two sides namely with the definition of the correct model substance and with the choice of its correct solid state reactivity.

In the silver-oxygen system it turned out that the neglect of an oxide phase on grounds of stability arguments[13] at thermodynamic standard conditions was not correct. For the oxidation of methanol to formaldehyde the selectively active surface species is oxygen embedded in the silver surface forming a unique crystal structure with strong silver-oxygen interactions at the thermodynamic conditions of the technical catalyst operation[14].

In the case of the ammonia synthesis reaction over iron it is still a matter of controversy how the non-equilibrium surface structure of iron (111) is formed in the technical reaction system[15 16]. In this case the material gap between „ammonia iron" (polycrystalline iron metal active as catalyst) and an iron (111) single crystal was closed qualitatively by structure-sensitive experiments[17]. A special iron-nitrogen sub-surface phase which is thermodynamically metastable at ambient conditions[18] is also known to be present in the operating catalyst system.

The ammonia synthesis process further serves as example for the operation of unexpected effects in the gas - interface reaction sequence causing the so-called pressure gap to open between UHV studies and conversion experiments under practical reaction conditions. It was found recently[19] that the chemisorption of molecular nitrogen, which is a crucial step in the overall reaction sequence[20] occurs under atmospheric pressure with a significant thermal activation barrier which was absent in low-pressure UHV studies[21]. This thermal activation step was not included in the extrapolation models[22 23 24] from low pressure UHV studies over the pressure gap into the range of practical application. Nevertheless, the models are surprisingly accurate in predicting technical reaction rates from UHV experimental parameters, qualifying the degree of scientific perception that can be gained from quantitative system descriptions. We note that ammonia synthesis is a chemically simple reaction and that generations[25] of scientists have worked on the elucidation of the reaction mechanism. The efforts on the ammonia synthesis mechanism[26 27] have contributed a great deal to the foundations of today's status of catalysis science and the results are a corner stone in the justification of the single crystal approach in understanding heterogeneous catalysis. Fundamental studies of the CO hydrogenation reaction[28 29] aimed at an understanding of the synthesis gas chemistry[30 31] as well as an attempt to combine experimentally within one apparatus surface physics and catalytic conversion experiments at elevated pressures[32] had a further significant influence on the co-operation between surface physics and catalysis science.

In order to overcome the problems of extrapolation between different types of experiments it seems essential to perform additional experiments aimed at the connections between the four fundamental sets of system properties. In Figure 1 four such additional properties are chosen as links which all require experiments under reacting conditions. These properties describe the reactivity of the catalytic system at different levels of resolution and complexity. Only because of the enormous progress in sharpening our diagnostic tools[4] it is possible to address problems like surface dynamics, solid state reactivity or microkinetics in the solid and at the gas-solid interface. Performing such experiments[5,6] is one front-end of contemporary catalysis science and the methods and concepts of surface science are required as foundation more than in the past, where both disciplines were developed to a large extent independently. Most relevant in this context are further the insights in mechanistic aspects and preparative techniques of solid state chemistry. Understanding and controlling defects and their mobility in catalytic solids and pathways to thermodynamically metastable catalyst precursors like metal-organic compounds, metallic glasses or template synthesis of hollow structures[33] complement the analytical efforts.

Specific Examples

In the second part of the paper we discuss two specific examples from our ongoing work. In the first case of a catalyst used for selective oxidation of isobutyric acid (IBA) to methacrylic acid (MAA) we address the problem of how the phase inventory of a given catalyst

changes under reaction conditions and brings about novel phases which have not been synthesised under conventional catalyst preparation conditions. This study serves as an example for the material gap problem.

The second case study is an old object of surface science, the system silver-oxygen. It was studied for many years[13,34] in the context of explaining the selective oxidation of ethylene to ethylene epoxide. Our group investigates the partial oxidation of methanol to formaldehyde which is carried out industrially under much more severe conditions[35] requiring hence an even more selective mode of operation than in the ethylene epoxidation with respect to a partially oxidised product relative to the thermodynamically favoured total oxidation. Here we illustrate that the pressure gap between UHV experiments and practically relevant oxygen partial

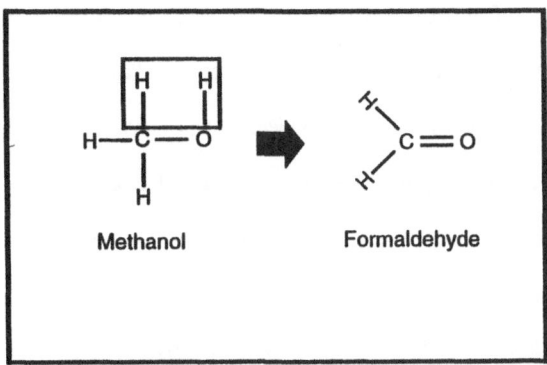

pressures of ca. 50 mbar[36] precludes the identification of the catalytically relevant phase which is formed in a thermally activated process only under conditions of „high pressure" synthesis. Thanks to its chemical stability it is possible to apply many tools of surface science and investigate the chemical bonding by electron spectroscopy, the localisation and the microstructural characterisation by STM imaging to derive a chemical portrait of this species.

The reader may ask the question about the common features of these apparently so different reactions. The reaction schemes 1 and 2 illustrate that it is the same elementary reaction, the dehydrogenation of two protons, which is required in the two processes. Both reactions are termed „selective oxidation" although the number of oxygen atoms remains constant in educts and products. It is the fictive oxidation number of the carbon skeletal atoms which is increased by these reactions.

Heteropolyacid Catalysts

In the IBA to MAA reaction a vanadium substituted Keggin-type $((PMo_{12}O_{40})^{3-})$ heteropolyacid (HPA) system was empirically found[37] to be a suitable catalyst with the only major problem of insufficient long-term stability (wanted are several years, found are several months)[38].

In the context of an extended study on the mode of operation of this catalyst we determined a set of X-ray crystal structures[39] and compared these data to diffraction patterns under in-situ reaction conditions[40] in order to identify the catalytically active phase[41]. This aim of using a bulk-sensitive techniques to approach a surface-derived property is generally to be considered as difficult as it is not possible to find a causality between bulk structures and catalytic performance. In the present case the problem is, however, much reduced as the catalyst is a molecular solid. The material consists of oligomeric anions of $(PVMo_{11}O_{40})^{4-}$ embedded in a complex network of water molecules[42]. It was common conjecture that this anion is isostructural with the parent Keggin anion[43] $(PMo_{12}O_{40})^{3-}$. It turned out, however, that powder diffraction patterns could not be indexed with the known structure of the Keggin

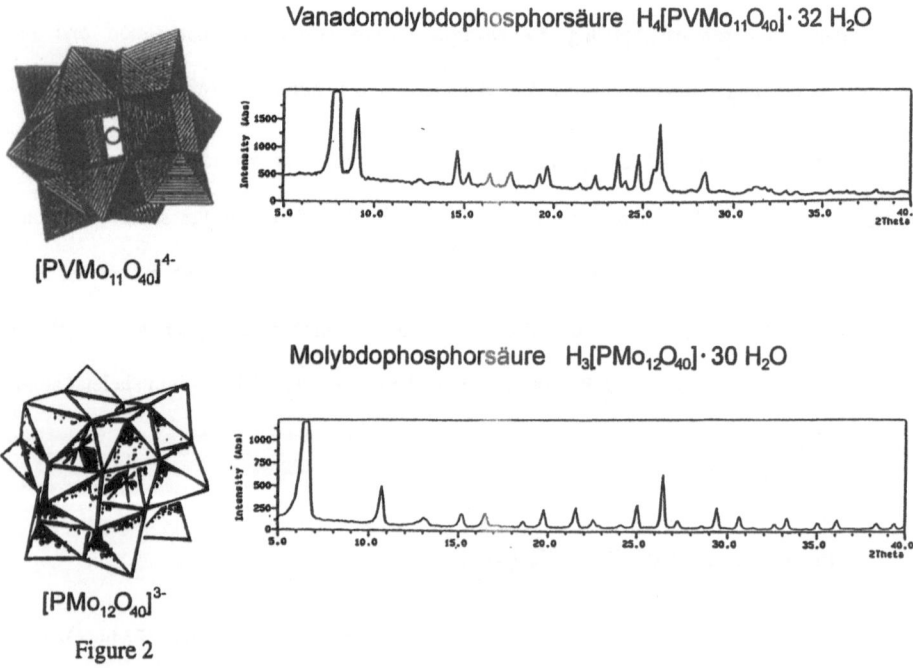

Vanadomolybdophosphorsäure $H_4[PVMo_{11}O_{40}] \cdot 32\ H_2O$

$[PVMo_{11}O_{40}]^{4-}$

Molybdophosphorsäure $H_3[PMo_{12}O_{40}] \cdot 30\ H_2O$

$[PMo_{12}O_{40}]^{3-}$

Figure 2

Polyhedra representation of the inverse Keggin (top) and of the normal Keggin structure. The corresponding powder patterns belong to phase-pure samples of the two materials and agree exactly in with theoretically derived patterns using the parameters of the corresponding single-crystal structures.

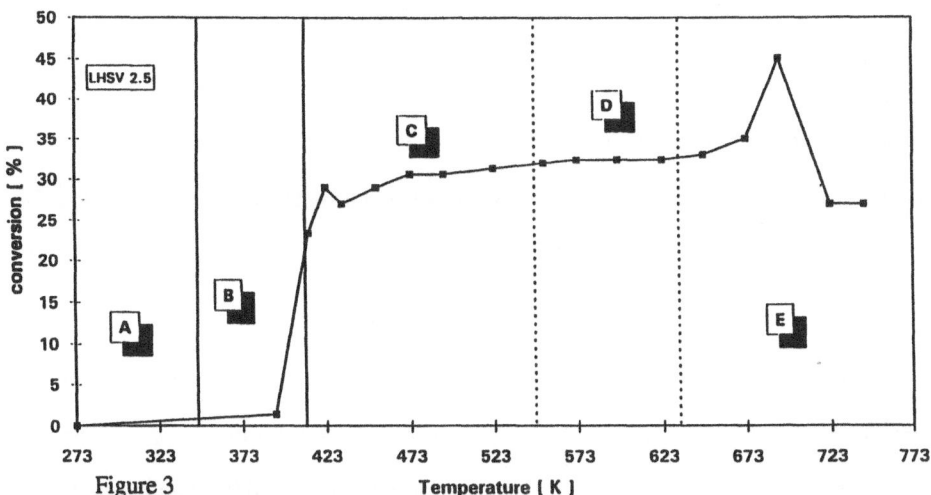

Figure 3

In-situ conversion experiment for the partial oxidation of IBA inside the X-ray diffraction camera. 40 mg of the $PVMo_{11}O_{40}$ catalyst in the form of the free acid were used. A feed of IBA/water 1:9 and of nitrogen/oxygen 40:80 ml/min yielded an absolute conversion of 0.44 mmol/h at 573 K. The indicated segments denote changes of the crystal structure (see text).

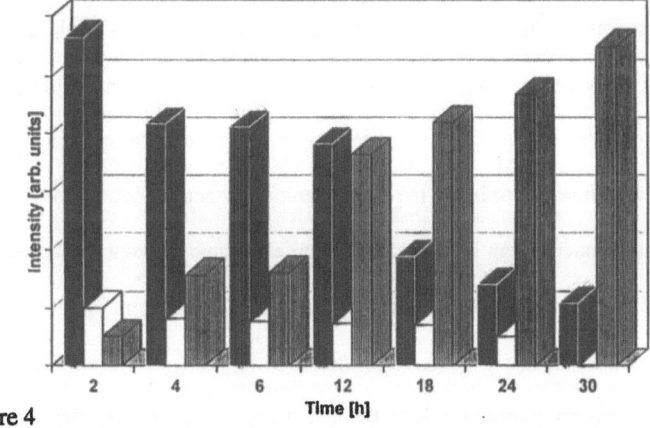

Figure 4

Temporal evolution of the phase inventory of the conversion experiment from Figure 3. The black bars denote the anhydride phase (C), the white bars indicate the abundance of the vanadyl salt (D) and the grey bars stand for MoO_3 as final decomposition product. The catalytic activity scales with the sum of the black and white bars[41].

material. A comparative single crystal structure of the catalytic material and the Keggin structure showed, that the structures were in fact different in the respect of octahedra interconnection[39,42]. The resulting slight difference in molecular shape caused drastic effects on the packing arrangement within the crystal and hence a quite different diffraction pattern. The results of the structure determination[39] and the corresponding powder patterns are shown in Figure 2.

With this firm basis we determined the crystal structure evolution from the catalyst precursor at ambient conditions to the active phase. Figure 3 shows a conversion-vs.-temperature diagram of a sample of $(PVMo_{11}O_{40})^{4-}$ inside our X-ray diffraction reactor[40]. The segments A-E in the Figure indicate the ranges of existence of a given crystal structure. Data analysis[41] showed that the parent structure used as the 24-hydrate (A1) converted via the 14 hydrate (A2) into the 6-hydrate (B) which were all three *catalytically fully inactive*. The catalytic activity rose only after formation of the „anhydride" structure (C), which is in fact the dihydrate derivative. The active catalyst proved, however, to be unstable under a wide variation of reaction conditions. It converted into a novel structure (D) which is the vanadyl salt of the HPA (a novel compound) and finally decomposed into a defective form of molybdenum trioxide and amorphous P-V oxides. It is significant that under reaction conditions the vanadium is not inside the Keggin anion but acts as counterion. For this reason it is unlikely that vanadium can be considered as an electronic promoter to enhance the redox properties of the parent anion as it was suggested from a quantum chemical study[44]. The temporal evolution of the active phases is illustrated in Figure 4 which indicates that the novel vanadyl salt is significantly more stable as the anhydride. It is noted that the decomposition is complete in this experiment after 30 hours whereas under technical conditions more than 200 days are required[38,43] for this process. An explanation for this apparent contradiction may be found in a self-regeneration ability of the decomposed material which is capable under the presence of sufficient water vapour (which we could not supply for technical reasons to our in-situ reactor) to self-organise into the starting compound[1].

These findings allow the following conclusions for our present context.

•There is no justification to assume that the parent catalyst is the active material, even in such a simple case as for a molecular solid.

•In-situ experiments require extensive studies into the temporal stability of the results.

•Great care must be taken in choosing a model system for mechanistic or theoretical studies without knowing the solid-state reactivity of the system under reaction conditions.

•The solid state reactivity can provide complex sequences of events under extended catalytic applications which lead to a steady state abundance of several crystalline phases (detected either in-situ or even in post-mortem studies) belonging all to the catalytic system. By no means, however, all of these phases must play a role in the desired substrate-to-product reaction sequence. The discrimination in catalyst-relevant phases and substrate-relevant phases becomes then important in order to define correctly the material responsible for the elementary steps of substrate conversion. Such an identified material requires then careful study by surface science methods in order to determine its catalytic mode of action.

One puzzling fact in the present story is the unexpected low stability of the dehydrated HPA material at moderate temperature (600K) in an oxygen-containing atmosphere. The catalytically inactive hydrated phases are perfectly stable under these environmental conditions. This allows to conclude that the solid state reactivity and the catalytic activity may have a common cause. This prompted us to study again the dehydration process which has been investigated in the past in great detail[42]. By the application of the novel detection technique of ion-molecule-reaction mass spectrometry (IMR-MS)[45] to a conventional fixed-bed dehydration

---

[1] The synthesis of the catalyst proceeds also via a hydrothermal treatment of binary oxides in an acidic medium. The re-organisation of the spent catalyst would thus be only a repetition of the synthesis sequence carried out under different kinetic boundary conditions.

experiment in nitrogen we were able to detect small amounts of molecular oxygen associated with the stepwise evolution of water.[1]

Figure 5 illustrates that in the temperature range of the formation of the anhydride and the vanadyl salt we observe small evolution of oxygen indicating a partial decomposition of the molybdate anions[2]. Such partly decomposed anions are called lacunary species[42] and have been characterised in great detail in the context of mechanistic studies on the formation of HPA materials. The finding allows to comprehend the reduced stability of the HPA after water abstraction which in fact also destroys the molecular integrity of the anions. In addition, the finding allows to speculate that the required dehydrogenation activity may be correlated with coordinatively undersaturated oxygen anions which form upon rupture of the octahedra network of the precursor HPA anions.

The Silver - Oxygen System

In a series of previously published papers [14,46 47 48 49] we analysed the interaction of oxygen with silver both at UHV and high pressure conditions. Based on TDS evidence[14] three atomic oxygen species labelled alpha (surface oxygen), beta (bulk oxygen, dissolved) and gamma oxygen (sub-surface, embedded in the topmost silver atomic layer) were identified and additionally characterised by Raman spectroscopy. The presence of molecular oxygen in any detectable abundance at or near reaction temperature could be clearly[47] ruled out.

As an example of a further spectroscopic study of the oxygenated surface we show in Figure 6 a sequence of He II UPS data of a Ag (111) single crystal after a high pressure

---

[1] The IMR-MS instrument allows to detect gas molecules without fragmentation and can be operated as a specific traceanalytical tool with detection limits of oxygen at the ppb level. The concentration of molecular oxygen in the present case was below 10 ppm.

[2] The resulting solid retained its yellow colour indicating the product must be a binary oxide like MoO_3. Careful phase analysis by XRD showed indications of very minor amounts of molybdenum oxide to be present in the anhydride/vanadyl salt mix.

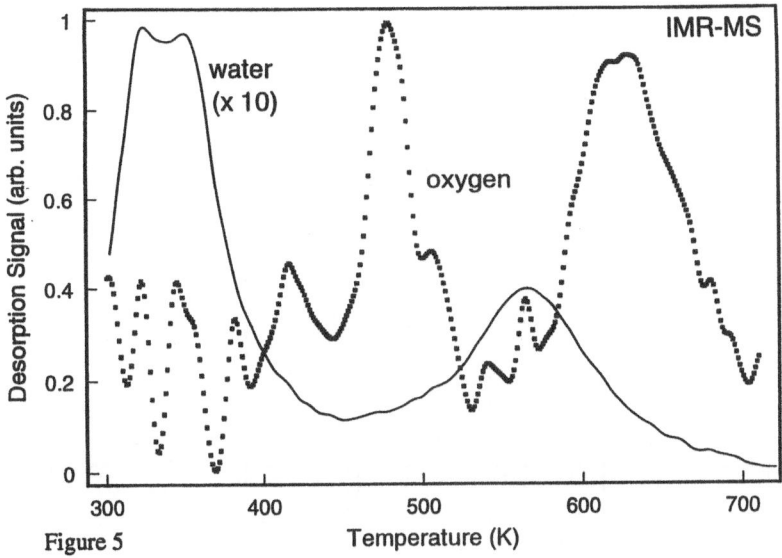

Figure 5

IMR-MS data for a dehydration experiment of the $PVMo_{11}O_{40}$ catalyst in a microreactor (1g diluted with inert SiC). A nitrogen carrier stream and a linear heating ramp of 3 K/min were applied. The reaction cross section for oxygen with Kr as primary reaction ion is about 6 times larger than that for molecular water; 1.5 % of the total gas evolved was molecular oxygen.

treatment at 780 K and 500 mbar oxygen for 1 and 13 hours respectively. The spectrum (b) is indicative of three oxygen species. All features contain significant contributions from oxygen 2p states as concluded from a comparison of the energy dependence of the cross sections (not shown). This analysis supports also the presence of three states. The high-energy doublet structure arises from OH/H_2O formed by the reaction of chemisorbed oxygen with hydrogen from the residual gas of the UHV system. Prolonged exposure reduces its surface abundance indicating that the molecular oxygen source is not contaminated with moisture.

The low energy features indicate the existence of two atomic oxygen species Similar data were found for silver oxide[50] allowing to conclude that the unusually low binding energy of the oxygen 2p emissions indicates indeed oxide-like species. Great care must, however, be

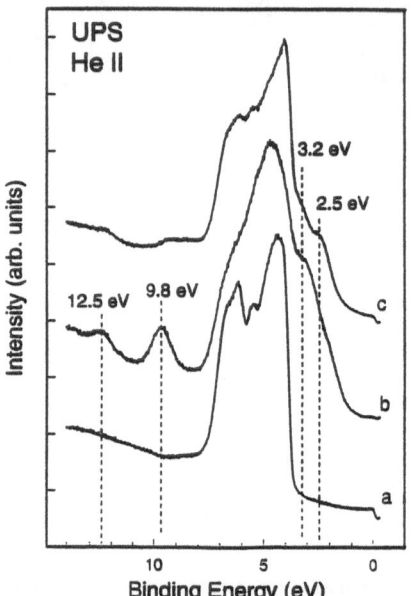

Figure 6

He II UPS data for an oxygenated Ag (111) crystal. Trace a) indicates the spectrum of the clean surface, b) after oxygen exposure at 750 K and atmospheric pressure for 30 min and trace c) after prolonged exposure for 7 hours.

taken in analysing the degree of charge transfer[51] as the low-energy cut-off of the valence band data indicates a drastic change in the work function of 1.5 eV upon oxygen treatment. The third atomic oxygen species found in the TDS data is located under the main Ag 4d band as it was demonstrated earlier[47] with polycrystalline silver foil. Prolonged exposure changes the spectrum again and reverts it partially back to the elemental silver pattern. This is attributed to the pronounced facetting[48] of the crystal which maximises the surface abundance of the gamma species and reduces the occurrence of unselectively adsorbed atomic alpha oxygen. This assignment classifies the lowest binding energy peak as to arise from gamma oxygen and the peak at 3.2 eV to be characteristic of alpha oxygen. The bulk-dissolved species causes the broad feature under the Ag 4d band.

This assignment should be reflected in corresponding XPS data. In Figure 7 oxygen 1s spectra and corresponding He ISS data are compared. This comparison is made to correct the

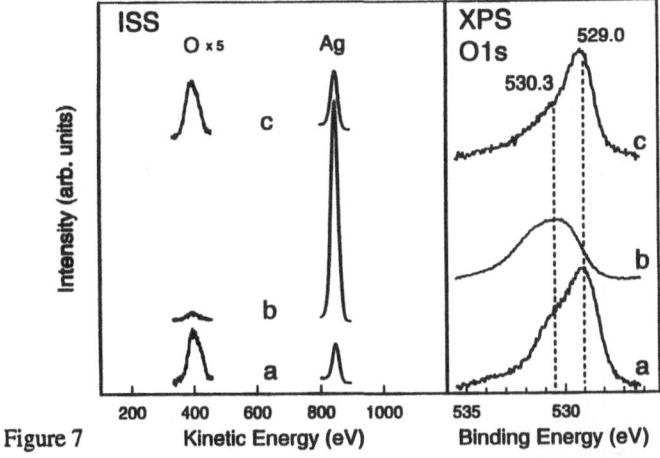

Figure 7   ISS   Kinetic Energy (eV)   Binding Energy (eV)

ISS and XPS data of an oxygenated Ag (111) crystal. Set a) was taken after a treatment as in Figure 6b, b) after sputtering at 300 K (He, 1000 eV, 10 mA) and c) after heating state b) in UHV to 750 K for 30 min.

assignment for the different information depth of XPS (not surface sensitive) and UPS or ISS respectively (surface sensitive).

The series of spectra (a) in Figure 7 corresponds to the state (b) of the oxygenated Ag (111) crystal. We detect two oxygen features in XPS and a pronounced shadowing of the silver by oxygen in ISS. Extensive sputtering with Ar (series (b)) removes the surface oxygen species (ISS) and the oxygen 1s structure at 529 eV. An intense oxygen signal characteristic of bulk-dissolved species above 530 eV remains. If the sample is heated in UHV at 780 K for 1 hour (series (c)) the bulk species segregates back to the surface and replenishes the state characterised by the 529 eV signal in XPS. These observations allow to conclude that gamma oxygen is the surface-embedded species with the low-binding energy XPS signal fully consistent with the low O 2p binding energy in UPS. The broad feature above 530 eV in XPS belongs to the bulk species which occurs in UPS as broad feature under the Ag 4 d band. The spectral parameters indicate a reduced charge transfer of the dissolved atomic oxygen relative to the gamma species. From low-temperature exposure experiments (not shown) we know[14,47]

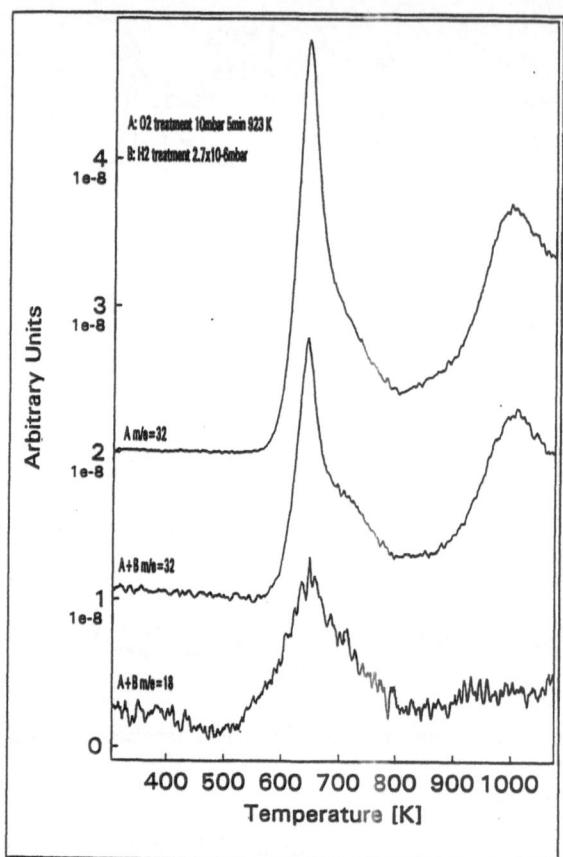

Figure 8

TPD and TPRS data of electrolytic silver grains after oxygen and hydrogen treatments in a quartz cell heated with a linear rate of 1K/s using external furnace and temperature control (calibrated for the gradient) in order to minimise the cell/sample holder background which is very intense after the severe treatment conditions.

that atomic oxygen chemisorbed on unreconstructed silver occurs in XPS at 530.5 eV and overlaps in the present experiment with the bulk oxygen signal. This reference experiment and comparison of the data in Figure 7 (b,c) indicate that bulk dissolved oxygen segregates under our experimental conditions as gamma oxygen to the surface and not as alpha species.

This detailed spectroscopic analysis may be questioned for its relevance to catalysis as the reaction conditions were quite static relative to a flow-through experiment done in kinetic studies. In order to bridge this experimental gap we analysed the chemical reactivity of the surface species by TPRS in the presence of hydrogen and methanol[52].

Figure 8 shows TDS results from a sample of electrolytically prepared silver powder (d= 0.4 mm) which is used in technical applications. The features are similar on single crystal surfaces. The exact peak shape changes from run to run, as any exposure of silver to oxygen above 600 K leads to a restructuring with the creation of facets and the incorporation of gamma oxygen. The kinetics of bulk dissolution of oxygen will, however, depend on the defect structure, grain boundaries and surface textures which are modified by the gamma oxygen formation as well as by the diffusion of previously incorporated beta oxygen.

The TD spectrum in Figure 8 (a) contains two well-resolved features and a shoulder at the sharp low-temperature peak. If the experiment is repeated as TPRS with molecular hydrogen we note from comparison of traces (b, c) in Figure 8 that only one species represented in the low-temperature peak reacts to OH/H$_2$O. The high temperature desorption occurs only in the molecular oxygen channel. The shape of the water desorption and the modifications in the oxygen profile of the TPRS imply that the two species of this peak interchange their location during heating. The high temperature species is also reduced in its abundance pointing again to an interconversion into reactive oxygen. This reactive oxygen is the alpha oxygen with the second feature in the low-temperature peak being attributed to beta oxygen as this feature cannot be saturated even at 500 mbar and 24 hours exposure. The high temperature peak of the nonreactive species is hence gamma oxygen.

We point out that the TPRS has clearly discriminated the chemical reactivity of the two surface oxygen species which is comprehensible from the spectral differences discussed above. The more weakly held alpha species is an oxidising agent and transferable from the surface to the substrate whereas the strongly held gamma oxygen has lost is oxidising ability. It is an electron-rich surface centre with strongly localised valence electrons leading to the „oxide-like" spectral properties.

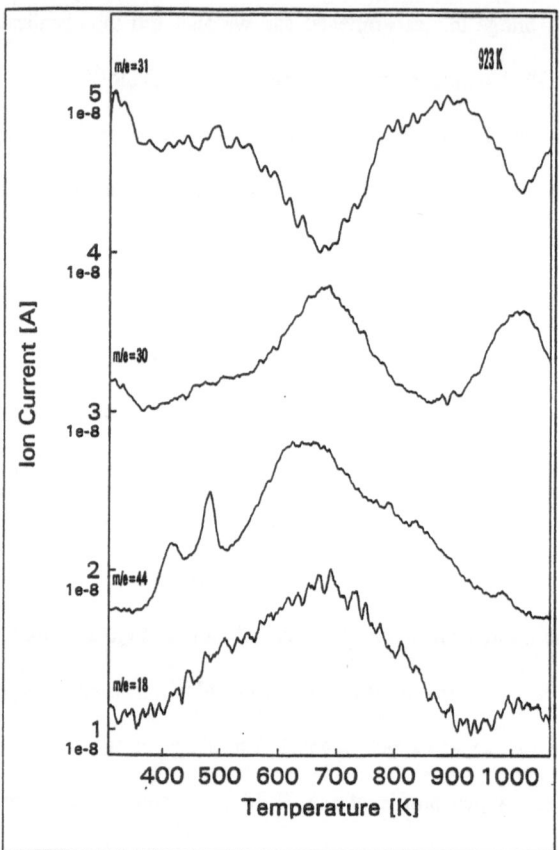

Figure 9

TPRS results of the reaction of oxygen loaded electrolytic silver with methanol. The catalyst was loaded at 923 K with 10 mbar oxygen for 5 min. A constant supply of methanol, amounting to a partial pressure of 2×10^{-6} mbar was used. All traces were simultaneously detected with a HIDEN 10 channel quadrupole mass analyser.

These differences are also reflected in the methanol TPRS experiment summarised in Figure 9. The top trace shows the consumption of methanol in two well separated temperature regions. The regions coincide exactly with the desorption features of molecular oxygen explaining evidently that the conversion of methanol is a surface reaction with both species involved being co-adsorbed on the silver surface. This is of importance as it was always impossible to detect chemisorbed methanol on silver at temperatures above 180 K.

Formaldehyde (m/e 30 trace) is formed also in the two regions of methanol consumption. Carbon dioxide (m/e 44 trace) -the total oxidation product- is, however, only formed in the low-temperature region. Also water, the second product of combustion of methanol occurs only at the low-temperature peak. These findings allow to conclude that the low-temperature oxygen is an oxidising agent which transfers itself onto the substrate and forms either formaldehyde and water or carbon dioxide and water with the selectivity depending on the sequence of elementary steps. If the abstraction of a proton can occur without simultaneous oxygen transfer onto the carbon atom then formaldehyde will result, if the proton exchange is simultaneous with the oxygen transfer total combustion will occur.

The high-temperature gamma oxygen, however, is not oxidising but acts as a Lewis base (or here Bijerrum base) and exclusively abstracts protons leading to a formaldehyde formation with 100% selectivity. This identification of two different functions of catalytically active atomic oxygen would have been impossible without all the spectroscopic evidence for two electronically different adsorbates summarised above.

We note the two sharp desorption features in the m/e 44 desorption trace and identify them as products of methanol consumption without formation of either water or formaldehyde. These peaks can be identified as chemisorbed methoxy and carbonate. Both species were investigated in single crystal low-pressure studies[13] as potential reaction intermediates to formaldehyde which is, however, not correct as can be seen from the present experiment. Formate and carbonate form at basic oxygen-containing surface sites[53] but remain firmly held until they are oxidatively decomposed. These surface sites are reactive towards methanol but bind the primary intermediates too strongly to the surface and are hence unproductive in steady state conversion. For their high reactivity towards methanol these surface sites were identified at the low pressures used in the single crystal studies as the only active sites under the model conditions.

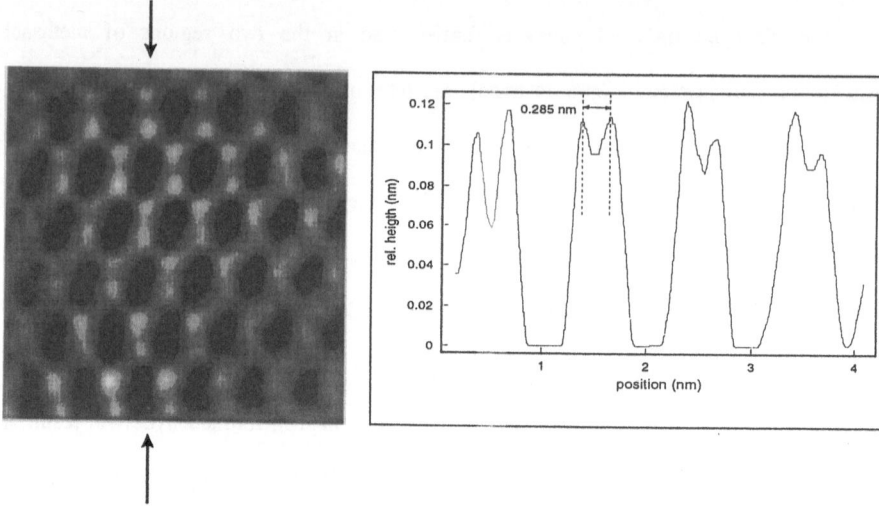

Figure 10

STM image and linescan of a electrolytic silver grain loaded in oxygen at 923 K for 3 h. The data were measured in dry air with a BURLEIGH ARIS 6100 instrument and a mechanically prepared Pt-IR tip in the constant current mode at 1 nA and 200 mV gap voltage. The raw data are shown after FFT smoothing but without any further corrections. The image was taken on the smooth side of a facet which had the same characteristic shape as (111) facets analysed on a silver sphere using the REM /RHEED technique[46].

The location of the selective species gamma oxygen within the top atomic layer of silver is unusual and requires further characterisation. The high chemical stability allowed to monitor loaded surfaces by STM in air. Transfer and evacuation to a UHV STM system led to contaminations which could not be removed without destroying the surface restructuring. A characteristic picture from a (111) facet on a grain of electrolytic silver powder loaded in oxygen at 900 K and at atmospheric pressure for 24 h prior to imaging is presented in Figure 10. The line scan along the two arrows shown indicates the silver atoms in the normal distance of 285 pm. Also the hexagonal pattern is the expected arrangement for a (111) close packed structure. The deep „holes" within the silver hexagons are attributed to the drastic variation in local work function caused by the presence of oxygen. A surface stoichiometry of Ag_2O is in full agreement with XPS data corrected for the contribution of beta oxygen. The images are of insufficient quality to allow the analysis of the exact location of the oxygen, in particular the determination of an off-centre position. In any case, a high co-ordination number will result for the oxygen atom allowing to rationalise the postulated strong interaction between silver and oxygen in terms of a d-sp hybridisation. This hybridisation is also indicated by the unusual low-energy shift of the Ag 3d photoemission spectrum of 367.3 eV vs. 368.0 eV for elemental Ag upon formation of the gamma oxygen species.

We note that the STM image is one of the first examples of an atomically resolved picture of an active species seen on a real catalyst particle.

The conclusions presented may be criticised as not to apply for high pressure conditions. As it is difficult to deduce mechanistic details from the observation of steady state conversion data we employed instationary kinetic measurements using the IMR-MS instrument as a time-resolved detector for the educt and all products. The instationary experiments were conducted after periods of 240 h of stationary operation to ensure a characteristic state of the catalyst. During this time a significant fraction of over 60% of the total conversion produced elemental hydrogen indicating that the oxydehydrogenation cannot be the main reaction pathway as it was postulated from low pressure single crystal studies[13,53]. The finding is full in accordance with earlier kinetic data assuming a mixed mechanism[54] with parallel reactions of oxidation and dehydrogenation.

In Figure 11 the response of the system is illustrated after loading the catalyst with oxygen without methanol feed and switching to only methanol without gas phase oxygen at time t=0. The catalyst being nominally elemental silver can store significant amounts of oxygen which exhibits, however, pronounced differences in reactivity. In the first 2 minutes 50% of the total reactivity occurs with a high but limited selectivity (compare traces of formaldehyde and carbon dioxide). After two minutes the production mechanism changes with a total drop of the carbon dioxide production. In terms of the above results the behaviour can be interpreted as follows. After loading the silver all three atomic species alpha, beta and gamma oxygen were

316

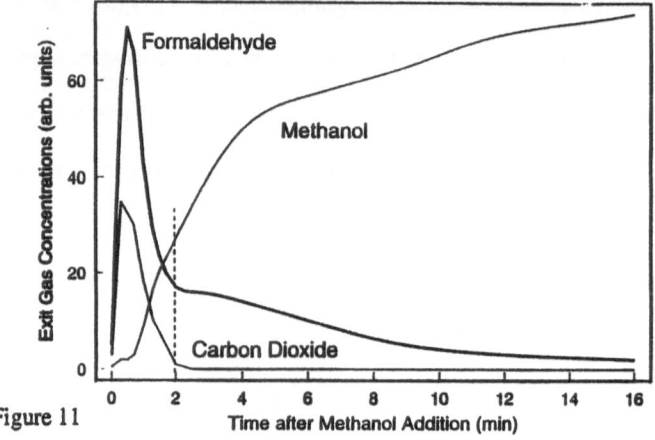

Figure 11

IMR-MS response to an instationary conversion experiment with a microreactor loaded with electrolytic silver and fed with a 1:1 water methanol mix at a load of 0.1 Kg/cm^2/h. Stoichiometric amounts of oxygen diluted in nitrogen at 400 ml/min were passed through the 0.8 cm thick catalyst bed. After steady state operation the feed was interrupted immediately before the experiment and the catalyst was loaded with the oxygen-nitrogen stream. After a nitrogen purge the methanol/water feed was switched on at t=0 without co-feeding oxygen. The high sensitivity of IMR-MS ensured that no adventitious oxygen source was present during the experiment. The analysis time for one set of conversion data was chosen to be 6 s.

present. Alpha and gamma oxygen reacted immediately after admission of methanol with the substrate leading to a rapid surface depletion terminated after 2 min. The steep part of the formaldehyde production curve is, however, superimposed on a broad exponential decay caused by the diffusion-controlled segregation of bulk beta oxygen to the surface where it is primarily converted to gamma oxygen. It reacts as basic dehydrogenation centre with methanol and is therefore not converted to oxidising alpha oxygen. The dehydrogenation activity is apparently of limited lifetime with the oxygen atoms ending most likely as water molecules. For this reason a catalytic amount of oxygen during steady state operation is insufficient.

One might be tempted to deduce from the present data intrinsic selectivities for the parallel occurring processes which could be used in a quantitative model of the reaction. Care must be taken in such derivations as the instationary boundary conditions will affect the surface abundancies of all species, in particular if possibly different kinetics for alpha and gamma oxygen formation are taken into account.

Conclusions

The two examples of apparently different catalyst and reaction systems have in common that the fundamental reaction step is in both cases a dehydrogenation step. In both cases two chemically different atomic oxygen species are involved. They differ greatly in their bonding relative to the central atom and may be discriminated in strongly bound oxygen (terminal oxygen in HPA or gamma oxygen in silver) and weakly bound oxygen (bridging oxygen in HPA or alpha oxygen in silver). This electronic difference is the basis of a different chemical function namely as dehydrogenation centre or as oxygen transfer centre. These ideas tie in with the concept of electrophilic and nucleophilic oxygen in catalysis which is described in detail in the book of *Bielanski and Haber.*[55]

Regarding the strategic side of the analysis of the mode of operation of a catalyst the data show that is essential to start with a thorough analytical approach from the beginning and to avoid a-priori *assumptions*. Both the material gap and the pressure gap cannot be bridged by linear extrapolations. In-situ experiments (as they are very expensive and far away from routine) should be aimed at crucial properties of the reaction system. They are prerequisites in order to justify the simplifications necessary on the material side in order to approach a detailed mechanistic study. A large number of experimental techniques including even methods not directed towards surface properties (such as diffraction experiments) are essential for the identification of the events on the surface which determine the steady state operation of the system. This includes not only the substrate-to-product cycle but also reactions of the catalyst solid with the gas phase or within its own phase inventory.

This a-priori *information* has to be built into a surface science study of the elementary reaction steps. Evaluation of the function of a heterogeneous catalyst cannot be achieved outside the „single crystal approach" This approach is, however, only applicable *after* a definition phase of the problem to be studies and requires confirmation of its results by in-situ (preferably kinetic) experiments.

This paper is focused onto the interplay between surface physics and solid state chemistry. The kinetic aspects concerning the educt-product cycle are of equal importance for the system description. As these aspects are more emphasised in the many reviews on the issue they were not covered in this text which does by no means indicate that they are of minor significance.

All experimental efforts should be paralleled by adequate theoretical studies including kinetic modelling, electronic structure calculations and investigations into the dynamic behaviour of the system. Only this integrated approach which is illustrated by the analysis of the oscillatory behaviour of the CO oxidation reaction over noble metals[56] can ensure that important system properties which are difficult to find by experiments (such as surface dynamics under realistic conditions) are not overlooked.

Catalysis science is now in the position to utilise all these tools and to perform the recommended integrated approach. For this reason there is good hope that the co-operation of the surface disciplines will produce in the near future much solid evidence about heterogeneous catalysis and fulfil the expectations which were put in the past just into one segment of the scientific strategy.

Acknowledgements

The author is grateful to all collaborators who performed the work mentioned here. M. Muhler, X. Bao, W. Bensch, B. Herzog, H. Schubert, U. Tegtmeyer, Th. Ilkenhans and U. Guntow contributed their work. Financial support came from BASF AG, Röhm AG, and the Bundesmisister für Forschung und Technologie.

References

[1] D.J. Dwyer, F.M. Hoffmann, „Surface Science of Catalysis", ACS Symposium Series, Vol. 482, 1992, p.1

[2] J.M. Thomas, K.I. Zamaraev, R. Schlögl, Angewandte Chemie, (106), 1994, 316

[3] R. Schlögl, Angewandte Chemie, (105), 1993, 402

[4] J.M. Thomas, Angewandte Chemie, (106), 1994, 963

[5] B.S. Clausen, L. Grabek, G. Steffenson, H. Topsoe, Catal. Lett., (20), 1993, 23, J.M. Thomas, G.N. Greaves, Catal. Lett., (20), 1993, 337,

[6] G.T. Moggridge, T. Rayment, R.M. Ormenrod, M. A. Morris, R.M. Lambert, Nature, (358), 1992, 658

[7] M. Muhler, Z. Paal, R. Schlögl, Appl. Surf. Sci., (47), 1991, 281, K. Noack, H. Zbinden, R. Schlögl, Catal. Lett., (4), 1990, 145

[8] B. Fastrup, H.N. Nielsen, Catal. Lett., (14), 1992, 233

[9] L. Riekert, Chem. Ing. Tech., (53),1981, 950

[10] G. Ertl, Catal. Rev. Sci. Eng., (21), 1980, 201

[11] G.A. Somorjai, „Chemistry in Two Dimensions", Cornell, 1981

[12] P. Stoltze, Physica Scr., (36), 1987, 824

[13] M.A. Barteau, R.J. Madix, in „The Chemical Physics of Solid Surfaces and Heterogeneous Catalysis", Ed.: D.A. King, P. Woodruff, Elsevier, Vol. 4, 1982, p. 95

[14] X. Bao, M. Muhler, B. Pettinger, R. Schlögl, G. Ertl, Catal. Lett., (22), 1993, 215

[15] R. Schlögl, in: „Catalytic Ammonia Synthesis", ed.: J.R. Jennings, Plenum, 1992, p.19

[16] J. Schütze, W. Mahdi, B. Herzog, R. Schlögl, Catal. Lett., 1994, to appear,

[17] R. Schlögl, R.C. Schoonmaker, M. Muhler, G. Ertl, Catal. Lett., (1), 1988, 237

[18] G. Ertl, M. Huber, N,. Thiele, Z. Naturforsch., (34a), 1979, 30

[19] M. Muhler, F. Rosowski, G. Ertl, Catal. Lett., (24), 1994, 317

[20] M. Boudart, G. Djega-Mariadassou, „Kinetics of Heterogeneous Catalytic Reactions", Princeton, 1984,

[21] F. Bozso, G. Ertl, M. Grunze, M. Weiss, J. Catal., (50), 1977, 519

[22] M. Bowker, I. Parker, K. Waugh, Surf. Sci., (197), 1988, L223

[23] P. Stoltze, J.K. Norskov, J. Catal., (110), 1988, 1

[24] J. A. Dumesic, A.A. Trevino, J. Catal., (116), 1989, 119

[25] see articles in ref. 15

[26] G. Ertl, J. Vac. Sci. Technol., (A1), 1983, 1247

[27] N.D. Spencer, R.C. Schoonmaker, G.A. Somorjai, J. Catal., (74), 1982, 129

[28] D.W. Goodman, R.D Kelley, T.E. Madey, J.T. Yates, J. Catal., (63), 1980, 226

320

[29] H.J. Krebs, H.P. Bonzel, Surf. Sci., (88), 1982, 269

[30] S.R. Craxford, E.K. Rideal, J. Chem. Soc., (1), 1939, 604

[31] C.S. Kellner, A.T. Bell, J. Catal., (70), 1981, 418

[32] D.W. Blakely, C.I. Kozak, B.A. Sexton, G.A. Somorjai, J. Vac. Sci. Technol., (13), 1976, 1091

[33] see e.g. „Solid State Chemistry, Compounds", ed. A.K. Cheetham, P. Day, Oxford, 1992

[34] R.A. van Santen, H.P.C.E. Kuipers, Advances Catal., (35), 1987, 265

[35] H. Sperber, Chem. Ing. Tech., (41), 1969, 962

[36] Ullmann, „Encyclopedia of Industrial Chemistry", Verlag Chemie, vol 21, 1982, 311

[37] M. Misono, Catal. Rev. Sci. Eng., (29), 1987, 269

[38] Th. Haeberle, G. Emig, Chem. Eng. Technol., (11), 1988, 392

[39] B. Herzog, W. Bensch, Th. Ilkenhans, R. Schlögl, Catal. Lett., (20), 1993, 203

[40] B. Herzog, Th. Ilkenhans, R. Schlögl, Fresen. J. Anal. Chem., 1994, in press

[41] Th. Ilkenhans, B. Herzog, Th. Braun, R. Schlögl, J. Catal., 1995 , in press

[42] M.T. Pope, A. Müller, Angewandte Chemie, (103), 1991, 56

[43] O. Watzenberger, G. Emig, D.T. Lynch, J. Catal., (124), 1991, 56

[44] K. Brückman, J.M. Tatibouet, M. Che, E. Serwicka, J. Haber, J. Catal., (139), 1993, 455

[45] U. Tegtmeyer, H.P. Weiss, R. Schlögl, Fresen. J. Anal. Chem., (347), 1993, 263

[46] X. Bao, J.V. Bart, G. Lehmpfuhl, R. Schuster, Y. Uchida, R. Schlögl, G. Ertl, Surf. Sci., (284), 1993, 14

[47] C. Rehren, M. Muhler, X. Bao, R. Schlögl, G. Ertl, Z. Phys. Chem., (174), 1991, 11

[48] X. Bao, G. Lehmpfuhl, G. Weinberg, R. Schlögl, G. Ertl, J. Chem. Soc., Faraday Trans., (88), 1992, 865

[49] X. Bao, B. Pettinger, G. Ertl, R. Schlögl, Ber. Bunsenges. Phys. Chem., (97), 1993, 322

[50] W. Segeth, J.H. Wjingaard, G.A. Sawatzky, Surf. Sci., (194), 1988, 615

[51] L.H. Tjeng, M.B.J. Meinders, J. van Elp, J. Ghijsen,G.A. Sawatzky, R.L. Johnson, Phys. Rev. B., (41), 1990, 3190

[52] H. Schubert, U. Tegtmeyer, R. Schlögl, Catal. Lett., 1994 in press

[53] M.A. Barteau, M. Bowker, R.J. Madix, Surf. Sci., (94), 1980, 303

[54] V.J. Atoschenko, I.P. Kusnharenko, Int.J. Chem. Eng., (4), 1964, 581

[55] "Oxygen in Catalysis", A. Bielanski, J. Haber, M. Dekker, 1991, p. 147

[56] G. Ertl, Adv. Catal., (37), 1990, 213, G. Ertl, Angewandte Chemie, (102), 1990, 1258

Characterization of Zeolite Catalysts and Related Materials by Multinuclear Solid-State NMR Spectroscopy

Günter Engelhardt

Institute of Chemical Technology I, University of Stuttgart
D-70550 Stuttgart, Germany

Abstract. The application of multinuclear solid-state NMR to zeolites and related microporous host-guest systems provides detailed information on the local structure of the host framework as well as on distribution and dynamics of the guest species and on the related host-guest interactions. Owing to the importance of silicon and aluminium as framework constituents in zeolites, ^{29}Si and ^{27}Al NMR play a dominant role but also other elements isomorphously substituted in the zeolite framework can be profitably studied by their respective NMR-active nuclides, e.g. ^{31}P, ^{11}B, and ^{71}Ga. Charge balancing cations in zeolites have been studied by e.g. ^{7}Li, ^{23}Na, ^{133}Cs, ^{139}La, and ^{205}Tl NMR, while ^{13}C and ^{1}H NMR is applied to study organic guest molecules present in the zeolite cavities. Finally, ^{1}H NMR also permits the characterization of catalytically active proton sites and terminal SiOH groups in zeolites. In this article a brief survey will be given on the basic principles and techniques of high-resolution solid-state NMR, and the progress achieved during recent years will be demonstrated by selected examples from the zeolite field.

1. Introduction

Since the phenomenon of nuclear magnetic resonance was discovered in 1946 by Bloch and Purcell, NMR spectroscopy has emerged in one of the most powerful tools for investigating structure and dynamics of molecular systems. Until recently, however, the application of high-resolution NMR spectroscopy was more or less restricted to liquids, the NMR spectra of which exhibit well resolved distinct lines for structurally different atoms in the molecule. Straightforward application of the conventional NMR techniques to microcrystalline or amorphous solids yields, in general, very broad and almost featureless spectral shapes without any resolution of distinct lines for resonance atoms in different structural environments. This line broadening originates from specific interactions of the nuclear spins tightly bound in the rigid lattice of solids. In liquids these interactions are averaged by rapid thermal motion of the molecules. However, the development of novel sophisticated NMR techniques starting in the early 1970's opened the possibility of measuring NMR spectra of solids with spectral resolution comparable to those of liquids.

Subsequently, high-resolution solid-state NMR has developed rapidly into an effective and widely used means for elucidating subtle details of the structural properties of different kinds of solid materials. Since high-resolution solid-state NMR gives, in general, information on the local order of the structure and can be applied to crystalline, microcrystalline and amorphous materials, it is a valuable complement to X-ray and other diffraction techniques which probe the long-range order of crystalline materials.

The first application of high-resolution solid-state NMR to zeolites was presented at a conference in 1979 [1], and a series of pioneering papers were published in 1981 (see e.g. [2-5]). In this early work mainly ^{29}Si NMR but also ^{27}Al NMR [6] were used to study the aluminosilicate framework of various zeolites. Subsequently, interest in the application of multinuclear solid-state NMR to zeolites has grown very rapidly and several thousend papers on the subject have now appeared. Progress in the field has been summarized in several reviews [7, 8] and a book [9].

In this article a brief overview will be given on the basic principles and experimental techniques of high-resolution solid-state NMR, followed by a concise survey on the application of multinuclear NMR to structural studies of zeolites and related materials.

2. Basic principles and methods of high-resolution NMR of solids

In condensed matter each nucleus is exposed to specific interactions with other nuclei and electric fields of electrons. In the external magnetic field B_0 each interaction effects a specific shift of the energy levels of the nuclear spin states, the differences of which are the measurable quantities in the NMR experiment. The critical differences between NMR of liquids and solids mentioned above arise from the directional dependence of the various spin interactions. The fixed orientations of nuclear environments in the solid lattice affect the shape and position of the NMR lines measured for different orientations of the sample with respect to the direction of the external magnetic field, B_0. In highly ordered single crystals having only a small number of symmetry-related orientations of nuclei, narrow lines are observed, the positions of which change with the orientation of the crystal relative to the B_0 direction. Microcrystalline or amorphous powders, however, are characterized by a random distribution of many different orientations and the observed NMR pattern is a broad superposition of the lines from randomly oriented individual nuclei.

2.1 Nuclear spin interactions and line broadening in NMR spectra of powders

There are three structure dependent interactions which preferably contribute to line broadening in NMR powder spectra:

The *dipolar interaction* between the magnetic moment of the observed nucleus and those of neighbouring nuclei. The dipolar interaction depends on the magnitude of the magnetic moments of the interacting nuclei, falls off rapidly with the nuclear distance ($1/r^3$), and is independent of B_o. It is the dominant source of line-broadening in the NMR spectra of ^1H and other spin-1/2 nuclei in proton-containing systems, and can range up to 100 kHz.

The *chemical shift anisotropy*, which follows from the spatial dependence of nuclear magnetic shielding, is determined by the electronic charge distribution around the NMR nucleus. The interaction increases linearly with B_o and its size ranges typically up to about 100 ppm but may be considerably larger for heavy nuclei.

The *quadrupolar interaction* for nuclei with spin quantum number $I > 1/2$. The nuclear quadrupole moment of these nuclei interacts with electric field gradients (EFG) at the nuclear site induced by a non-spherical charge distribution around the nucleus. The quadrupolar interaction can range up to several MHz and dominates very often the NMR line shapes of quadrupolar nuclei. In general, the quadrupolar powder patterns of nuclei with non-integer spin I are mainly affected by the quadrupolar interaction to second order which decreases with increasing B_o.

2.2 Experimental techniques for line narrowing in NMR powder spectra

Measurements of high-resolution NMR spectra of powders characterized by individual resonance lines for structurally distinct nuclei require the application of special techniques to remove or at least to reduce substantially the line broadening interactions considered above. The following techniques are of particular importance:

Dipolar decoupling (DD) removes the heteronuclear dipolar interaction by irradiating a strong rf field at the resonance frequency of the nucleus giving rise to the dipolar broadening, while observing the nucleus under study. Dipolar decoupling is preferably applied to remove line broadening effects due to dipolar interactions with protons in the NMR spectra of other nuclei.

Magic-angle spinning (MAS), i.e. fast mechanical sample rotation about an axis inclined at the "magic" angle $\theta = 54°44'$ to the direction of B_o, also eliminates line broadenings from dipolar interactions, averages chemical shift anisotropy to the isotropic value, and reduces second order quadrupolar line broadening. The magic angle follows from the condition $(3\cos^2\theta - 1) = 0$, where the left-side expression is a geometric term governing the dipolar and shift anisotropy interaction. If the spinning rate in MAS experiments exceeds the static line width (in Hz), just one narrow line is observed. In practice, however, this is often not the case (though spinning frequencies of more than 20 kHz can be attained) and the central line is flanked by a comb of spinning side bands.

Double rotation (DOR) [10] and *dynamic-angle spinning (DAS)* [11] are applied to eliminate second-order quadrupolar line broadening which cannot be removed by MAS. Since the second-order quadrupolar interaction depends on geometric factors like $\sin\theta$, $\cos\theta$, $\sin^2\theta$, and $\cos^2\theta$ terms, the sample must be spun around more than one axis. In the DOR experiment the sample is rotated simultaneously around two axes, while DAS involves successive rotation around two different axes by switching the direction of the spinning axis in subsequent time periods.

High magnetic field strengths B_o (up to 17.6 T) further reduce the linewidth dominated by second-order quadrupolar interactions.

Two-dimensional quadrupole nutation MAS NMR [12] permits the separation of lines characterized by weak and strong quadrupolar interactions.

Application of these techniques to microcrystalline powder samples results in substantial line-narrowing and, therefore, considerable improvement of the spectral resolution which will be demonstrated below by selected examples.

2.3 Survey of NMR parameters and related structural information

Provided that well resolved signals appear in the NMR spectrum (or may be separated by spectra simulation and line decomposition), the number of signals in the spectrum gives immediately the number of nuclei in different structural environments present in the sample. In addition, specific parameters of each line can be extracted from the spectra which are related to the structural surroundings of the resonance atoms. These parameters are in particular:

The *isotropic chemical shift, δ_{iso},* which, for spin-1/2 nuclei, is the position of the line in the spectrum relative to that of a standard sample, but must be corrected for quadrupolar shift contributions in the case of quadrupolar nuclei. δ_{iso} provides information on the chemical environment of the resonance atom, such as number and type of neighbouring atoms, coordination number, and bonding geometry.

The *quadrupole coupling constant, QCC,* and *asymmetry parameter of the EFG, η,* ($0\leq\eta\leq1$ with $\eta = 0$ for axialsymmetric EFG) for quadrupolar nuclei which may be determined from computer simulations of the quadrupolar MAS line shape and/or from MAS or DOR experiments at different B_o field strenghts. Valuable information on the charge distribution around the quadrupolar nucleus, i.e. on distortions of the coordination symmetry, may be derived from QCC and η.

The *relative line intensities, I,* which correspond to the relative populations of the structurally distinct sites. For spin-1/2 nuclei the intensities can directly be determined from spectra integration, while for non-integer spin quadrupolar nuclei the experimental line intensities must be carefully corrected for contributions from central and satellite transitions and/or spinning side bands [13].

3. Application to zeolites and related materials

Zeolites are microporous aluminosilicates with a complex three-dimensional framework of SiO_4 and AlO_4 tetrahedra forming a regular system of interconnected cavities and channels of molecular dimensions. The negative charge of the framework introduced by the four-coordinated aluminium is compensated by metal cations or protons located in the zeolite cavities which may further contain neutral guest species such as water or organic molecules. In principle, each of the three basic atomic constituents of the aluminosilicate framework - silicon, aluminium and oxygen - are amenable to NMR measurements by their naturally occurring isotopes ^{29}Si, ^{27}Al and ^{17}O. However, ^{17}O has a very low natural abundance (0.037%) and is a quadrupolar nucleus ($I = 5/2$) rendering the application of ^{17}O NMR difficult without expensive isotopic enrichment. Nevertheless, some interesting and promising ^{17}O NMR studies of ^{17}O enriched zeolites have been published [14]. ^{27}Al is also a quadrupolar nucleus ($I = 5/2$) but is 100% abundant, and the ^{29}Si isotope, although of low natural abundance (4.7%), has $I = 1/2$, i.e. no quadrupole moment, and thus gives rise to narrow resonance lines. Both nuclei are well suited for NMR experiments and play the dominant role in NMR studies of aluminosilicate zeolites. In addition, ^1H NMR is extensively used to study protons in bridging Si(OH)Al and terminal SiOH or AlOH groups or proton-containing guest species (e.g., H_2O, NH_3, organic sorbates, etc.) in zeolites [15]. More recently, NMR of the quadrupolar ^{23}Na nucleus ($I = 3/2$) has been applied for detailed investigations of location and migration of Na^+ cations in zeolites [16, 17]. Several other elements that replace silicon or aluminium in tetrahedral sites of the zeolite framework (e.g., B, P, Ga, Ge, Be), or are present in non-framework constituents can also be profitably studied by their respective NMR-active nuclides (e.g., ^{11}B, ^{31}P, $^{69,71}Ga$, ^9Be, ^{73}Ge, ^{13}C, $^{14,15}N$, ^{133}Cs, ^{205}Tl, ^{207}Pb).

In what follows, a short survey will be given on important applications of ^{29}Si, ^{27}Al, and ^{23}Na NMR to structural studies of zeolites. NMR studies of these nuclei provide detailed insights into the framework structure and host-guest interactions, and cover at least 80% of the published NMR work in the field.

3.1 ^{29}Si NMR

The ^{29}Si chemical shifts of zeolites depend sensitively on number and type of T-atoms (T = Si, Al) connected with a given SiO_4 tetrahedron. In framework aluminosilicates like zeolites up to five different silicon-centered structural units of type $Si(OSi)_{4-n}(OAl)_n$ with n = 0-4 may exist which are conventionally denoted by Si(nAl). Neglecting the presence of crystallographically inequivalent Si(nAl) sites (vide infra), the ^{29}Si NMR spectrum of a zeolite may thus consist of one to five peaks corresponding to the five possible Si(nAl) environments in the zeolite framework. This is demonstrated in Figure 1 by the ^{29}Si MAS NMR spectra of a series of zeolites of type X and Y with different numbers of Si and Al atoms in the framework given by the atomic Si/Al ratio [3]. Up to five lines appear in the

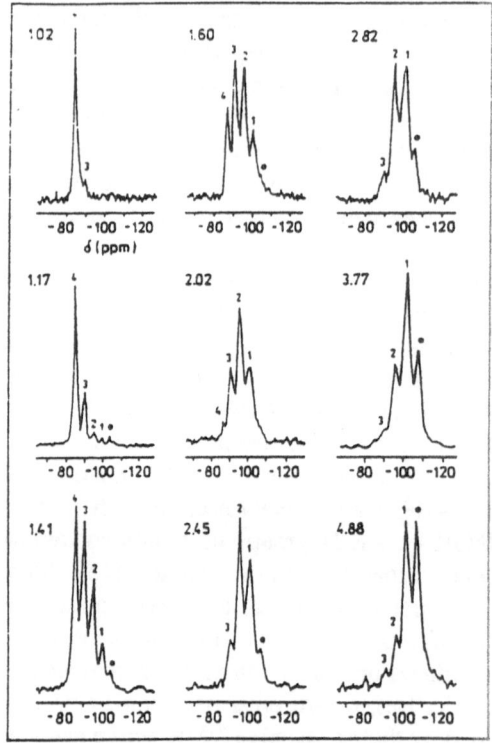

Fig. 1. ^{29}Si MAS NMR spectra of zeolites X and Y with different Si/Al ratios

spectra which can be attributed to the five distinct Si(nAl) environments (indicated at the lines by the number n) by their chemical shifts. As the number n of Al atoms increases, the lines are systematically shifted to low field by steps of about 5 ppm for each Al. From the intensities I_n of the distinct lines, the Si/Al ratio of the framework can easily be determined by the equation Si/Al = $\Sigma I_n / \Sigma (0.25 n \, I_n)$ which is independent of the specific structure of the zeolite. Moreover, since the intensities correspond to the populations of the various Si(nAl) sites in the structure, information on the topological distribution of Si and Al atoms on the T-sites of the framework (Si,Al ordering) can be deduced from comparison of the relative intensities with relative Si(nAl) populations derived from the distribution patterns of appropriate structure models. This procedure has been successfully applied to study the Si,Al ordering in zeolites X and Y of different Si/Al ratios [3, 4] and in several other zeolites.

In addition to the chemical environment, i.e. the number of Si and Al atoms in the Si(nAl) sites, the ^{29}Si chemical shift is further affected by the bonding geometry around the Si atom, i.e. by SiOT bond angles and SiO bond lengths.

Therefore, chemically equivalent but crystallographically inequivalent Si sites may have different chemical shifts. Though these effects may complicate the interpretation of the spectra in terms of the Si(nAl) units, they provide valuable information on the number, population and local geometry of crystallographically distinct Si sites in the zeolite framework. A case in point is the aluminium-free variant of zeolite ZSM-5 which shows 20 well resolved lines in its ^{29}Si MAS NMR spectrum at 295 K representing 24 crystallographically distinct Si sites of the monoclinic form (space group $P2_1/n$), while at 393 K only 10 lines appear in the spectrum, indicating the temperature induced transformation to the orthorhombic form containing 12 distinct Si positions (space group Pnma) [18]. Empirical correlations have been established between ^{29}Si chemical shift and mean SiOT bond angles of the Si(OT)$_4$ units, and subsequently rationalized by theoretical considerations (see [19] and references therein), which may be used to estimate SiOT angles from chemical shifts and, vice versa, to calculate shift values from crystal structure data.

3.2 ^{27}Al NMR

The ^{27}Al NMR spectra of zeolites are, in general, much simpler than their ^{29}Si NMR counterparts since according to Loewenstein's rule [20] (which forbids AlOAl pairing) only one tetrahedral Al environment, namely Al(OSi)$_4$, exists in the aluminosilicate framework. Comparatively narrow lines at about 60±5 ppm (standard: aqueous Al(NO$_3$)$_3$ solution) are usually observed for tetrahedral framework Al in hydrated zeolites, indicating only weak to moderate quadrupolar interactions owing to small deviations from tetrahedral symmetry of the AlO$_4$ environments. Similarly to ^{29}Si NMR, a linear correlation has been found between ^{27}Al chemical shifts and mean AlOSi bond angles [21], provided the shifts were corrected for quadrupolar contributions. In contrast to the hydrated cation-containing zeolites, very strong quadrupolar interactions were observed for ^{27}Al nuclei in Al-OH-Si sites of dehydrated hydrogen forms of zeolites [22], rendering these Al atoms "NMR invisible" in conventional ^{27}Al MAS NMR spectra due to extensive line broadening. Large quadrupolar interactions have also been detected in several aluminate sodalites the framework of which is built up entirely from AlO$_4$ tetrahedra. While the MAS NMR spectra exhibits broad humps of several superimposed quadrupolar line shapes, up to seven narrow lines for differently distorted AlO$_4$ environments could be resolved in the ^{27}Al DOR NMR spectra of these materials [23].

An important application of ^{27}Al NMR to zeolites is the detection and characterization of non-framework aluminium species formed e.g. by various thermal or hydrothermal treatments of zeolites as usually applied in the preparation of zeolite catalysts and their practical use. Non-framework aluminium in zeolites shows preferably six-fold AlO$_6$ coordination and gives rise to a line at about

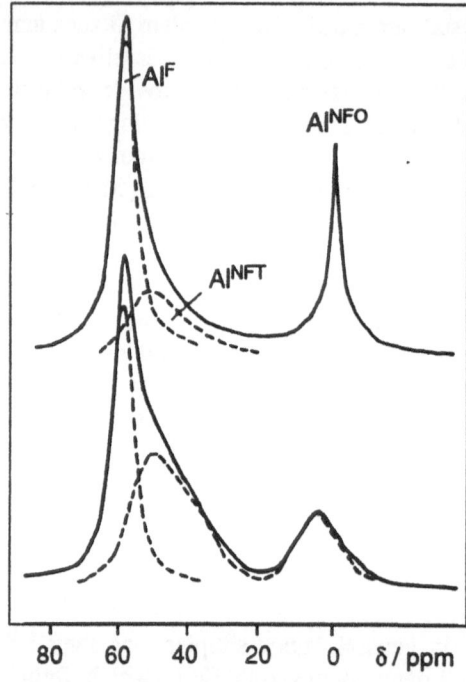

Fig. 2. ^{27}Al MAS NMR spectra of two hydrothermally treated zeolites Y

0 ppm, i.e. well separated from four-coordinated framework Al. However, quadrupolar line broadening due to major distortions of the octahedral symmetry of the AlO_6 environments may complicate the quantitative interpretation of the spectra. In addition, four-coordinated non-framework Al in strongly distorted AlO_4 environments has been detected in hydrothermally treated zeolites by broad signals at lower field which heavily overlap with the lines of tetrahedral framework Al [24, 25]. As an example, Figure 2 shows the ^{27}Al MAS NMR spectra (measured at 130.3 MHz) of two under different conditions hydrothermally treated samples of zeolite Y. The three signals of six-fold coordinated non-framework aluminium, Al^{NFO}, four-coordinated non-framework aluminium, Al^{NFT}, and four-coordinated framework aluminium, Al^F, are indicated, their specific shapes were derived from two-dimensional nutation MAS NMR spectra [24].

3.3 ^{23}Na NMR

Solid state ^{23}Na NMR offers considerable potential for determining the location of sodium cations present in the zeolite cavities. However, complex spectral patterns are generally observed due to superposition of the quadrupolar line shapes of different sodium sites, and an array of NMR techniques has to be applied for line separation. These include MAS, DOR, two-dimensional nutation MAS, experiments at different magnetic field strenghts, and, most important, computer simulation and line decomposition of the spectra. In addition, theoretical calculations of the EFG at the ^{23}Na nucleus based on crystal structure data or sensible structure models [26] are very helpful for spectra interpretation.

Detailed ^{23}Na NMR investigations on sodium location in dehydrated zeolite NaY and related materials have been performed applying all the methods mentioned above [16, 17]. Figure 3 displays the ^{23}Na MAS NMR spectrum (measured at 105.8 MHz) of dehydrated zeolite NaY and its simulation by four

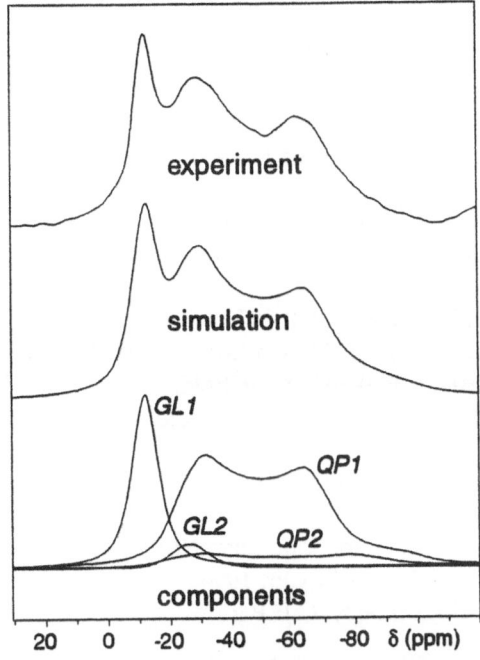

Fig. 3. Experimental and simulated ^{23}Na MAS NMR spectra of dehydrated zeolite NaY

components. The two quadrupolar patterns QP1 and QP2 are attributed to sodium sites SII and SI', located above the centers of the six-ring windows in the supercages and ß-cages, respectively, while the two gaussian lines GL1 and GL2 originate from sodium in the hexagonal prisms (GL1) and from small amounts of rehydrated sodium cations (GL2). From the corrected line intensities [13], the specific populations of the distinct sites are obtained which agree well with those derived from X-ray diffraction studies. It should be noted that the simulation and assignment of the MAS spectrum given in Fig.3 is greatly supported by the information derived from the corresponding DOR, 2D nutation and MAS NMR spectra measured at up to five different magnetic field strengths (7.0 to 17.6 T), and calculations of the EFG by a point charge model.

References

1 G.Engelhardt, D.Kunath, M.Mägi, A.Samoson, M.Tarmak, E.Lippmaa, Workshop on Adsorption of Hydrocarbons in Zeolites, Berlin-Adlershof, 1979

2 E.Lippmaa, M.Mägi, A.Samoson, M.Tarmak, G.Engelhardt, J.Am.Chem.Soc. 103 (1981) 4992

3 G.Engelhardt, U.Lohse, E.Lippmaa, M.Tarmak, M.Mägi, Z.anorg.allg.Chem. 482 (1981) 49

4 S.Ramdas, J.M.Thomas, J.Klinowski, C.A.Fyfe, J.S.Hartman, Nature 292 (1981) 228

5 J.Klinowski, J.M.Thomas, C.A.Fyfe, J.S.Hartman, J.Phys.Chem. 85 (1981) 2590

6 D.Freude, H.J.Behrens, Cryst.Res.Technol. 16 (1981) 1236

7 G.Engelhardt, Stud.Surf.Sci.Catal. 58 (1991) 285

8 J.Klinowski, Anal.Chim.Acta 283 (1993) 929

9 G.Engelhardt, D.Michel, High-Resolution Solid-State NMR of Silicates and Zeolites, Wiley 1987

10 A.Samoson, E.Lippmaa, A.Pines, Mol.Phys. 65 (1989) 1013

11 K.T.Mueller, B.Q.Sun, G.C.Chingas, J.W.Zwanziger, T.Terao, A.Pines, J.Magn.Reson. 86 (1990) 470

12 A.Samoson, E.Lippmaa, Chem.Phys.Lett. 100 (1983) 205

13 D.Massiot, C.Bessada, J.P.Coutures, F.Taulelle, J.Magn.Reson. 90 (1990) 231

14 H.K.C.Timken, G.L.Turner, J.P.Gilson, L.B.Welsh, E.Oldfield, J.Am.Chem.Soc. 108 (1986) 7231, S.Yang, K.D.Park, E.Oldfield, J.Am.Chem.Soc. 111 (1989) 7278

15 H.Pfeifer, in: NMR - Basic Principles and Progress, Vol.31, p.31, Springer-Verlag 1994

16 M.Hunger, G.Engelhardt, H.Koller, J.Weitkamp, Solid State NMR 2 (1993) 111

17 G.Engelhardt, Proc. 10th Intern.Zeolite Conf., Elsevier, 1994

18 C.A.Fyfe, H.Strobl, G.T.Kokotailo, G.J.Kennedy, G.E.Barlow, J.Am.Chem.Soc. 110 (1988) 3373

19 G.Engelhardt, Stud.Surf.Sci.Catal. 52 (1989) 151

20 W.Loewenstein, Am.Mineral. 39 (1954) 92

21 E.Lippmaa, A.Samoson, M.Mägi, J.Am.Chem.Soc. 108 (1986) 1730

22 H.Ernst, D.Freude, I.Wolf, Chem.Phys.Lett. 212 (1993) 588

23 G.Engelhardt, H.Koller, P.Sieger, W.Depmeier, A.Samoson, Solid State NMR 1 (1992) 127

24 A.Samoson, E.Lippmaa, G.Engelhardt, U.Lohse, H.-G.Jerschkewitz, Chem.Phys.Lett. 134 (1987) 589

25 G.J.Ray, A.Samoson, Zeolites 13 (1993) 410

26 H.Koller, G.Engelhardt, A.P.M.Kentgens, J.Sauer, J.Phys.Chem. 98 (1994)

Spectroscopic Studies of Zeolite Single Crystals

K.T. Jackson and R.F. Howe

Department of Physical Chemistry, University of New South Wales,

P.O. Box 1, Kensington NSW 2033, Australia

Abstract. This paper describes spectroscopic studies of large single crystals of zeolite ZSM-5, illustrating the single crystal surface science approach to zeolite chemistry. Single crystal XPS studies reveal extreme zoning of aluminium in the outer surface of the zeolite, confirmed by electron microprobe and chemical etching experiments. The non uniform aluminium distribution in the large crystals is also evident from NMR measurements of xenon physically adsorbed in the zeolite pores. FTIR microscopy has been used to study the oligomerization of ethene in ZSM-5 single crystals. Polarization studies allow the orientations of oligomer chains to be determined. Raman microscopy on cleaved single crystals has sufficient spatial resolution to study the distribution of adsorbed species within different regions of the crystal.

1. Introduction

Zeolites are widely used industrially as catalysts, adsorbents and ion-exchangers. The crystalline microporous environment offered by these materials has also prompted more recent interest in their use as hosts for nano-scale molecular engineering. Many different physical and spectroscopic techniques have been applied to characterize zeolites and to study molecules adsorbed within zeolite pores. From a surface science viewpoint, the well defined internal surface of zeolites provides an opportunity to undertake surface science in three dimensions rather than two.

Such studies may involve "traditional" surface science techniques, although the ability of these to penetrate beyond the external surface can be limited. Alternatively, methods more usually associated with bulk materials characterization can become surface sensitive in the case of three dimensional surfaces.

Until recently, all spectroscopic studies of zeolites were undertaken with polycrystalline powders. To fully exploit the well defined surface structures offered by zeolites however, it is highly desirable to undertake single crystal studies, and several reports have now appeared of microspectroscopic measurements on zeolite single crystals [1-5]. Single crystal zeolite experiments are the 3 dimensional analogs of the more usual single crystal surface science experiment, and offer opportunities for determining structure, location and orientation of adsorbed molecules in a manner not possible with polycrystalline samples.

In this article, we review some recent work we have carried out with single crystals of the zeolite ZSM-5. This particular zeolite was chosen because of its importance as a catalyst, and because it proved to be relatively straightforward to grow crystals large enough for a number of different microspectroscopic experiments. More detailed descriptions of the individual experiments will be presented elsewhere.

2. Experimental

ZSM-5 single crystals were synthesized according to the process described by Kornatowski [6], using tetrapropylammonium bromide as the template, fumed silica (Aerosil 200) and aluminium hydroxide. The as-synthesized crystals were calcined to remove the template at 600°C for 24 hours. The hydrogen form of the zeolite was prepared by ammonium ion exchange, followed by calcination in air at 500°C for 5 days. Some experiments were also undertaken with crystals kindly provided by Dr Kornatowski [6]; these had been prepared by the same method but using different silica and alumina sources.

X-ray photoelectron spectroscopy used a Kratos XSAM800 instrument fitted with an AXIS lens system, using Mg Kα x-rays. Zeolite crystals were pressed into indium foil mounted on a sample stub, and outgassed at a base pressure of 10^{-9} torr before analysis. Photoelectrons were collected and analyzed from a sample area of ca. 100 microns square.

Argon ion etching used a 5kV ion beam at a current density of $10\mu A\ cm^{-2}$. Calibration experiments with tantalum oxide films of known thickness indicated an etch rate of ca. 10 Å s^{-1} for tantalum oxide under these conditions.

Scanning electron microscopy and electron microprobe analyses were obtained with a Cambridge scan 360 instrument.

FTIR microscopy employed a Spectra-Tech IR Plan microscopy coupled to a Bomem MB100 series spectrometer. Infrared spectra were measured in transmission mode over the range 4000 to 700 cm^{-1}. Raman spectra were measured with a Dilor Microprobe (CCD detector), using the 514.5 nm line of an argon ion laser (300 mW). Solid state NMR spectra (^{13}C, ^{29}Si and ^{27}Al) were measured with magic angle spinning (and cross-polarization in the case of ^{13}C) on a Bruker MSL300 instrument. Xenon-129 spectra were obtained with a static glass high vacuum cell in a Bruker AC300P instrument.

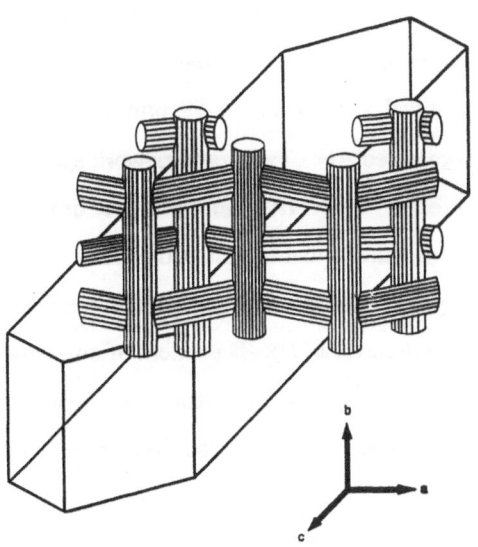

Figure 1 Schematic of crystal morphology and pore geometry in ZSM-5

3. Aluminium Zoning in Large ZSM-5 Crystals

Figure 1 shows a schematic of the two dimensional pore structure in the zeolite ZSM-5. Linear channels of approximately circular cross-section (actually 5.3 × 5.6 Å) run parallel to the crystal b axis, and intersect sinusoidal channels of elliptical cross-section (5.1 × 5.5Å) running parallel to the crystal a axis.

The synthesis method used here yielded uniform large crystals ca. 150 microns along the c axis, and ca. 40 microns along the a and b axes, with morphology similar to that shown in Figure 1. ZSM-5, after calcination to remove the template and conversion to the hydrogen exchanged form, has the ideal unit cell composition H_x Si_{96-x} Al_x O_{192} $y.H_2O$, where x (the number of aluminium ions substituted for silicon in the zeolite lattice) may vary between 0 and 8.

Bulk analysis of the crystals indicated an average Si:Al ratio of 27 (i.e. x = 3.4). Not all of the aluminium in a zeolite is necessarily substituted into the aluminosilicate lattice, however. So-called "extraframework aluminium species" may be present within the zeolite pores, such as free Al^{3+} cations, or polymeric oxy- and hydroxyaluminium species, which modify the reactivity of the zeolite. The presence of non-lattice aluminium can be sometimes detected by ^{27}Al NMR spectroscopy. Aluminium in tetrahedral coordination, as in the zeolite lattice, gives an NMR signal at a chemical shift of about 58 ppm (relative to $Al(H_2O)_6^{3+}$), whereas octahedrally coordinated Al^{3+}, as extraframework species, gives a signal at close to 0 ppm. Conventional ^{27}Al NMR cannot however quantitatively determine extraframework aluminium, since quadrupolar broadening may make aluminium species in lower symmetry coordinations NMR invisible (methods to overcome this problem have recently been described [7]). The extent of non-lattice aluminium can however be deduced from the ^{29}Si NMR spectra of zeolites, since silicon atoms bonded via oxygen to one, two, three or four aluminiums give ^{29}Si NMR resonances at distinctly different chemical shifts from that of silicon with four silicon next nearest neighbours. Quantification of the ^{29}Si spectra of the large ZSM-5 crystals studied here in this manner indicated that less than 25% of the aluminium content

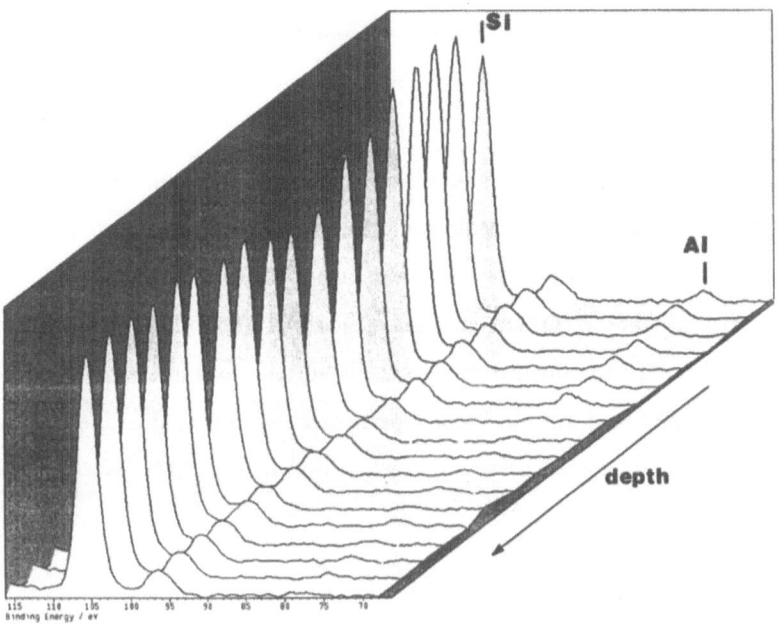

Figure 2 XPS depth profile of a highly zoned ZSM-5 single crystal ((010) face)

determined by bulk analysis is present in the zeolite lattice. Qualitatively, this conclusion was supported by the ^{27}Al NMR spectra, which showed approximately equal intensities for signals at 58 and 0 ppm.

Figure 2 shows a set of XPS spectra measured in a depth profile experiment on a single crystal of ZSM-5. The crystal was oriented such that the x-ray beam was incident on the (010) crystal face. The first spectrum, at the back of the Figure, is that of the fresh crystal face. The three features observed in the binding energy range shown are Si 2p at 106 eV, a Si 2p satellite at 97 eV, and Al 2p at 75 eV (all binding energies uncorrected for sample charging). Quantification of the Si 2p and Al 2p peak intensities gives a surface Si:Al ratio of 12, indicating a dramatic enrichment of aluminium at the outer surface relative to the bulk. The remaining spectra in Figure 2 were measured sequentially following argon ion bombardment of the crystal. The Si:Al ratio remains constant for approximately 30 minutes of argon ion etching, then increases dramatically to a value greater than 35 (the Al 2p peak has insufficient intensity to allow accurate determination of Si:Al ratios in the etched crystal).

Similar depth profiles were measured for a number of different ZSM-5 crystals, indicating that the apparent segregation of aluminium to the outer surface of the crystals is a general phenomenon for this particular synthesis method.

Confirmation that the surface enrichment of aluminium observed by XPS is a real effect and not a consequence of selective sputtering or other artefacts of ion bombardment comes from two different experiments. Figure 3 shows the results of an electron microprobe scan across a cross-section of a ZSM-5 crystal cleaved

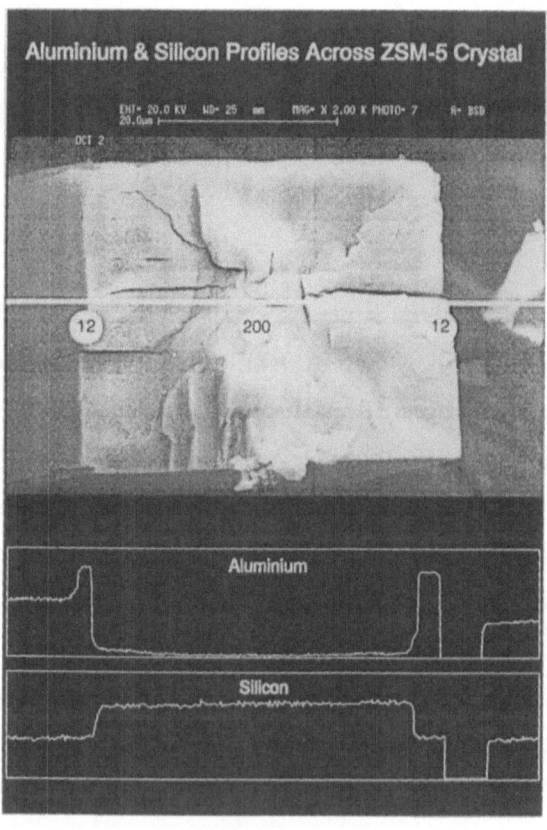

Figure 3 Electron microprobe Si and Al profiles across (001) face of a cleaved crystal

perpendicular to the c-axis. This shows clearly that the outer 1-2 micron rim of the crystal contains a much higher concentration of aluminium than the interior. A Si:Al ratio of 12 estimated for this outer rim agrees well with that found by XPS. The aluminium concentration in the interior of the crystal is too low to be accurately determined by EDX, but the Si:Al ratio in the centre of the crystal is greater than 100.

The second experiment confirming the presence of an outer 1-2 micron high aluminium rim on the zeolite crystals involved chemical etching. Mobil workers [8] have reported that high-silica ZSM-5 is soluble in refluxing sodium carbonate solution, whereas low silica zeolites are not. Treatment of the large ZSM-5 crystals with sodium carbonate completely dissolved the interior, leaving hollow shells with wall thicknesses of 1-2 microns. X-ray diffraction and NMR spectroscopy confirmed that the surviving zeolite shells were crystalline ZSM-5 with an Si:Al ratio of about 12 and negligible concentrations of extraframework aluminium.

Heterogeneous aluminium distributions have been previously reported from studies on polycrystalline samples, and attributed to particular synthesis conditions [9]. The single crystal studies described here show however a particularly striking example of aluminium zoning during synthesis. A silica rich core is initially formed preferentially, and high aluminium ZSM-5 forms only at the concluding stages of crystal growth when solution silica species are depleted.

The extent of aluminium zoning is critically dependent on the precise details of the crystal synthesis. We have recently undertaken similar measurements on single crystals of ZSM-5 provided by Dr Kornatowski [6]. These were prepared by a nominally identical method, although different sources of silica and alumina were used. The composition of the outer surface of this second batch of crystals as measured by XPS studies of single crystals was identical within experimental error to the bulk composition, and no significant variation of the Si:Al ratio with argon ion sputtering was observed in depth profile experiments. Electron microprobe scans across cleaved crystal cross-sections could not detect Al in any part of the crystal and the Si profile remained constant.

The origins of aluminium zoning in ZSM-5 crystals have been the subject of much speculation [9]. Aluminium enrichment at the outer surface has been attributed to specific interactions between aluminosilicate and template or alkali metal cations in the synthesis mixture. Effects similar to those described here have recently been reported by Schuth and Althoff [4], although their crystals also showed some increase in aluminium concentration towards the centre. The spatial distribution of aluminium and hence catalytic sites within zeolite crystals has a strong influence on catalytic performance. It is clear from this work and that of Schuth [4,10] that single crystal studies can provide a great deal more detailed information about lattice aluminium and extraframework aluminium distributions than previous studies of polycrystalline powders. Further studies of this type on model single crystals will lead to improved understanding of how synthesis and treatment conditions determine the location and distribution of active sites in zeolite catalysts.

4. NMR spectroscopy of Adsorbed Xenon

Xenon is physically adsorbed in zeolite pores at room temperature, and the ^{129}Xe NMR spectrum of the physically adsorbed xenon has been widely used as a probe of the environment within the pores [11]. The method relies on the fact that the ^{129}Xe chemical shift of the large polarizable xenon atom is extremely sensitive to electrostatic fields, such as those present within zeolites due to the negatively charged lattice and the charge balancing cations. In xenon gas, the NMR chemical shift is a linear function of the gas density (at least at lower densities); this has been explained in terms of xenon-xenon interactions perturbing the magnetic shielding at the nucleus from the outer electrons [12].

For xenon physically adsorbed in many zeolites, the chemical shift of the xenon is a linear function of the amount of xenon adsorbed. The usual explanation [11] for such linear variation is that xenon-xenon interactions within the zeolite similar to those in the gas phase determine the chemical shift. Extrapolation to zero coverage of xenon yields a chemical shift which is determined by xenon-zeolite interactions; these may include both xenon-lattice and xenon-cation interactions.

We have shown previously [13] that xenon adsorbed in a series of well defined microcrystalline ZSM-5 zeolites which were homogeneous in composition and contained no extra-framework aluminium showed a linear variation of chemical shift with coverage. The gradient was independent of aluminium content but the intercept increased with increasing aluminium content of the zeolite, consistent with increasing magnitude of electrostatic fields within the zeolite pores.

The environment probed by physically adsorbed xenon in the NMR experiment is an average of that experienced by the xenon atom on the NMR time scale as it diffuses through the zeolite pores. Measurements of xenon diffusion coefficients

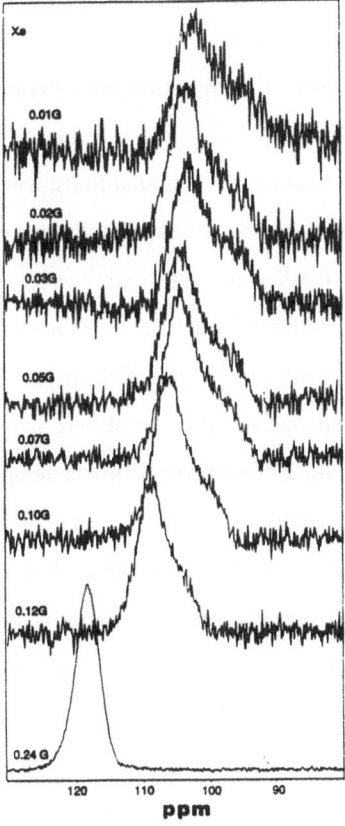

Figure 4 Xenon-129 NMR spectra of xenon adsorbed in ZSM-5 crystals at room temperature as a function of xenon coverage

indicate that xenon may diffuse over a distance of several microns during the microsecond time scale of the NMR measurement [14]. Thus, experiments on microcrystalline samples yield an average environment for the entire crystal or for several crystals, since inter-crystalline diffusion may also be rapid.

The extreme aluminium zoning in the large ZSM-5 crystals described above provided us with the opportunity to investigate the behaviour of xenon adsorbed in a system where much longer diffusion paths are available.

Figure 4 shows NMR spectra of xenon adsorbed in the large single crystals of ZSM-5 as a function of xenon coverage   At high coverages, a single symmetrical line was observed. Comparison with the microcrystalline experiments indicates that the chemical shift of this line, for a given xenon coverage, is typical of ZSM-5 containing about 1 Al per unit cell. At lower xenon coverages, however a second component can be observed in the signal as a shoulder toward lower chemical shift. The chemical shift of this second component, for a given xenon coverage, is typical of silicalite, a ZSM-5 zeolite containing no aluminium at all.

Our interpretation of these results is that at low coverage and room temperature, we are resolving two different environments for xenon in the zeolite crystals; a lattice aluminium-free, silicalite-like environment at the centre of the crystals, and a low lattice aluminium environment which is an average of the outer high aluminium rim and the low aluminium interior immediately beneath the outer 1-2 microns. At high xenon coverage, the single line observed is an average of all of the environments in the crystal. Experiments at lower temperatures (190K) and very low xenon coverage gave spectra in which a third component signal became evident as a poorly resolved shoulder at higher chemical shift.. The position of this third component corresponds to that expected for xenon adsorbed in a high aluminium ZSM-5; we thus attribute it to xenon in the outer rim of the zeolite crystals.

This interpretation is supported by measurements of the xenon-129 spin lattice relaxation times as a function of coverage. In microcrystalline ZSM-5 samples T_1 depends strongly on the aluminium content of the zeolite [15]. The average T_1 value for the large single crystals lies in the range (0.5 - 1 s) typical of a low aluminium ZSM-5, but not as high as that of silicalite (up to 5 s). At low coverage, the lower chemical shift component signal was found to relax more slowly than the main signal, consistent with our assignment of this signal to the silicalite core of the crystals.

5. Ethene Oligomerization in Single Crystals of ZSM-5

Reactions of alkenes over acid zeolite catalysts have been extensively investigated, and commercial processes such as the Mobil Olefins to Gasoline and Distillate (MOGD) process exploit the ability of acid zeolites to catalyze alkene oligomerization reactions. Numerous spectroscopic studies have been reported of the interaction of alkenes with the acid (proton exchanged) form of ZSM-5 (16,17]. Protonation of an alkene adsorbed in the zeolite forms an alkyl carbenium cation which can then undergo further reaction to form oligomers which may either crack to give higher alkene products or cyclize and lose hydrogen to form aromatic products. Propene and higher alkenes readily form branched oligomeric species within the zeolite pores at room temperature [16]. Ethene is less reactive, but is reported to form linear oligomers in ZSM-5 at low temperatures which undergo chain branching at higher temperatures [18].

All previous spectroscopic studies of alkenes in ZSM-5 have used polycrystalline samples. We have recently shown however that the techniques of infrared and Raman microscopy can be applied to single crystals to give information about location and orientation of the adsorbed species [5]. The zoned crystals may be regarded as microreactors in which the outer 2 micron shell contains a high density of reactive sites, while the interior contains a low density.

The ability of the zoned crystals to catalyze ethene oligomerization was established by carbon-13 NMR studies of a sample of the crystals exposed to ethene at various temperatures. At room temperature ethene reacted very slowly over a period of many hours; the NMR signal characteristic of physically adsorbed ethene

(121 ppm) was gradually replaced by a signal at 32 ppm due to -CH_2- groups in a linear oligomer species. The oligomerization reaction occurred more rapidly on heating in ethene at 100°C, and reached completion within 1 hour (studies on microcrystalline samples have shown that the low temperature oligomerization reaction stops when the zeolite pores are completely filled with hydrocarbon [19]. The NMR spectrum of the adsorbed oligomer after heating in ethene at 100°C or 200°C continued to be dominated by the signal due to -CH_2- groups, although additional weak signals due to CH_3- groups could be detected. Heating above 200°C however caused a dramatic change in the NMR spectrum; signals due to CH_3- groups and -CH_2- groups bonded to CH_3- groups grew in intensity, while that due to -CH_2- groups in linear oligomer decreased.

The oligomerization reaction and the changes occurring in the structure of the oligomer on heating can be seen very clearly in FTIR microspectroscopic experiments with single crystals of HZSM-5. Figure 5 shows spectra recorded in the ν(CH) region of a single crystal exposed to ethene at successively higher temperatures. These spectra were recorded in transmission mode; the unpolarized infrared beam was focussed onto the (010) face of a single crystal with a spot size of ca. 50 microns.

At low temperatures, the spectra are dominated by a pair of ν(CH) bands at 2934 cm^{-1} and 2860 cm^{-1}, which correspond to the asymmetric and symmetric stretching modes respectively of -CH_2- groups. Other weak shoulders at 2960 cm^{-1} and 2880 cm^{-1} are the corresponding vibrations of CH_3- groups, while a fifth component at about 2900 cm may be due to $\overset{\backslash}{\diagup}$ CH groups.

At 300°C, the -CH_2- bands are overtaken by those due to CH_3 groups, and at higher temperatures all bands decreased progressively in intensity. These infrared spectra confirm the conclusions drawn from the NMR data concerning the initial formation of a linear oligomer which becomes more branched at higher temperatures.

Measurements using polarized infrared light provide further information about the orientation of the adsorbed oligomer within the zeolite crystal. Figure 6 shows a

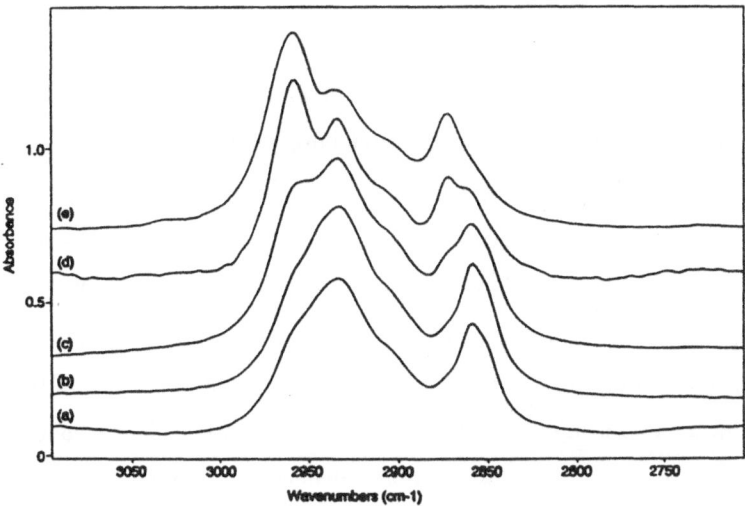

Figure 5 IR spectra (unpolarized) of a single crystal of HZSM-5 after exposure to ethene at a), 25°C; b), 100°C; c), 200°C; d), 300°C; e), 400°C

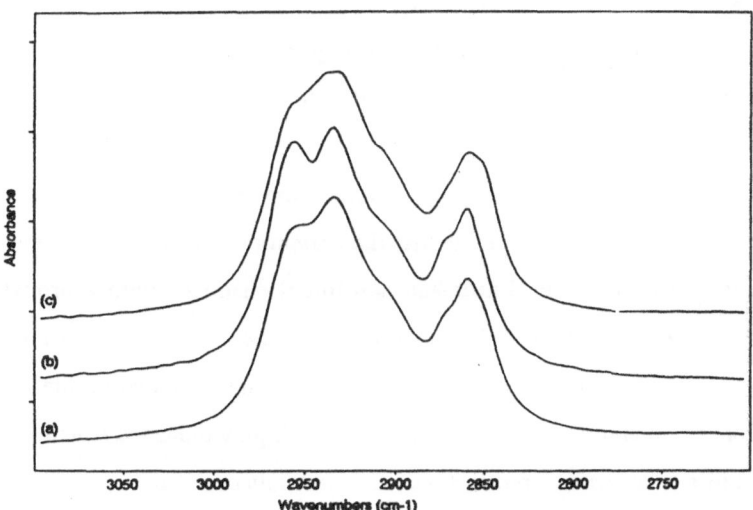

Figure 6 IR spectra of HZSM-5 single crystal after exposure to ethene at 200°C (a), unpolarized; (b) polarized parallel to a axis; (c) polarized parallel to c axis

set of spectra for ethene in ZSM-5 after heating to 200°C obtained with respectively unpolarized light incident on the (010) face, light polarized parallel to the crystal a axis, and light polarized parallel to the c axis. Polarization parallel to the crystal a axis enhances relatively the asymmetric stretching mode of CH_3 groups, whereas rotation through 90°C diminishes (but does not completely remove) the CH_3 bands. Crystals examined with polarized light incident on the (100) face showed similar effects; polarization parallel to the b axis enhanced the CH_3 contribution to the spectrum, relative to polarization parallel to the c axis.

The polarization effects for light incident on both the (100) and (010) faces indicate that the linear oligomer is formed in both the linear zeolite channels (which run parallel to the b axis) and the sinusoidal channels (parallel to the a axis). The stretching modes of CH_2 groups in a linear polyethylene oligomer will be seen most strongly when the plane of polarization of the infrared radiation is perpendicular to the linear axis of the oligomer, whereas for terminal CH_3 groups on such a linear oligomer little difference would be expected between parallel and perpendicular polarization.

Crystals heated in ethene to 300°C or higher showed no variation in relative intensities of the CH_3- and $-CH_2-$ bands when the polarization direction was varied relative to the crystal axes. A similar absence of polarization effects was found in the infrared spectra of the tetrapropyl ammonium cation initially present in the zeolite following synthesis. In this case the tetrapropylammonium cation is located at the intersections of the linear and sinusoidal channels, with a propyl group extending down each of the 4 channel directions. The infrared spectra of the ethene oligomer after heating to 300°C are closely similar to those of the occluded tetrapropylammonium cation, suggesting that the highly branched ethene oligomers are also located at channel intersections, and randomly oriented.

The chemistry of ethene oligomerization in HZSM-5 may be summarized simply by the following scheme

$$C_2H_4 + H^+ \rightarrow C_2H_5^+$$

$$C_2H_5{}^+ + nC_2H_4 \xrightarrow[\longrightarrow]{\leq 200°C} C\,H_3(CH_2)_{2n}\,CH_2{}^+$$

$$CH_3(CH_2)_{2n}\,CH_2^+ \longrightarrow (CH_3(CH_2)_{\underset{3}{2n\text{-}2}})_3\,C^+$$

where n=4 would give a tri n-propyl species after high temperature oligomerization.

The distribution of oligomer between the high and low aluminium regions of the zoned crystals cannot be determined in the infrared microscope, since the spatial resolution is limited to about 25 microns at best. Variations in the density of acid sites are expected however to influence strongly the oligomerization chemistry described above. Higher spatial resolution can be achieved using Raman microscopy, since a visible laser can be focussed to a spot size of about one micron. Figure 7 shows spectra of ethene oligomerized in a ZSM-5 crystal at 200°C measured with a Raman microscope (the corresponding infrared spectrum is shown for comparison also). The zeolite crystals, after heating in ethene to 200°C, was cleaved across the c-axis, and spectra were recorded from either the outer edge or the centre of the cleaved face. The samples fluoresced strongly, and we experienced

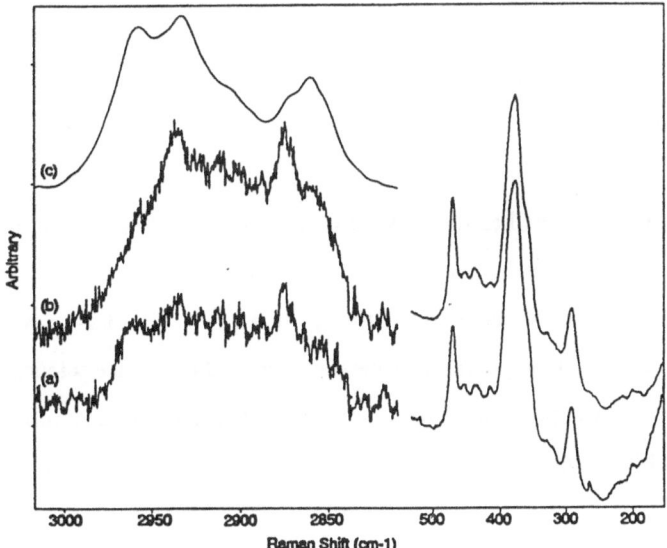

Figure 7 Raman spectra of cleaved crystals after exposure to ethene at 200°C (a), edge; (b), centre; (c), IR spectrum of whole crystal

extreme difficulty in obtaining acceptable signal to noise ratios, particularly in the $\nu(CH)$ region. The initial results illustrated in Figure 7 indicate that the oligomer in the centre of the crystal is less branched than that at the outer rim (compare Figures 7(a) and (b)), since the $-CH_2-$ bands are more intense both in relative and absolute terms (the low frequency bands due to the zeolite lattice provide an internal intensity standard). Similar differences between the edges and centres of ethene loaded crystals were found for all crystals examined. Unfortunately sample fluorescence has so far prevented Raman spectra from being recorded from crystals heated in ethene at 300°C or higher, where the infrared spectra indicate the onset of extensive chain branching. Nevertheless it appears that the extent of chain branching does depend on the density of acid sites in the crystal. Conclusions drawn from infrared or NMR spectra which average over the entire crystal may not tell the entire story.

It is particularly interesting that high concentrations of oligomer are found in the interior of the crystals where very few acid sites are present. Further study of this system is underway, but is clear that Raman microscopy will be an invaluable complement to infrared microscopy in studying intrazeolite chemistry in single crystals, particularly if fluorescence problems can be overcome (by using near infrared laser excitation, for example).

6. Future outlook

The work presented here has illustrated for one particular zeolite how single crystal studies can provide much more detailed information about zeolite composition, structure and chemistry occurring within zeolite pores than has previously been available from work with polycrystalline powders. Other surface science techniques which have been applied to zeolite single crystals include scanning tunneling microscopy [20], atomic force microscopy [21] and thermal desorption spectroscopy [1]. Single crystal studies on many of the new larger pore zeolite materials involving elements other than silicon and aluminium will be particularly important as these materials are more appropriate hosts for conducting polymers or nano-scale metal clusters or semiconductors.

References

[1] R.F. Howe in "Surface Science : Principles and Applications" Springer Proc. Physics 73 (1992) 242

[2] M. Nowotny, J. Lercher and H. Kessler, Zeolites 11 (1991) 454

[3] F. Schüth, J.Phys.Chem. 96 (1992)

[4] F. Schüth and R. Althoff, J.Catal. 143 (1993) 338

[5] K.T. Jackson and R.F. Howe, Proc.International Symposium on Zeolites and Microporous Crystals, Nagoya, 1993, in press

[6] J. Komatowski, Zeolites 8 (1988) 77

[7] E W Wooten, K T Mueller and A Pines, Acc.Chem.Res. 25(1992) 209

[8] R.M. Dessau, E.W. Valyocsik and N.H. Goeke, Zeolites 12 (1992) 776

[9] P.A. Jacobs and J.A. Martens, "Synthesis of High Silica Aluminosilicate Zeolites" Stud.Surf.Sci.Catal. 33 (1987) 91

[10] R. Althoff, B. Schulz-Dobrick, F. Schüth and K. Unger, Microporous Materials 1 (1993) 207

[11] J. Fraissard and T. Ito, Zeolites 8 (1988) 350

[12] A.K. Jameson, C.J. Jameson and H.S. Gutowsky, J.Chem.Phys. 53 (1970) 2310

[13] S.M. Alexander, J.M. Coddington and R.F. Howe, Zeolites II (1991) 368

[14] W Heink, J Karger,H Pfeiffer and F Stallmach, J.Am.Chem.Soc. 112(1990) 2175

[15] K.T. Jackson, R.F. Howe, J. Hook and L. Van Gorkom, manuscript in preparation

[16] A.K. Ghosh and R.A. Kydd, J.Catal. 100 (1986) 185

[17] J.F. Haw, B.R. Richardson, I.S. Oshiro, N.D. Lazo and J.A. Speed, J.Am.Chem.Soc. 111 (1989) 2052

[18] J.P. Van den Berg, J.P. Walthuizen, A.D.H. Claque, G.R. Hays, R. Huis and C. Van Hoof, J. Catal. 80 (1983) 130.

[19] Y. Liu, J.B. Metson and R.F. Howe, manuscript in preparation

[20] J.C. Jansen, J. Schoonman, H. van Bekkum and V. Pinet, Zeolites 11 (1991) 306.

[21] J E MacDougall et al, Zeolites 11(1991) 429

This work is supported by research grants from the Australian Research Council.

Analysis of Alumina-Supported Catalysts by XPS

B.G. Baker and M. Jasieniak

School of Physical Sciences, Flinders University,
G.P.O. Box 2100, Adelaide, SA 5001, Australia

Abstract. The application of X-ray photoelectron spectroscopy for determining the chemical state of supported catalysts is discussed with particular reference to the system iron-praseodymium-alumina. The XPS studies are complicated by the variable valency, intermediate stoichiometries and by the variety of structure possibilities in the iron-praseodymium-alumina system. Techniques to cope with differential charging of the sample are described and results presented for the surface characterization of hydrogen conditioned catalysts. Extensive reaction between the catalyst components is shown to occur. It is shown that praseodymium is incorporated in the alumina support and that the reaction of iron at this surface can be distinguished from the reaction in the unpromoted iron-alumina catalyst.

1. Introduction

Catalysts containing iron have activity for the Fischer-Tropsch synthesis of hydrocarbons from carbon monoxide and hydrogen. Other components of such catalysts are structural stabilisers or supports and promotors. Under the reducing conditions of the reaction some iron will be present in the zero valence state but some may react with the support or promotor. The resulting compounds, likely to be semiconductors, could influence the catalytic behaviour of the metallic iron.

It has been shown that control of the chain growth process in Fischer-Tropsch synthesis and selectivity to the formation of alkenes can be achieved by lightly loaded alumina-supported iron catalysts [1]. In this type of catalyst iron is deposited on alumina which has been treated with praseodymium. There are a number of possible reactions involving alumina, praseodymium and iron which could occur either during catalyst preparation or under the reducing conditions of the reaction. The objective of the present work is to determine the extent to which such reactions occur in the activated catalyst.

Analysis by X-ray photoelectron spectroscopy (XPS) provides elemental information of the catalyst surface. Valence states can be obtained from chemical shifts provided that the effects of electrostatic charging can be accounted for. Alumina is an insulator but the catalyst surface may contain semiconductors. Differential charging needs to be detected and corrected. Methods to overcome this problem have been described previously [2].

Because the active components are highly dispersed and of low concentration, these supported catalysts are difficult to analyse by XPS. An alternative approach, previously used to investigate the reactivity of alumina towards iron, involved the direct introduction of metal to the external oxide surface by vacuum evaporation. XPS analyses of such samples showed that iron reacted with alumina and with praseodymium oxide under reducing conditions [3]. The conditions (hydrogen at 600 K) were similar to the conditions for catalyst activation. The same treatment was shown to reduce iron oxide to metal in the absence of other oxides.

Other structural and spectroscopic studies of the iron-alumina interface have shown that various reactions occur depending on the conditions [4,5]. Model catalysts prepared as films do not reproduce either the support characteristics or conditions of preparation of the real catalyst. The present series of experiments is designed to study samples of the separate catalyst components and of catalysts under conditions which closely match the activation treatment of the supported iron catalyst.

2. Experimental

The iron-praseodymium-alumina catalysts were prepared by pore volume impregnation with aqueous solutions of appropriate nitrates. The catalysts were prepared by a two-step procedure with Pr being introduced first. After each impregnation the system was agitated in ultrasonic bath to aid in uniform wetting of the support, followed by microwave drying and calcination in air at 723 K for 2 h.

The catalyst support was derived from a γ-alumina containing about 0.2 wt% Na_2O (Merck). The fraction of the alumina (125-150 μ) was preheated at 1073 K for 15 min and then heat treated at 1473 K for 30 min. X-ray diffraction showed that the product of this treatment was predominantly α-alumina. A surface area of this support was around 20 m^2g^{-1}.

The praseodymium oxide, Pr_6O_{11} (SPEX, 99.99 %) was the starting material for the preparation of catalysts, as well as other praseodymium oxides investigated in this study. Praseodymium sesquioxide, Pr_2O_3 was obtained by reducing Pr_6O_{11} *in situ* in a stream of high purity hydrogen at 600 K. Praseodymium dioxide, PrO_2 was prepared by continuous stirring of a suspension of Pr_6O_{11} in a 5 % aqueous solution of acetic acid at room temperature for 21 days [6].

Ferric nitrate was of high purity (Ajax Chemicals).

X-ray photoelectron spectroscopy was performed using a modified Leybold LHS10 spectrometer, fitted with a sample preparation chamber [3]. The sample stage can be heated during analysis if required.

The mounting of powder or granular samples for analysis must allow for good thermal and electrical contact with the holder and ensure mechanical stability so that sample is not spilled during evacuation, thermal treatment or reduction. Alumina and alumina-supported samples were mounted as pellets into small stainless steel cups. Praseodymium oxides were pressed as very thin wafers onto annealed copper disks. This method is very successful for samples undergoing contraction during reduction.

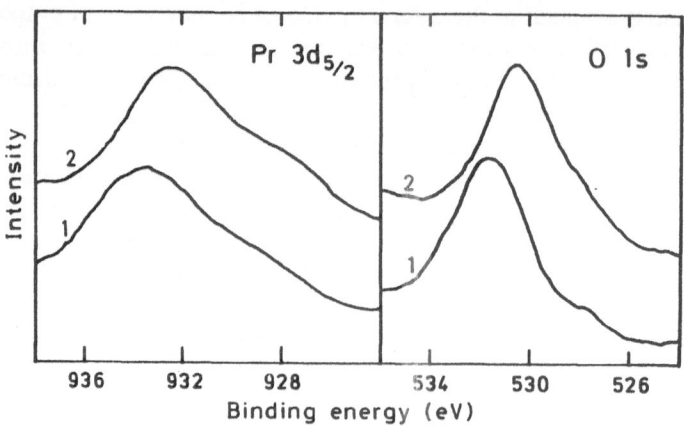

Figure 1. XPS spectra of the Pr $3d_{5/2}$ and O $1s$ levels for Pr_6O_{11}. 1. standard referencing to Au $4f_{7/2}$ (without flood gun); 2. biased referencing to Au $4f_{7/2}$.

The principle objective in XPS is the identification of chemical states. A comparison of core levels and their displacement due to modification of the local environment yields significant information about various ionic states of an atom. When a non-conductive sample is investigated by XPS, the surface acquires a positive charge which decreases the observed kinetic energy of the photoelectrons. This shift is in the same direction as a chemical shift where the element has a more positive valence state. Specimens frequently show both effects and in addition often exhibit differential charging.

Iron-praseodymium-alumina is a very complex system comprising iron and praseodymium species at the insulating alumina surface. As the analysed specimens were subjected to a variety of surface destructive treatments (high temperature, reduction) which could induce reactions with the alumina support, utilisation of some charge correction techniques was excluded in advance (eg. referencing to the C $1s$ and Al $2p$ peaks).

Preliminary experiments have shown that referencing to the Au $4f_{7/2}$ peak (thin gold layer) is unreliable. For example, spectra for Pr_6O_{11} exhibited differential charging (Figure 1). It can be seen that differential charging not only makes the determination of the true O $1s$ and Pr $3d_{5/2}$ binding energies impossible but it changes the shape of spectra. Consequently, in order to avoid differential charging effects and to achieve a higher degree of confidence in estimation of BE's in this study we adopted a method based on the biased referencing technique [2,7,8]. In this method a 2 mm gold spot is affixed to a sample by means of vacuum deposition. The sample is flooded with thermal electrons. If an insulating sample exhibiting differential surface charging is bombarded with low energy electrons the differential potential undergoes equlilization as positive sites on the surface preferentially attract electrons. The value of the surface potential at equilibrium depends upon the biased voltage applied to the surface and the flux of electrons the sample is exposed to. While in principle a complete neutralisation of the surface charge can be achieved, it is necessary only to operate the flood gun at a level which homogenises the surface charge such that the Au $4f_{7/2}$ peak performs a stable energy over an extended time. The difference between the measured and the true BE for gold is then applied as the charge correction for all other peaks.

3. Results

3.1 Alumina

The catalyst support prepared by heating γ-alumina in air at 1473 K has been described previously [2]. The effect of the heat treatment is to destroy a surface impurity phase of β-alumina containing sodium and to change the alumina structure to α-alumina. By limiting the time and temperature the extent of the conversion is limited so as to retain the porous character of a catalyst support.

Alumina samples were subject to electrostatic charging in XPS analysis. Binding energies for O $1s$ and Al $2p$ corrected by the biased referencing method are included in Table 1.

3.2 Praseodymium Oxides

The most common oxides of praseodymium, Pr_6O_{11}, Pr_2O_3 and PrO_2, were studied in view of their possible presence at the surface of the catalyst support. Characterization of these compounds was based on the analysis of the Pr $3d_{5/2}$ and O $1s$ regions.

The $3d$ final state configuration of praseodymium oxides has a complex, multiplet structure. Two main lines $3d_{5/2}$ and $3d_{3/2}$ are due to the large $3d$ spin-orbit splitting. Upon photoionisation, a $3d$ core electron is removed and the created positive hole forces the O $2p \rightarrow Pr$ $4f$ transition [9,10]. A strong coupling between the core electron hole and the Pr $4f$ orbital gives rise to the distinct satellite structure present on the low binding energy side of each of the main $3d$ peaks. It has been suggested that in the case of PrO_2 each spin-orbit splitting component comprises 3 lines [11]. The main line of the $3d_{5/2}$ inner-core level overlaps with the lowest binding energy satellite present in the $3d_{3/2}$ outer-core level. For the purpose of analysis it would be impracticable to use this line as the energy reference for the Pr $3d_{5/2}$ region. In this study, which is fully consistent with other authors [12], the Pr $3d_{5/2}$ region was characterized by the most intensive peak present in this level.

Pr_6O_{11} has focused our special attention as it was used as a precursor of the other praseodymium oxides and the praseodymium promoted iron catalyst.

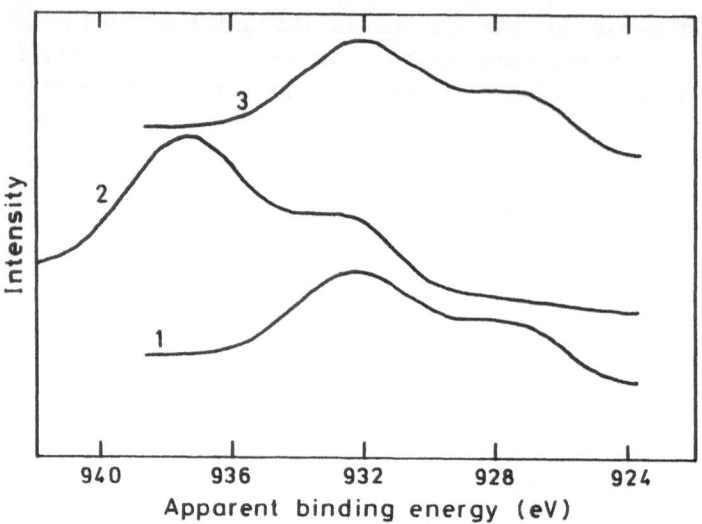

Figure 2. Effect of reduction of Pr_6O_{11} on the Pr $3d_{5/2}$ spectra. 1. Pr_6O_{11} outgassed in vacuum at 600 K for 1 h (uncharged at 293 K); 2. reduced oxide (H_2, 600 K, 2 h), spectrum recorded at 293 K→charged sample; 3. reduced oxide (H_2, 600 K, 2 h), spectrum recorded at 600 K→uncharged sample.

The effect of reduction of Pr_6O_{11} on the $3d_{5/2}$ spectra is shown in Figure 2. The outgassed praseodymium oxide does not charge and the $3d_{5/2}$ region comprises two overlapping peaks. Reduction of this oxide in a stream of hydrogen forms insulating Pr_2O_3. The temperature dependence of the conductivity of Pr_2O_3 can be clearly seen from the comparison of traces 2 and 3. It is sufficient for this oxide to be heated to 600 K in order to become conducting.

The reduction-oxidation behaviour of Pr_6O_{11} is illustrated in Figure 3. Short exposure of the insulating praseodymium sesquioxide to air forms a conducting mixed-valence state compound (traces 2 and 3). Surprisingly, the Pr $3d_{5/2}$ electron binding energies recorded for Pr_6O_{11}, Pr_2O_3 and the re-oxidized product were almost identical (see Figure 4 and Table 1).

In order to determine the effect of the oxidation state on the Pr $3d_{5/2}$ electron binding energy we prepared and analysed the tetravalent praseodymium oxide. The spectra of Pr_6O_{11}, Pr_2O_3 and PrO_2, corrected for charging effects, are given in Figure 5. It can be clearly seen that the Pr $3d_{5/2}$ electron binding energy in PrO_2 was only slightly higher than that recorded for the other oxygen containing samples. This behaviour indicates that the valence characterization of various praseodymium oxides by XPS is impossible. The question arises as to the reason for the observed phenomenon. The composition of praseodymium oxide depends upon the temperature and oxygen pressure. In vacuum, the higher praseodymium oxides tend to loose oxygen from the surface layer. This process is accelerated by

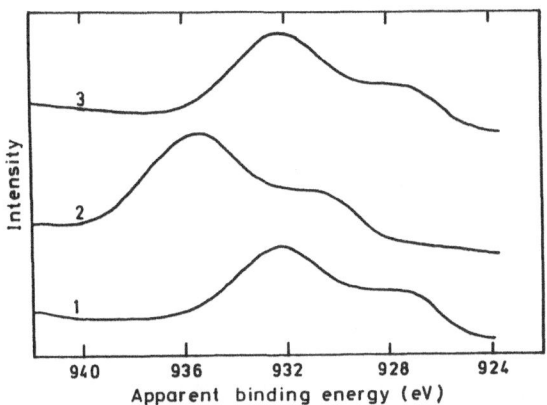

Figure 3. (left) Reduction-oxidation behaviour of Pr_6O_{11}. XPS spectra of Pr $3d_{5/2}$. 1. Pr_6O_{11} outgassed in vacuum at 600 K for 1 h (uncharged at 293 K); 2. reduced oxide (H_2, 600 K, 2 h), spectrum recorded at 293 K→charged sample; 3. re-oxidized sample (air, 293 K, 1 h), spectrum recorded at 293 K→uncharged sample.

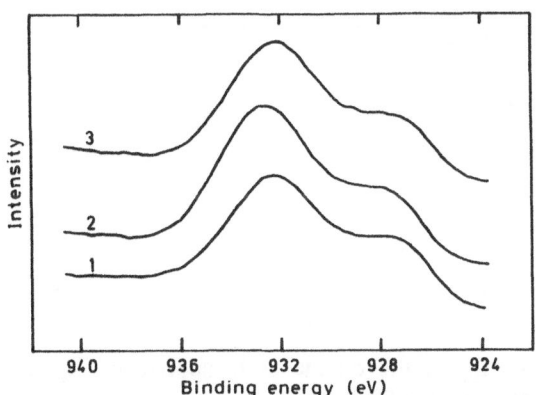

Figure 4. (right) Reduction-oxidation behaviour of Pr_6O_{11}. XPS spectra of Pr $3d_{5/2}$ corrected for electrostatic charging. 1. Pr_6O_{11} outgassed in vacuum at 600 K for 1 h; 2. reduced oxide (H_2, 600 K, 2 h), spectrum recorded at 293 K; 3. re-oxidized sample (air, 293 K, 1 h), spectrum recorded at 293 K.

temperature and lasts until a composition of $PrO_{1.5}$ is reached. Regarding our results, only Pr_2O_3 was present at the outermost surface of all analysed oxides due to their instability under the measurement conditions. However, the bulk oxide was responsible for the electrical behaviour of the surface praseodymium species.

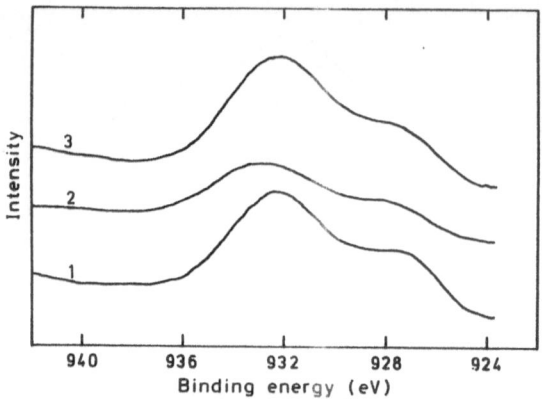

Figure 5. XPS spectra of the Pr $3d_{5/2}$ core level for various praseodymium oxides. 1. Pr_6O_{11}; 2. PrO_2; 3. Pr_2O_3.

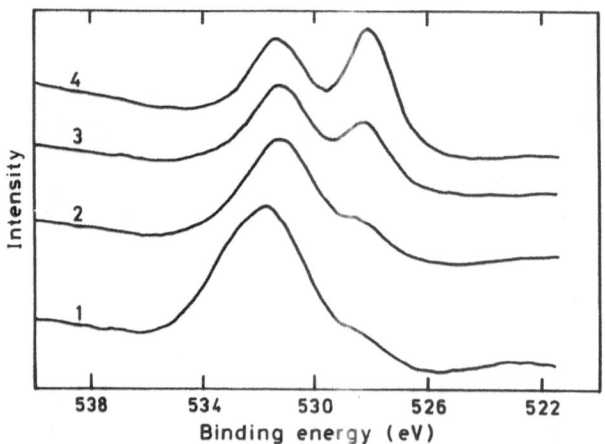

Figure 6. Effect of temperature on the O $1s$ spectra of Pr_6O_{11}. 1. 293 K; 2-4. increase in temperature up to 600 K (trace 4).

The effect of thermal treatment of Pr_6O_{11} on the O $1s$ spectra is shown in Figure 6. Trace 1, recorded for the "as loaded" sample, exhibited a very broad feature with a small shoulder on its low binding energy side. Increase in temperature resulted in the evolution of the low binding energy peak which was accompanied by a significant decrease in the intensity of the main feature. A quadrupole mass spectrometer attached to the main chamber of the instrument showed an evolution of carbon dioxide and water due to this treatment. The rate at which CO_2 and H_2O were formed steadily decreased on heating of the sample at constant temperature. The resultant spectrum comprised two uncharged peaks.

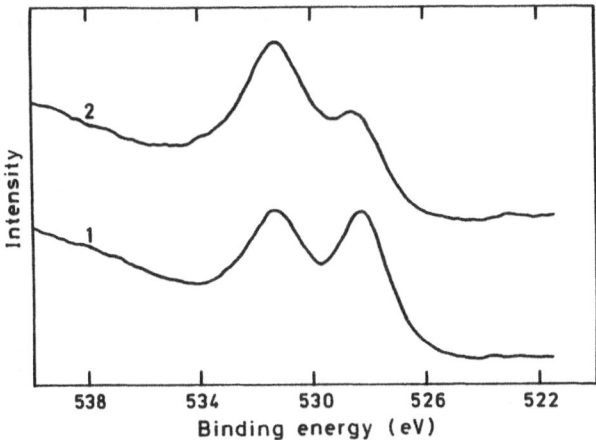

Figure 7. (left) Effect of CO_2 on the O $1s$ spectra of Pr_6O_{11}. 1. Pr_6O_{11} outgassed in vacuum at 600 K for 1 h; 2. outgassed sample exposed to CO_2 (293K, 5 min).

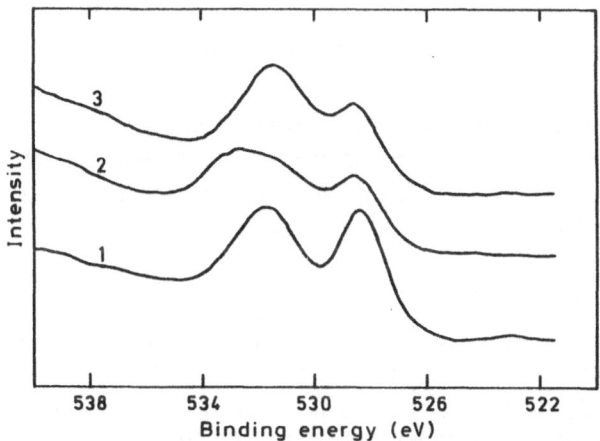

Figure 8. (right) Effect of water vapour on the O $1s$ spectra of Pr_6O_{11}. 1. Pr_6O_{11} outgassed in vacuum at 600 K for 1 h; 2. outgassed sample exposed to argon saturated with water vapour (293K, 5 min).

In order to characterize the chemistry of the Pr_6O_{11} surface and to identify components present in the O $1s$ spectrum (Figure 6, trace 1), a sequence of experiments has been performed. A sample of the oxide, outgassed at 600 K, was subjected to a variety of treatments, including exposure to carbon dioxide and water vapour. The effect of CO_2 on the O $1s$ region is shown in Figure 7. The change in the intensity ratio between two components present in the spectrum clearly indicates that the high binding energy peak corresponds to the praseodymium carbonate (or other related compounds). The low binding energy peak is attributed to oxygen from the praseodymium oxide lattice. The effect of water on the outgassed praseodymium oxide is illustrated in Figure 8. A short-time exposure of

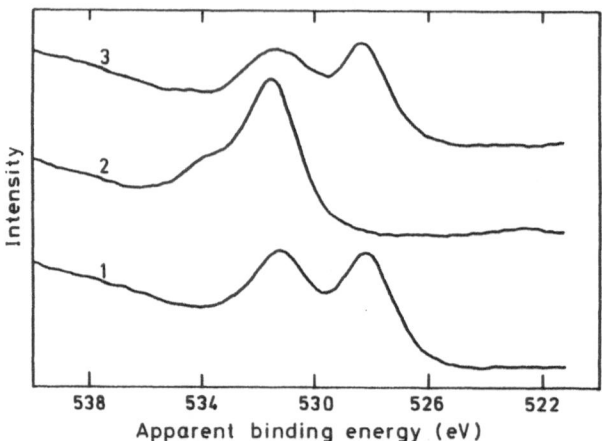

Figure 9. Effect of reduction of Pr_6O_{11} on the O $1s$ spectra. 1. Pr_6O_{11} outgassed in vacuum at 600 K for 1 h (uncharged at 293 K); 2. reduced oxide (H_2, 600 K, 2 h), spectrum recorded at 293 K→differentially charged sample; 3. reduced oxide (H_2, 600 K, 2 h), spectrum recorded at 600 K→uncharged sample.

the sample to water vapour significantly broadened the high energy component and changed its position and symmetry (see trace 2 recorded without flood gun). However, the well-defined shape of this peak and its original position were recovered by applying thermal electrons (trace 3). This behaviour suggests that differential charging observed for untreated Pr_6O_{11} (chapter 2) was mainly due to the adsorbed water. The final conclusion regarding the complex O $1s$ region of untreated Pr_6O_{11} is that there are at least three oxygen-containing species at the surface of this oxide. The low binding energy component is assigned to praseodymium oxide. The peak of the high binding energy corresponds to praseodymium carbonate and weakly bonded molecules of water responsible for differential charging of the sample. It is highly likely that praseodymium hydroxide contributes to this feature.

The effect of reduction and oxidation on the O $1s$ spectra of Pr_6O_{11} is shown in Figure 9. The reduced sample is an insulator. Treatment with hydrogen diminishes the energy separation between the components of the spectrum (see traces 1 and 2) and introduces differential charging. Comparison of the traces indicates that reduction-oxidation processes are fully reversible and a short-time exposure to air is sufficient to re-oxidized the sample.

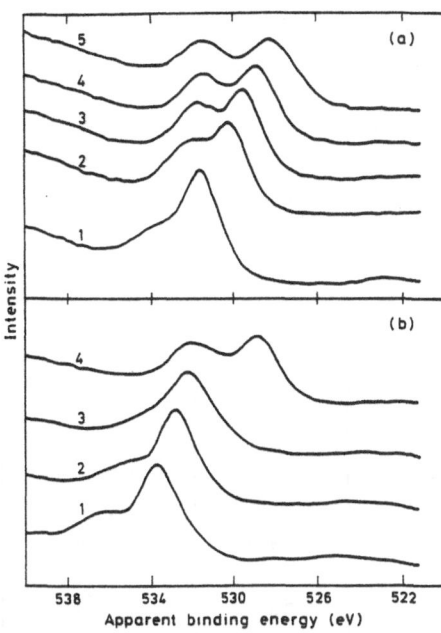

Figure 10. Effect of flood gun (a) and temperature (b) on the O $1s$ spectra of Pr_2O_3. (a) traces 1-5 were recorded at increasing flux of electrons. (b) traces 1-4 were recorded at increasing temperature of the sample (trace 1 - 293 K; trace 4 - 600 K).

Pr_2O_3 is a p-type semiconductor. The effect of temperature on the O $1s$ spectra is shown in Figure 10(b). The conductivity of the praseodymium sesquioxide increases with temperature and reaches its maximum below 600 K. At this temperature the lines present in the spectrum were stable. The analysed sample was not homogenous and exhibited differential charging in regard to the oxide and the carbonate. An increase in the energy separation between these two lines with temperature indicates gradual equlilization of the surface potential due to the increase in the conductivity of the sample. Figure 10(a) shows that the same goal can be achieved using a flux of low energy electrons. This experiment confirms the biased referencing method can be successfully applied to XPS analysis of praseodymium oxide and other compounds exhibiting differential charging.

The conductivity of Pr_2O_3 and PrO_2 at room temperature was low and a positive charge was built on these samples upon photoionisation. However, both oxides showed a substantial increase in conductivity as the temperature raised, as would be expected for semiconductors [13].

3.3 Praseodymium on Alumina

In the first stage of the catalyst preparation alumina is impregnated by a Pr(III) nitrate solution. Either the initial adsorption or the subsequent heating in air to 720 K could possibly result in incorporation of praseodymium at the alumina surface. If this does not occur then the surface would be expected to show the characteristic of a praseodymium oxide, the product of the thermal decomposition of the nitrate.

Sample	Electron Binding Energy, eV			Intensity Ratio $\left(\dfrac{Pr\ 3d_{5/2}}{Al\ 2p}\right)$
	Pr $3d_{5/2}$	O $1s$	Al $2p$	
alumina	-	531.8	75.0	-
Pr_6O_{11}	932.4	528.4	-	-
Pr_2O_3	932.3	528.7	-	-
2% Pr-alumina	934.2	531.4	74.7	2.08
6% Pr-alumina	934.3	531.5	74.7	2.56
10% Pr-alumina	934.5	531.5	74.9	2.69
20% Pr-alumina	934.4	531.3	74.7	3.32

Table 1. XPS characteristics of some praseodymium-alumina systems and reference compounds

The XPS analyses are in Table 1. At 2 percent loading of praseodymium, the Pr $3d_{5/2}$ BE is 1.8 eV higher than for praseodymium oxide. The O $1s$ BE is 3.0 eV higher and is close to the value for alumina.

Higher loadings of praseodymium, exceeding those used in catalysts, showed the same pattern of binding energies. Even at 20 percent there is no evidence for a separate praseodymium oxide phase. The relatively small increase in the intensity ratio for Pr/Al is consistent with incorporation of Pr into the alumina.

3.4 Iron on Alumina

A sample of iron on alumina was analysed to provide a comparison with the promoted catalyst. The binding energies for O $1s$ and Al $2p$ in Table 2 are higher than for alumina. This suggests that a compound involving these elements and iron has formed at the surface. The Fe $2p_{3/2}$ spectrum in Figure 11 (trace 1) is consistent with Fe(III). Treatment in hydrogen at 600 K results only in partial reduction of iron. The broad feature in the spectrum (trace 2) shows evidence of Fe(III), Fe(II) and Fe(0) in the conditioned catalyst.

Sample	Electron Binding Energy, eV			Intensity Ratio $\left(\dfrac{Fe\ 2p_{3/2}}{Al\ 2p}\right)$
	Pr $3d_{5/2}$	O $1s$	Al $2p$	
1.6%Fe-alumina	-	532.0	75.2	0.7
4.8%Fe-alumina	-	532.1	75.3	1.0
1.6%Fe-2%Pr-alumina	934.4	531.2	74.6	1.5
4.8%Fe-2%Pr-alumina	934.2	531.3	74.8	2.6

Table 2. XPS characteristics of iron-alumina and iron-praseodymium-alumina systems

Sample	Intensity Change, %			
	O $1s$	Al $2p$	Fe $2p_{3/2}$	Pr $3d_{5/2}$
1.6%Fe-alumina	9	15	-54	-
4.8%Fe-alumina	10	21	-10	-
1.6%Fe-2%Pr-alumina	-8	14	-42	47
4.8%Fe-2%Pr-alumina	-8	35	-23	118

Table 3. Relative changes in intensities of spectral lines due to treatment with hydrogen (600 K, 12 h)

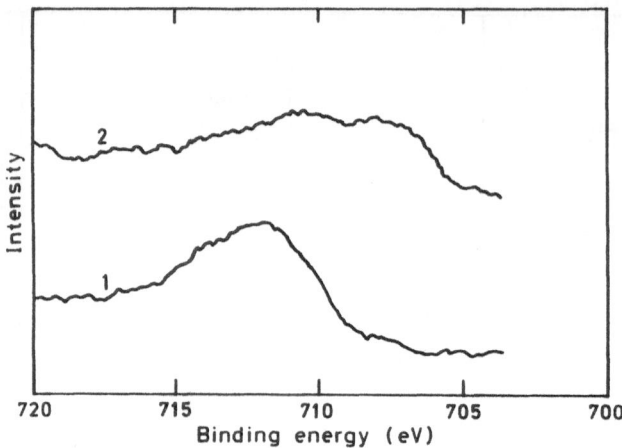

Figure 11. Effect of reduction of the iron-alumina system (4.8% Fe) on the Fe $2p_{3/2}$ spectra. 1. sample "as loaded"; 2. reduced sample (H_2, 600 K, 12 h).

3.5 Iron-Praseodymium on Alumina

Samples of catalysts with two different loadings were analysed by XPS. The binding energies for O $1s$, Al $2p$ and Pr $3d_{5/2}$ shown in Table 2 are equal to those observed for the praseodymium-alumina system. The intensity ratios Fe $2p_{3/2}$/Al $2p$ for these promoted catalysts are more than twice those measured for the unpromoted catalyst of equal loading.

Spectra for Fe $2p_{3/2}$ are shown in Figure 12. Treatment in hydrogen resulted in partial reduction of Fe(III). The amount of reduced iron in the 4.8 % Fe sample is much greater than for the unpromoted catalyst of the same loading.

The reduction process also caused marked changes in the XPS intensities. In Table 3 the percent changes for the Fe $2p_{3/2}$, all negative, indicate reaction of iron into the support. Reduction of the promoted catalyst resulted in increased praseodymium and decreased oxygen. This last effect is the reverse of that observed for the unpromoted catalyst.

4. Discussion

Referring to the binding energies in Tables 1 and 2 it is noted that Pr $3d_{5/2}$ = 932.3 eV and O $1s$ = 528.7 eV, found for Pr_2O_3, are not the values observed for praseodymium-alumina or the promoted catalyst. This shows that the alumina-supported materials do not contain free praseodymium oxide.

The most likely compound at the alumina surface is $PrAlO_3$ which has Pr(III) in a perovskite lattice. The observed Pr $3d_{5/2}$ BE = 934.3 ± 0.2 eV is 2.0 eV above the value for Pr_2O_3. This shift is attributed to the difference in local charge

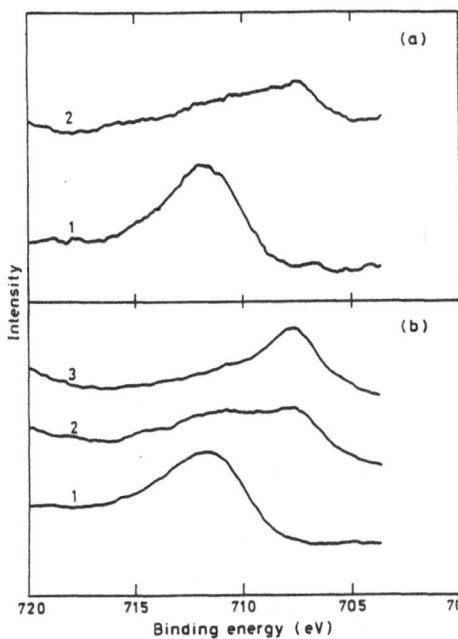

Figure 12. Effect of reduction of the iron-praseodymium-alumina catalyst on the Fe $2p_{3/2}$ spectra. (a) 1.6%Fe-2.0%Pr-alumina. 1. sample "as loaded"; 2. reduced sample (H_2, 600 K, 12 h). (b) 4.8%Fe-2.0%Pr-alumina. 1. sample "as loaded"; 2. reduced sample (H_2, 600 K, 4 h); 3. reduced sample (H_2, 600 K, 12 h).

environment in the perovskite structure. The Al $2p$ level = 74.7± 0.2 eV is almost the same as for alumina. Both α-alumina and $PrAlO_3$ have aluminium in octahedral coordination.

Binding energies for unpromoted iron on alumina catalysts were Al $2p$ = 75.3 eV and O $1s$ = 532.1 eV, higher than for alumina. There are numerous possible compounds and ranges of solid solutions in the iron-alumina system [4]. It is most likely that spinel-type structures with some Al(III) in tetrahedral coordination are formed.

For the praseodymium-promoted catalysts these higher values are not observed. The O $1s$ level, 531.4 ± 0.2 eV, was found for both the praseodymium-alumina support and the catalysts containing iron. A phase study of the Pr-Fe-O system found that $PrAlO_3$ was a constituent over a wide range of compositions [14].

The extent to which any surface compound could form is limited by the low concentration. If distribution over the alumina support is uniform, the loadings 2% Pr and 1.6% Fe represent less than one monolayer. The XPS analyses are sensitive to surface structural relationships which would not be detected by powder diffraction techniques.

The differences between unpromoted and promoted catalysts can be summarized as follows:
- iron on alumina reacts to form a surface phase in which most iron remains in an oxidized state after treatment with hydrogen. Very low activity results.
- praseodymium on alumina forms a surface phase in which praseodymium is combined with aluminium and oxygen.

- iron on the praseodymium treated support is combined in a form which is more readily reduced. Superior catalytic activity, alkene selectivity and methane suppression are the characteristics of the catalyst.

References

1. B.G. Baker and N.J. Clark, Studies in Surface Science and Catalysis, **31**, 455 (1987).
2. B.G. Baker and M. Jasieniak, in *Surface Science: Principles and Applications*, in Springer Proceed. in Physics 73, ed. by R.F. Howe, R.N. Lamb and K. Wandelt, (Springer-Verlag 1993) pp. 279-289.
3. B.G. Baker, in *Surface Analysis Methods in Material Science*, Springer Ser. in Surface Sci. 23, ed by D.J. O'Connor, B.A. Sexton and R.St.C. Smart, (Springer-Verlag 1992) pp. 337-351.
4. I. Sushumna and E. Ruckenstein, J. Catal., **94**, 239 (1985).
5. M.L. Colainni, P.J. Chen and J.T. Yates, Surf. Sci., **238**, 13 (1990).
6. R.L.N. Sastry, P.N. Mehrotra and C.N.R. Rao, J. Inorg. Nucl. Chem., **28**, 1579 (1966).
7. D.A. Stevenson and N.J. Binkowski, J. Non-Cryst. Solids, **22**, 399 (1976).
8. W.J. Landis and J.R. Martin, J. Vacuum Sci. Technol., A2, 1108 (1984).
9. A. Kotani and H. Ogasawara, J. Electron Spectrosc. Relat. Phenom., **60**, 257 (1992).
10. P. Burroughs, A. Hamnett, A.F. Orchard and G. Thornton, J. Chem. Soc., Dalton Trans., **17**, 1686 (1976).
11. A. Bianconi, K. Kotani, K. Okada, R. Giorgi, A. Gargano, A. Marcelli and T. Miyahara, Phys. Rev. B, **38**, 3433 (1988).
12. *Handbook of X-ray Photoelectron Spectroscopy*, ed. by J. Chastain, (Perkin-Elmer Corporation, Physical Electronics Division 1992).
13. L. Erying, in *Handbook of the Physics and Chemistry of Rare Earth*, Vol. 3, ed. by K.A. Gschneidner and Erying, (North-Holland Publishing Company 1979) pp. 337-399.
14. Yu.D. Tretyakov, V.V. Sorokin and A.P. Erastova, J. Solid State Chem., **18**, 263 (1976).

Precious Metal-Ceria Interactions in Car Exhaust Catalysts

D.L. Trimm, C. Padeste*, D.J. Pettigrew & B. Whittington

*University of New South Wales, School of Chemical Engineering & Industrial Chemistry
PO Box 1, Kensington, NSW 2033, Australia*

&

N.W. Cant
Macquarie University, School of Chemistry, NSW 2109, Australia

*Present address:
Micro- and Nanostructures Lab, Paul-Scherrer-Institute, CH-5253 Villigen, PSI, Switzerland*

Abstract

The importance of precious metal-ceria interactions in car exhaust catalysis has been examined using steam reforming and water gas shift as test reactions. The water gas shift reaction is catalysed most effectively by rhodium and platinum, and the catalytic activity is enhanced by the presence of ceria.

Cerium dioxide was found to be hard to reduce, but the presence of 2% Rh reduced the temperature at which reduction was initiated. Reduction was found to lead to Ce(III) on the surface and, in the presence of carbon oxides, to adsorbed carbon. Re-oxidation was facile, being possible using water, carbon oxides or nitric oxide as oxidants.

Interaction of rhodium and ceria to favour the production of surface Ce(III) is found to be an important effect in controlling catalytic activity.

Introduction

Although car exhaust catalysts are now regarded as commodity chemicals, understanding of all of the processes occurring in such systems is incomplete. The active components are platinum and palladium, added to promote the oxidation of unburnt hydrocarbons and carbon monoxide, and rhodium, which selectively reduces nitric oxide to nitrogen [1,2]. The precious metals are distributed in a washcoat, consisting mainly of thermally stabilised alumina [3,4] suspended on a ceramic or metallic monolith. Significant quantities of ceria (ca 15-25%) are also added to the washcoat [2].

The role of ceria is open to considerable question. The oxide was originally added as an oxygen storage component [5] which could be reduced when the exhaust gas ran rich.

$$CeO_2 + x \text{ reductant} \rightarrow CeO_{2-x} + \text{oxidised reductant}$$

and re-oxidised in a lean gas

$$CeO_{2-x} + x/2\ O_2 \rightarrow CeO_2$$

Examination of the thermodynamics of the Ce-O system shows that this is unlikely (6,7). CeO_2 crystallises in a fluorite structure in which Ce(IV) is stabilised. Reduction is only possible under strongly reducing conditions, rarely achieved in car exhaust gases.

Since the addition of ceria is known to produce more active catalysts, attention has been focused on examination of the role of the oxide. Ceria imparts some short term thermal stability to alumina (4) and promotes noble metal dispersion (8) in a reduced state (9). However, the most plausible explanation of the role of ceria is based on metal-support interaction effects (10). Such interactions appear to be important in promoting water gas shift and steam reforming reactions, as well as nitric oxide reduction over the catalysts (10,11). Similar observations have been reported for nickel-ceria catalysts used to promote conventional steam reforming (12).

As part of a systematic investigation of the preparation and application of ceria containing catalysts, studies have now been completed on reactions that can occur in car exhaust catalysts and their effect on the surface chemistry of the catalysts.

A typical exhaust gas can contain ca 0.5% hydrogen, 1.5% carbon monoxide, 13.4% carbon dioxide, 0.15% hydrocarbon, 15% water 0.1-0.2% nitric oxide and ca 70% nitrogen/oxygen. It was decided to use the steam reforming of propane and the associated reverse water gas shift reactions as test systems.

$$C_3H_8 + 3H_2O \rightarrow 3CO + 7H_2 \qquad\qquad 1$$

$$CO_2 + H_2 \rightarrow CO + H_2O \qquad\qquad 2$$

As a result, a test mixture was prepared as above, using propane as the hydrocarbon and omitting oxygen and nitrogen oxides from the mixture.

Experimental

Cerium (IV) oxide (Aldrich 99.9%) and cerium (III) carbonate hydrate (Aldrich 99.9%) were used as starting materials. The preparation of cerium oxide from the carbonate has been previously described (7). Impregnation with Rh was carried out at room temperature using aqueous $RhCl_3$ solutions to obtain Rh loadings of 2% and 5% by mole (1.2% and 3% by weight). After stirring for 30 minutes at room temperature, the temperature was increased to 90°C to evaporate the solvent. After drying overnight at 150°C, the solids were calcined in air for 4 hours at 600°C.

For the temperature programmed reduction and reoxidation studies, the powder samples were pressed to pellets, crushed with a razor blade, and sieved to 100-400 μm. The reactions were carried out in a flow system in which gas mixtures were passed at 40ml/min over 100 to 300 mg samples fixed in a glass tube using glass wool. The tube was placed in a furnace which was temperature programmed at a rate of 10°C/min up to 550°C. Changes in the composition of the product gas were determined using a VG SX300 quadruple mass spectrometer.

Catalyst testing involved a conventional continuous flow system operated at one atmosphere pressure. Reactant and product analysis was carried out using a Gow Mac Series 550 TCD gas chromatograph fitted with a silica gel column. The reverse water gas shift test mixture contained 10% CO_2, 10%H_2 and 80% He at a flow rate of 25cm^3/min passed over ca 0.4g catalyst.

Surface areas were determined by single point nitrogen adsorption at liquid nitrogen temperature from a 30%N_2/He mixture in a flow system equipped with a thermal conductivity detector.

The XPS spectra were recorded on a Kratos XSAM AXIS 800pci spectrometer equipped with a concentric hemispherical analyser which was run in the fixed transmission mode at pass energy 40 eV. A Mg Kα X-Ray source was used at 180 W. The base pressure of the system was less than 10^{-9} torr. The samples were ground to fine powders and suspended in acetone. The resulting slurry was deposited on a stainless steel sample holder. After evaporation of the solvent, the powder was bound to the holder sufficiently strongly to be introduced in the vacuum system. The energy scale of the spectra of non-conducting samples was calibrated against the C1s line of adventitious carbon at E_B=285 eV. Some samples were thermally pretreated in vacuum (less than 10^{-4} torr) or in reactive gas at ambient pressure in a chamber attached to the spectrometer and subsequently transferred into the main chamber of the instrument without contact with air. A dosing system allowed admission of water at pressures up to $5.x10^{-4}$ torr into a turbo-pumped preparation chamber with a base pressure of less than $5x10^{-7}$ torr.

Results and Discussion

The activity of the three way catalyst was first examined using the vehicle exhaust test mixture (Figure 1). The catalyst began to promote both steam reforming and water gas shift at temperatures greater than ca 400°C. The use of single component catalysts showed that the water gas shift reaction was catalysed mainly by Rh and Pt, and that the presence of ceria enhanced the activity significantly. Ceria itself was catalytically inactive.

The largest effect of ceria occurred with a palladium based catalyst, and the hydrogenation of carbon dioxide (reverse water gas shift) was used as a test reaction to examine activity. The turnover numbers (Table 1) were found to be increased by a factor of approximately 10 times on addition of ceria to the catalyst.

Hydrogen is known to be stored in the lattice of both palladium and ceria (13). It seemed possible that reduction of ceria by hydrogen was occurring followed by reoxidation of the ceria by carbon dioxide. With this concept in mind, the surface chemistry of the systems was examined, with particular reference to the effect of reactants on the Ce(III)/Ce(IV) ratio.

The 3d spectra of Ce(III) and Ce(IV) exhibit complicated features due to shake-up and shake-down processes which have been extensively investigated both theoretically and experimentally (13-16). For qualitative determination of Ce(III) versus Ce(IV) it is sufficient to recognise that strong peaks at E_B=886.2eV and 904.7eV (labelled with v′ and u′ respectively) are typical for Ce(III) while the main features of Ce(IV) are at E_B= 889.2eV (v″), 899.1eV (v‴), 908.2eV (u″), and 917.3eV (u‴). Peaks at E_B=883.2eV(v) and 901.2eV(u) arise from both Ce(III) and Ce(IV).

Previous studies have examined the oxidative decomposition of cerium carbonate to ceria (7). The Ce 3d spectrum of the carbonate clearly shows the presence of Ce(III) together with a

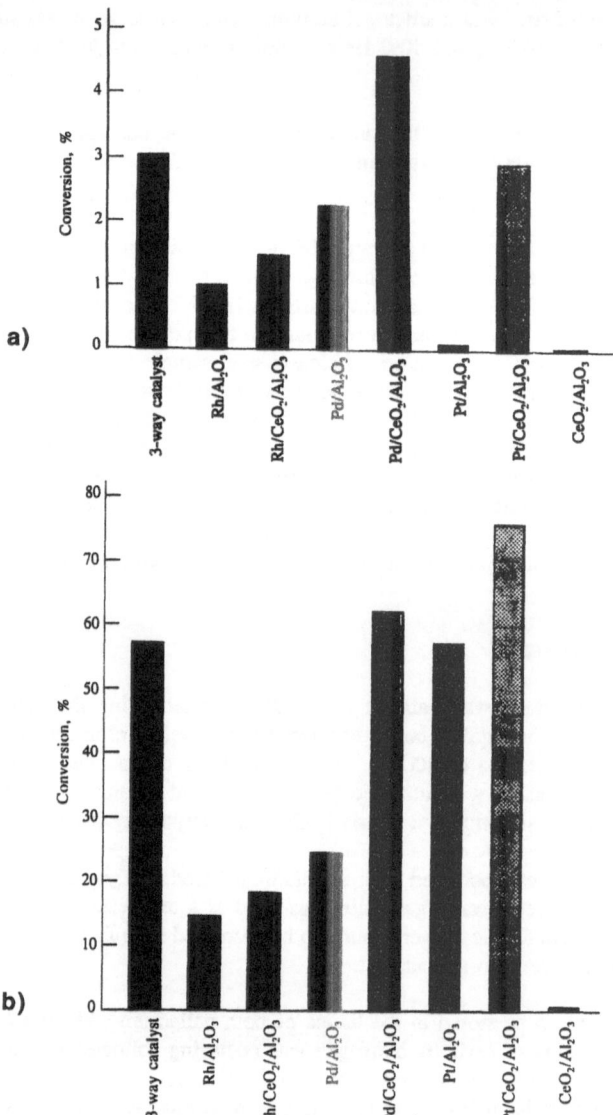

<u>Figure 1</u>   Catalytic activities for steam reforming and water gas shift at 400°C
Activities are assessed in terms of -

(a)   % conversion of propane (steam reforming)

(b)   % conversion expressed as (water gas shift)

$$\frac{CO_{exit}}{CO_{exit} + 3 \times C_3H_{8_{exit}}} \times 100$$

The standard test mixture was used.

VI.

Characterization of Catalysts

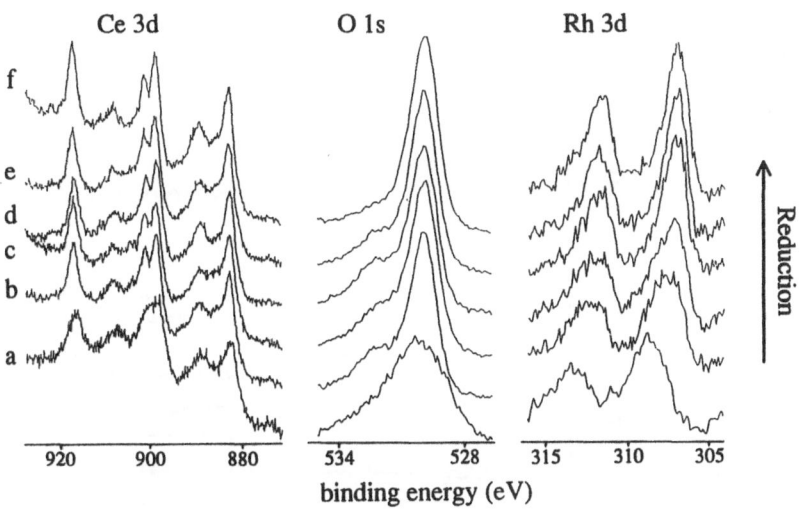

Figure 4   Ce 3d, O 1s and Rh 3d XPS spectra obtained on reduction of 2% Rh/CeO$_2$ in 10% H$_2$/N$_2$.

(a)	5 min	70°C
(b)	5 min	100°C
(c)	5 min	125°C
(d)	5 min	150°C
(e)	5 min	200°C
(f)	10 min	400°C: pure hydrogen

partly, the reduction of ceria. In Figure 4 the Ce 3d, O 1s and Rh 3d spectra of a sample after stepwise reduction in the preparation cell of the XPS in a 10%H$_2$/N$_2$ mixture are shown. Treatment at 70°C did not produce any signs of reduction of rhodium or of ceria support. After treatment at 100°C, metallic rhodium becomes clearly apparent, but some reduction of CeO$_2$ is also indicated by the narrower and better resolved Ce3d and O1s peaks due to the increased conductivity of oxygen deficient ceria (17). After treatment at temperatures at 150°C and above, rhodium exists predominantly in the metallic state. Reduction at 400°C (10 min) showed some evidence of Ce(III) production, based on better resolved Ce3d peaks (figure 4). This would infer that the onset of additional hydrogen consumption and water production above 400°C (Figure 3) is associated with Ce (IV) reduction. If this is the case, then the reducibility cycle of ceria has been significantly enhanced by the presence of Rh. The extent of reduction is difficult to quantify but the temperature of reduction has been substantially decreased.

Reduction of the ceria with carbon monoxide was then examined. Simultaneous measurements of carbon monoxide removed or carbon dioxide produced were in good agreement, the extent of reduction being measured both by the CO$_2$ produced and by the oxygen consumed on reoxidation at 400°C in a 2%O$_2$/He gas stream.

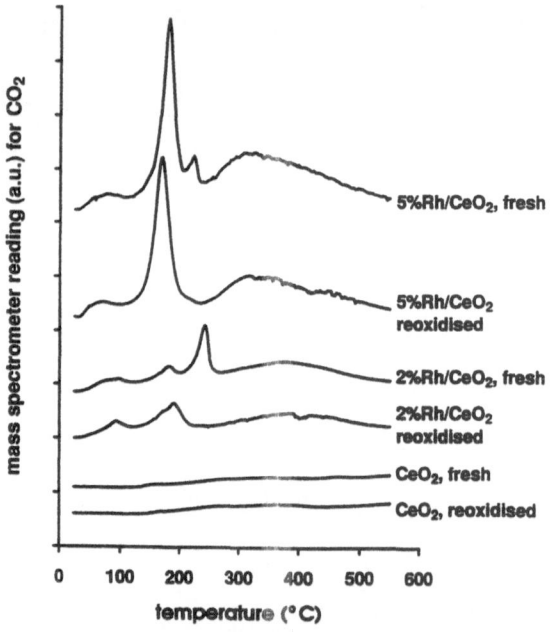

Figure 5   Temperature Programmed Reduction of samples using 2% CO/He

Fresh   = samples after calcination
reoxidised   = samples that had been reduced in the gas mixture up to 550°C
             and their reoxidised in 2% O₂/He for 10 min at 400°C.

Two samples were examined. More crystalline ceria A and less crystalline ceria B produced very different results, Ceria A showed at most 1-2% reduction (figure 5). Ceria B showed 10-15% reduction, beginning at ca 300°C and peaking at ca 550°C. The influence of crystallinity on the reactivity of ceria (7) was re-emphasised.

Ceria A was coated with 2 or 5% Rh, and subjected to the same temperature programmed reduction. Calcined samples were reduced in CO up to 550°C, cooled to 400°C and reoxidised in a 2% O₂/He gas stream. The samples were then cooled to room temperature and a second reduction up to 550°C was performed. Reoxidation was carried out using 2%O₂/He.

Rhodium impregnated samples show the first sign of reduction at ca 40°C and a maximum in reduction at ca 180°C (figure 5). This reduction is assigned to reduction of rhodium oxide.

A second peak was formed at ca 250°C but this disappeared after the first reduction/oxidation cycle (figure 5). This is thought to originate from surface carbonates.

Finally a broad peak was observed in the 300-400°C region (Figure 5). Thus was ascribed to the reduction of ceria. Surprisingly, however, reoxidation of the reduced sample at 400°C showed the presence of carbon oxides, inferring the formation of carbonaceous deposits on the surface during reduction. The extent of reduction and the formation of carbon oxides during reoxidation both increased with the Rh loading (Figure 5).

The reoxidation of Ce(III) in the absence and presence of rhodium was then examined. Reaction with oxygen was very fast, and oxidation with other reagents was attempted. In the first case, samples that had been reduced at 500°C were flushed with helium and then exposed to various gases. The first gas used was helium saturated with water at 0°C. Oxidation of pure ceria occurred, with hydrogen being detected in the product gas.

$$CeO_{2-x} + xH_2O \rightarrow CeO_2 + xH_2$$

The initial reaction was very rapid (100-200 secs) but the reaction was slower once the top layers of ceria had been oxidised.

Pure ceria that had been reduced at 550°C could also be reoxidised by passing nitric oxide over the sample. Both nitrogen and nitrous oxide were produced. Because of the importance of the nitrogen oxides reduction process in car exhaust catalysts, a more detailed study was carried out using 2% Rh/CeO_2.

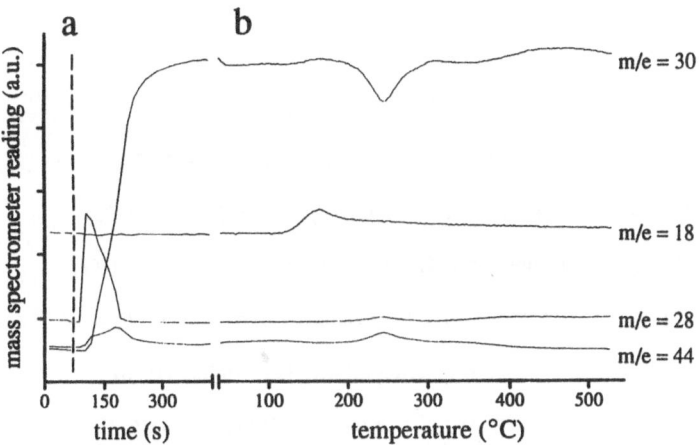

Figure 6   Temperature programmed oxidation of 2% Rh/CeO_2 by 1% NO/He.

   The sample had been reduced in 2% H_2/Ar up to 550°C, cooled to room temperature in He, and the oxidised mixture was then admitted at constant temperature. After ca 40s, the temperature was increased.

   The loss of NO (m/e=30) and the production of H_2O (m/e=18), N_2 (m/e=28) and N_2O (m/e=44) are shown.

The Rh/CeO_2 catalyst was reduced with hydrogen at 550°C and cooled to room temperature under helium. A stream of 1% NO in helium was passed over the catalyst at this temperature (Figure 6a).

Nitrogen was produced in the gas phase together with small quantities of N_2O. At least part of the nitrogen came from oxidation of Ce(III) since XPS examination of a sample that had been treated similarly showed that complete oxidation had occurred.

The catalyst was then subjected to temperature programmed heating under the same gas mixture (figure 6b). Both TPO and XPS spectra showed that the peak at ca 200-300°C was associated with oxidation of most of the rhodium. The major gaseous product from the oxidation was nitrous oxide.

Reduction and reoxidation of ceria by nitric oxide is one of the desired reactions in car exhaust catalysts. Synergistic effects between rhodium and ceria allow easier and (probably) more extensive reduction of ceria. Reoxidation by nitric oxide is facile, even at room temperature.

Acknowledgments

Thanks are due to the Australian Research Council for support of this project and the Swiss National Science Foundation for support of CP.

References

1. B. Harrison, B.J. Cooper and A.J. Wilkins, Platinum Metals Rev. **25** 14 (1981)

2. H.C. Yao and Y.F. Yu Yao, J. Catal. **86** 254 (1984)

3. S. Subramanian, M.S. Chattha and C.R. Peters, J. Mol. Cat. **69** 235 (1991)

4. J.S. Church, N.W. Cant and D.L. Trimm, Appl. Catal. **101** 105 (1993)

5. E.C. Su, C.N. Montreuil and W.G. Rothschild, Appl. Catal. **16** 75 (1985)

6. G. Brauer, K.A. Gingerich and K. Holtschmidt, J. Inorg. Nucl. Chem. **16** 77 (1960)

7. C. Padeste, N.W. Cant and D.L. Trimm, Catal. Lett. **18** 305 (1993)

8. F.J. Sergeys, J.M. Masellei and M.V. Ernest, US Patent 3,903,020 (1974)

9. A.F. Diwell, R.R. Rajaram, H.A. Shaw and T.V. Truex, Crucq A (ed) Catalysis & Automotive Pollution Control II **139** Elsevier, Amsterdam (1991).

10. B. Harrison, A.F. Diwell and C. Hallett, Platinum Metals Rev. **32** 73 (1988)

11. G. Kim, Ind. Eng. Chem. Prod. Res. Dev. **21** 267 (1982)

12. J. Barrault, A. Chafik and P. Gallezot, Appl. Catal. **67** 257 (1991)

13. J.L.G. Fierro, J. Soria, J. Sanz and J.M. Rojo, Solid State Chem **66** 154 (1987)

14. A. Fujimori, Phys. Rev. B28 (4) 2281

15. G. Praline B.E. Koel, R.L. Hance. H.I. Lee and J.M. White, J. Electron Spectros. **21** 21 (1980)

16. F. Le Normand, J. El Fallah, L. Hilaire, P. Legare, A. Kotain and J.C. Parlebas, Solid State Commion. **71** 885 (1989)

17. C. Padeste, N.W. Cant and D.L. Trimm, Catal. Lett. (in press)